Photovoltaic Power Generation

Solar Energy R&D in the European Community

Series C
Photovoltaic Power Generation

Volume 3

Publication arrangements: G. BRESLIN, D. NICOLAY

Solar Energy R&D
in the European Community

Series C   Volume 3

Photovoltaic Power Generation

Proceedings of the
EC Contractors' Meeting held in
Brussels, 16-17 November 1982

edited by

R. VAN OVERSTRAETEN
Katholieke Universiteit Leuven, Belgium

and

W. PALZ
Commission of the European Communities

D. REIDEL PUBLISHING COMPANY

Dordrecht, Holland / Boston, U.S.A. / London, England

for the Commission of the European Communities

Library of Congress Cataloging in Publication Data

EC Contractors' Meeting (1982 : Brussels, Belgium)
  Photovoltaic power generation.

    (Solar energy R&D in the European community. Series C ; v. 3)
    1. Photovoltaic power generation—Congresses. I. Overstraeten,
R. van. II. Palz, W. (Wolfgang), 1937–   III. Commission of the
European Communities. IV. Title. V. Series.
TK2960.E3   1982   621.31'244   83-3159
ISBN-13:978-94-009-7138-7   e-ISBN-13:978-94-009-7136-3
DOI: 10.1007/978-94-009-7136-3

Organization of the Contractors meeting by
Commission of the European Communities
Directorate-General Science, Research and Development, Brussels

Publication arrangements by
Commission of the European Communities
Directorate-General Information Market and Innovation, Luxembourg

EUR 8399
Copyright © 1983, ECSC, EEC, EAEC, Brussels and Luxembourg
Softcover reprint of the hardcover 1st edition 1983

Published by D. Reidel Publishing Company
P.O. Box 17, 3300 AA Dordrecht, Holland

Sold and distributed in the U.S.A. and Canada
by Kluwer Boston Inc.,
190 Old Derby Street, Hingham, MA 02043, U.S.A.

In all other countries, sold and distributed
by Kluwer Academic Publishers Group,
P.O. Box 322, 3300 AH Dordrecht, Holland

D. Reidel Publishing Company is a member of the Kluwer Group

INTRODUCTION

The EC programme on Photovoltaic Power Generation has two distinct parts :
Pilot Projects and R&D contracts on cells, processes, materials, modules
and systems. The pilot projects are managed in a totally different way
from the R&D contracts. Emphasis is put on all aspects of the system
such as power conditioning, environmental factors, structures, etc. The
aim is to examine all different system aspects, to improve the performance
and the reliability of the overall photovoltaic system and to reduce its
cost. A comprehensive overview of the PV pilot programme of the European
Communities was given in Volume 1 of this Series C.

The R&D contracts, on the other hand, are intended to advance our knowledge on all components in order to reduce the cost, mainly of the
photovoltaic modules. This book reports on the results presented at the
sixth contractors' meeting, held about six months before the end of the
current R&D programme. Substantial progress can be observed in several
areas :

- solar cell processing : screenprinting can now be used for junction
  formation and for metallization on single crystalline and polycrystalline material; the work on cold junction formation shows promise;

- low cost silicon wafers : the process of unidirectional crystallization
  in a crucible gives ingots 20 kg in weight; the potential of a new
  pendant drop growth method is discussed; continuous growth was demonstrated on a RAD puller up to 30 m at a pulling rate of 10 cm/min ;

- encapsulation : new materials with cast resin, with laminated resin and
  with acrylate polymer and mineral concrete are investigated;

- amorphous silicon : it is shown that geminate recombination does not
  impose any fundamental limitation in photovoltaic applications; magnetron sputtering produces a-Si:H of the same quality as the best glow-
  discharge; the use of a double ion-gun sputtering apparatus is investigated;

- Cu$_2$S-CdS solar cells : cells fabricated in a pilot plant framework by
  a spray technique give a statistical mean efficiency of 6%, the possibility of forming the cuprous sulfide by cathodic reduction is investigated as is the use of electrophoresis as a deposition technique;

- a simple MIS technology for polycrystalline CdSe films gives efficiencies up to 7%;

- special collectors : a 1 m^2 fluorescent collector has been built;
  phase volume holograms are studied as dispersive concentrators; a
  hybrid àar energy concentratio collector is built;

- system aspects, including a microcomputer for the control of a photovoltaic power station and a new highly efficient inverter.

At the end of the meeting, contractors met in 3 parallel group meetings to
discuss in detail the state of the art in their particular field of interest
and future R&D needs. The chairmen of these meetings agreed to prepare
short summaries which are published at the beginning of this book.

This book gives an excellent review of the state of the art of photovoltaics in Europe and we hope that it will contribute to the preparation of
the next R&D programme on energy.

                                        Roger VAN OVERSTRAETEN
                                        Project leader

C O N T E N T S

REVIEW PAPERS

Review paper on silicon cells
    R. VAN OVERSTRAETEN and E. FABRE

Review paper on thin film solar cells
    W. BLOSS and D. BONNET

Review paper on photovoltaic module technology, system
studies and concentrator devices
    K. KREBS and V. MAKIOS

REVIEW PAPER ON SILICON SOLAR CELLS

R. VAN OVERSTRAETEN and E. FABRE

The cost of silicon solar cells did not come down as fast as predicted by the D.O.E. in the United States. It is still between 7 and 10 ECU per peak Watt and is made up of 50% for the wafer cost, 30% for the processing and 20% for the interconnection and encapsulation.

There are several promising ways to reduce the cost of the silicon wafers :

- production of a solar grade quality silicon, which is expected on the market around 1986;

- the use of several new or improved methods to grow the crystals.

The most important ones are :

- casting of silicon. Casted polycrystalline silicon has been investigated for several years now. The cells give reasonable efficiencies. The cost can be reduced by using larger casting units and by using a continuous process. A major price drop can be expected around 1987;

- advanced Czochralsky pulling. The crystal growth is optimized, but the price can still be reduced by continuous feeding and by increasing the diameter;

- epitaxial growth of a 10-20 micron thick Si layer on unidirectionally solidified upgraded metallurgical grade silicon substrates. Efficiencies close to 10% have been obtained;

- the R.A.D. (ribbon against drop) method. Substantial progress has been made and industrial production can be expected around 1986.

Except for the last method, slicing in wafers has to be done and remains an expensive process. It should be mentioned that ribbons become more economic, the higher the cost of the feedstock.

In cell processing, we have two important steps : the junction formation and the deposition of metal contacts. The state of the art is the following :

- the diffusion is done at higher temperatures (either a conventional gaseous diffusion or a deposition of the impurities using screen printing followed by a diffusion in a conveyor furnace);

- the screen-printing of the contacts on the back and on the front has to be considered as an important achievement.

A strong effort is now devoted to cold junction processing (implantation and beam annealing). Problems to be studied include laser homogeneity, spot size and overlapping and the solid phase regrowth versus liquid regrowth. The use of the excimer laser seems interesting but could be too expensive. Incoherent light annealing should receive more attention.

It is also important to consider the compatibility between different processing steps as for example between the use of cold junction processing and screen printing, between the use of lower grade silicon and diffusion, or between the use of brittle materials and some processing steps. The overall production yield is an important factor.

REVIEW PAPER ON THIN FILM SOLAR CELLS

W. BLOSS and D. BONNET

The current R&D programme on alternative cells contains eight projects on three thin film cells which have been reported on at the contractors' meeting. They are conducted in industrial and university laboratories in Great Britain, France and Germany. In the following, first the present status of the three cell-types is briefly described, followed by the recommendations of the discussion group.

Curent situation

Amorphous silicon is being studied in four laboratories. One project has been terminated in 1981. Two are concerned with the relatively new technique of sputtering (magnetron and ion gun) and one is concerned with glow discharge deposition and with intensive characterisation of the material. The projects are at this time mainly aimed at improving the technology of film formation and diode preparation or film characterisation. Due to the limited budget and time the sputtering projects are not yet at a stage to produce cells with efficiencies of more than 3 percent.

The Cu_2S/CdS thun film cell is being studied within three contracts. One project, in which four French institutes are participating, is concerned with the technology of spray deposition of CdS-film preparation and has achieved encouraging efficiencies of up to 8 percent on areas of 1 cm^2. First stability measurements have been conducted. The importance of good encapsulation has again been evident. A second project from an industrial laboratory in Great Britain is concerned with electrophoretic deposition of CdS layers and post-treatment steps as a way to cheaper production technologies. A French Institute of Learning is conducting an interesting study on Cu_2S film formation as well as on new grid contact techniques.

The CdSe cell is being developed as a purely European effort by a German institution. Application of the MIS technology to evaporated thin CdSe films has led to efficiencies of 7 percent on 0.6 cm^2 active area with relatively little effort.

Future needs

After discussing the state of these thin film cells, the group tried to define future needs in order to establish a basis for an efficient programme for the future.

REVIEW PAPER ON PHOTOVOLTAIC MODULE TECHNOLOGY, SYSTEM STUDIES AND

CONCENTRATOR DEVICES

K. KREBS and V. MAKIOS

Modules :

It is felt that the development of conventional silicon modules has
reached a level where further improvements may be left to manufacturers.
Remaining problems are now more often related to poor or insufficient
workmanship control than to basic design errors. The use of new encapsulation materials could lead to further cost reductions but the choice of
these technologies should probably remain that of the producers. Support
should be given to really novel concepts and in particular to the development of modules for roof integration should be encouraged. In addition,
feasibility studies of encapsulation concepts for thin film devices
should be initiated.

Systems :

It is recommended to install and investigate prototype residential
systems of both stand-alone and grid-connected versions. Within entire
systems special attention should be given to the use of appliances
with low energy consumption. Photovoltaic generators should be combined
with passive solar systems. The efficient execution of such projects
requires close collaboration between scientists, design engineers and
architects, adequate emphasize should also be given to non-technical
obstacles.

It has also been stated that it would be worthwhile to implement a
network of specified identical systems below 10 kW at climatologically
different sites in the Community. This action should be supplemented by
the setting up of reference systems for specific applications which
consist of components of different technical design. For both cases it
would be essential to provide efficient data acquisition and data storage
possibilities.

Two other subjects which would benefit from contract research are concerned
with more economical support structures and with the introduction of
total energy systems, especially those which would include the production
of storable fuels.

With reference to amorphous silicon solar cells, the group came to the conclusion that the present programme of essentially three projects is too little and too small on an absolute scale with regard to the world-wide state of the art. More detailed investigations on processes and materials are felt to be needed. To this regard cooperation between different institutions and laboratories should be solicited. With respect to stability problems the future programme should put emphasis on the examination of potential degradation mechanisms and ways for their future elimination. Besides trying to make cells approaching the international standard, some of the future effort should be dedicated to evaluating and characterising the material itself in order to achieve films with good and well-defined electronic properties.

In the case of CdS/Cu_2S cells the future programme should strive to achieve prototypes of larger areas and greater number in order to get a good evaluation of the processes employed. Alternative, cost effective technologies for parts of the cells, such as CdS deposition, Cu_2S formation or deposition and grid generation should also be of future interest. Efforts to inherently avoid stability problems or to create cost efficient encapsulation techniques should not be disregarded.

Work on the CdSe cell as a more recent achievement should be continued with the aim of demonstrating efficiencies of around 10 percent. As in the case of the CdS/Cu_2S cell alternative deposition methods would greatly help to reach high cost-effectiveness.

The group strongly stated the recommendation, that the future programme should be held open for new cells and new concepts. No fundamental limitations should be posed on studies of the feasibility and development of new cell types.

Concentrators :

Classic concentrator systems are no longer considered to need further support, since it appears that enough experience has been acquired worldwide. To reach costs below 2-3 $/Wp seems to be very questionable. However, novel concepts deserve further backing, especially if they lead to new approaches or to a better understanding of how to collect and/or concentrate light. Much interest still exists for multiple cell arrangements, spectral control devices and for very high efficiency concentrator cells. In all these cases economic aspects should provide the essential guideline for further development.

SILICON CELLS

a) Cell process development

Development of new techniques for silicon solar cell fabrication

Study of mono- or polycrystalline solar cell process using screen printing technology

Technology of amorphous silicon thin films for solar cells and applications to power systems

b) Ion implantation

Ion implantation without mass analysis and laser annealing for fabrication of silicon solar cells

Investigation on the potentiality offered by ion implantation and electron beam annealing to obtain high efficiency solar cells

An experimental investigation of ion implantation combined with laser and incoherent-light annealing and of laser-induced diffusion for the production of solar cells

Optimization of polycrystalline silicon solar cells produced by ion-implantation and pulsed laser annealing

Low cost implantation into silicon

SILICON CELLS

(continued)

c) Material

Design, construction and optimization on the industrial prototype scale of a furnace able to produce polycrystalline silicon ingots as material for solar cells

Classification of crystal defects in solar base material with diamond lattice

Implementation of low cost semicrystalline silicon solar cells

Optimization of processing conditions of solar cells versus physical properties of relatively low cost silicon

Continuous production of photovoltaic silicon ribbons by the new pendant drop growth (PDG) method

Growth and solar cell aspects in relation to polycrystalline silicon ribbons grown by the RAD process

d) Modules

Studies relating to new encapsulation materials

Encapsulation of photovoltaic solar cell modules

R&D work on the encapsulation of solar cells with improved potting and cover materials

DEVELOPMENT OF NEW TECHNIQUES FOR
SILICON SOLAR CELL FABRICATION

Authors   : R. MERTENS, G. CHEEK, R. JANSSENS, M. LEEMPOELS, M. HONORE

Contract number : ESC-R-018-B

Duration   : 36 months   1 July 1980 - 30 June 1983

Total budget   : 16 MBF, CEC contribution 50%, Belgian Government 50%

Head of project : Prof. R. VAN OVERSTRAETEN, University of Leuven, Belgium

Contractor   : ESAT Laboratory, University of Leuven

Address   : ESAT Laboratory, Kardinaal Mercierlaan 94
                        3030 Heverlee - Belgium

Summary

The aim of this project is the development of new techniques for silicon solar cell fabrication. Most emphasis has been put on the use of
screen printing for junction formation, metallization and for the
realization of the antireflection coating. Through process optimization and improved understanding of the front contact formation
efficiencies exceeding 12% on single crystal, using the integral printing process, have been achieved. Solar cells made on single crystal
silicon in small batch quantities (400 wafers), using the printing
technique for junction formation and electrode deposition and a spin-
on ARC, have mean efficiencies over 11% with good yields and reproducibility. It has been found that the glass frit is very critical for
the optimization of the front electrode metallization as well as for
the formation of a BSF. Cell processing that is optimized for single
crystal silicon is not directly transferable to polycrystalline silicon
materials. The initial surface preparation and the silver ink firing
temperature were more critical for polycrystalline materials. A reduction in cell efficiency of about 2 to 3% is found when implementing
the screen printing fabrication technique on polycrystalline materials.
This loss factor is similar to that when other more conventional fabrication technologies are implemented on the same materials. Initial
results on epitaxially deposited thin films on upgraded metalurgically
grade silicon obtained with the screen printing technology are encouraging since efficiencies of 9% have been reached.

1. Introduction

The use of thick film technology for the fabrication of silicon solar cells offers several advantages. The capital investment for the equipment is relatively small and the processing is low-cost, due to the high materials yield, the high throughput and the possibilities of automation. Printing techniques can be used for the formation of the junction, the front and back metallization, including a back surface field, and for the antireflection coating.

Today screenprinting is accepted as the most cost effective method for cell metallization in a production environment since most solar cell manufacturers use this technique for their production.

This paper describes the progress that has been made during the last one and half year in the use of screen printing for silicon solar cell fabrication. First the general characteristics of the integral screen printing process will be reviewed, then the specific progress made and problems encountered during the last eighteen months will be discussed.

2. Description of the process

Starting material is p-type as cut silicon, 300-500 μm thick. After having etched the wafer, a home-made diffusion paste is printed and the diffusion occurs in a belt-furnace. The parasitic junction at the edges is etched using dry plasma etching technique. The antireflection layer is deposited by spin-on (TiO_2) or by printing (Ta_2O_5). For the metallization a silver layer is printed at the front and fired through the antireflection coating. At the backside an aluminium paste is printed and dried; a contact pad in silver-palladium is printed onto the aluminium and cofired with the aluminium. This process is characterized by an excellent yield and reasonable efficiencies. The best cell efficiency, achieved to date, is 13.3% with a FF = 75%, a J_{sc} = 30 mA/cm^2 and a V_{oc} = 596 mV.

3. Front side metallization

The front side metallization, using a modified ESL5964 silver ink, is the most critical part of the process. During the last 18 months considerable progress has been made in the understanding of the actual contact formation. Fig. 1 shows the influence of the peak firing temperature on the open circuit voltage and the fill-factor in the case of an ink with frit. One observes that for the non modified ink the open circuit voltage decreases with the firing temperature T_f (Fig. 1, dashed curve). However, the effect can be eliminated by modifying the ink, in which case the open circuit voltage does not decrease and the fill-factors are also good.

To explain these curves one must understand the role of the glass frit in the inks. Figure 2 shows V_{oc} and FF versus T_f for a fritless paste. Clearly in this case the FF is very poor but almost no V_{oc} degradation occurs at higher firing temperatures. Independent series resistance measurements show that this poor FF results from a large series resistance. Comparison of Figs. 1 and 2 show that the glass frit is necessary for good ohmic contact formation and adhesion, but at the same time that it is responsible for the V_{oc} degradation at higher firing temperatures. During firing, the glass frit, becoming a liquid, dissolves part of the silicon that, during cooling, will recrystallize. This recrystallized layer has poor electronic properties and serious V_{oc} degradation will occur when large fractions of the emitter will be dissolved by the frit and recrystallize. The frit containing paste ESL 5964 however, can be modified to compensate the effects of the interaction of the molten frit with the underlying silicon, at the same time the diffusion of Ag in Si can be reduced.

The achievement of sufficiently low series resistance, using printed silver inks, is extremely important for solar cell applications and depends on a number of parameters like the fact whether or not the Ag paste is fired through the ARC, the crystallographic orientation, the eventual presence of an HF dip after firing and the nature of the silver paste. We will now describe these different parameters.

3.1. Firing through the ARC

The idea of firing through the ARC has been proposed several years ago [1]. In our case this procedure consistently gives better results than the process in which the ARC is deposited after the front metal pattern. A systematic investigation has shown that the antireflection coating not only acts as a diffusion barrier against impurities during the firing but that it aids in the ohmic contact formation. The series resistance is typically a factor of two lower in the case of fired-through contacts. A good explanation of this phenomenon has not been found yet. It could be due to the interaction of the TiO_x ARC with the glass frit resulting in a better ohmic contact or to a silicide formation between the titanium and the silicon. The presence of oxygen in the ARC, however, makes this second mechanism less likely.

3.2. The influence of crystal orientation

We also found that the fill-factors that are obtained with the fired-through process are considerably larger on ⟨111⟩ than on ⟨100⟩ material. The major reason for the smaller fill-factor on ⟨100⟩ is related to the higher contact resistance. Measured series resistances on ⟨111⟩ are on the average twice as small as on ⟨100⟩. This can be explained by the fact that the glass frits act differently on ⟨100⟩. This statement will be validated when the formation of a BSF with the Aluminium screen printing technology will be discussed.

3.3. The influence of an HF dip

It has been reported [2] that dipping of a cell with a screen printed front electrode in a diluted HF solution results in a larger cell fill factor caused by a lower series resistance. We have systematically investigated the use of such an HF dipping method and came to the following conclusions :
- HF dipping only results in a lower R_s, no other cell parameters are affected.
- only cells with a high series resistance (e.g. printed with a non optimized Ag ink) will improve. Cells that already have a low series resistance will not become better. Surprisingly and contrary to what has been reported [2], the improvement also occurs on pastes without glass frit. Most likely the etching removes the oxides that can grow during firing.
- the improvement is not stable : cells with a large R_s will become poor after a few weeks; cells that were good before dipping will remain good.
The temporary improvement, resulting from an HF dip, is probably caused by the formation of electrolytes that fill the gaps created by the removal of the oxide. After some time the electrolytes evaporate and a high series resistance will be present.

3.4. The influence of the nature of the Ag paste

The best results have been obtained with a modified ESL 5964 ink; these results have been described in chapter 2.

The results obtained with other inks are given in Table I.

AG Paste	Performance (maximum efficiency)
Engelhard A4345	printing through ARC : 11.2% without ARC : 10.4%
Engelhard A4162	printing through ARC : 11.2% without ARC : 10.7%
Demetron 6190-0438	printing without ARC : 10.1% very poor if fired through
Demetron 6190-0492	in both cases $<$ 10%
TFS 3347	printing through ARC : 10% without ARC : 10%
ESL 5964 (= reference)	printing through ARC : 13.3% without ARC : 10.8%

Table I : Performance of different silver pastes when used as front
electrode.

The following conclusions can be drawn :
1. almost all cells perform better when the firing occurs through
   the ARC
2. the ESL 5964 is by far the best Ag ink; it should be realized,
   however, that this is true for the process, described in this paper.
   Some of the other inks may perform much better if they are modified
   and optimized with respect to a particular process.
3. The inks, considered in Table I, mainly differ in the composition
   of the glass-frit. Therefore, the nature of the glass-frit is
   extremely important when screenprinted contacts on solar cells
   are considered.

4. Back side metallization

4.1. Back surface field formation

The back surface field is formed by the screenprinting and firing of
aluminium paste. This occurs since part of the silicon dissolves in the
molten ink and will recrystallize being doped with the p-type dopant
aluminium up to the solid solubility limit (= 10^{19} cm^{-3}). We have determined the parameters influencing the formation of the back surface field.
Fig. 3 shows the sheet resistence of the p$^+$ layer as a function of the
Aluminium peak firing temperature for a $\langle 111 \rangle$ and $\langle 100 \rangle$ silicon substrate,
when using an ink with frit (Engelhard T2497) and one without frit. Most
of the results have been obtained with a 165 meshnumber screen to obtain
thick enough layers and low enough sheet resistances, the lower the sheet
resistance, the stronger the BSF will be, the thickness of the printed
aluminium layer is very important.

From Figs. 3 the following conclusions can be drawn :
- the glassfrit strongly helps to dissolve the silicon, that will recrystal-
  lyze, saturated with Aluminium. The nature of the glassfrit therefore
  will be of direct importance for the strength of the BSF.
- the dissolving power of the printed Aluminium layer is much less for
  $\langle 100 \rangle$ than for $\langle 111 \rangle$ oriented silicon. This observation is, most likely,
  related to the difference in contact resistance observed at the frontside
  between $\langle 111 \rangle$ and $\langle 100 \rangle$ silicon.

4.2. Elimination of the formation of bubbles at the backside

Sometimes during firing the Aluminium ink, which has been printed on the back side, tends to create bubbles. These bubbles effect the flatness of the cell and can cause problems during further cell processing and panel manufacturing. It has been found that the creation of bubbles can be attributed to high surface tension at the liquid-vapor interface.

To solve this problem elements such as Pb and Mg which reduce the surface tension can be added to the aluminium ink.

Table II describes results obtained with aluminium to which 7% (weight percentage) Pb has been added. All results refer to average results obtained on 18 (1.8 x 1.8) cm^2 cells.

	temp profile (5 zones)	$\overline{V}_{oc}$ mV	$\overline{J}_{sc}$ mA/cm^2	$\overline{FF}$ %	$\overline{\eta}$ %
no Pb	450-690 (3X)-450	584	26.6	72.9	11.4
with Pb	450-690 (3X)-450	580	26.5	71.7	11.1

Table II : Results for Al-metallization to which Pb has been added.

It can be concluded from table II that the electrical performance of the cells with an aluminum metallization to which Pb has been added is only slightly degraded. The problem however is that the thermal expansion coefficient of the Aluminum layers changes with the addition of Pb. Sometimes the aluminum layers to which Pb has been added tend to shear off from the silicon substrate. Therefore, experiments with Mg have been performed. Table III shows some typical results obtained as an average over 18 cells (1.8 x 1.8 cm^2)

	temp. profile (5 zones)	$\overline{V}_{oc}$ mV	$\overline{J}_{sc}$ mA/cm^2	$\overline{FF}$ %	$\overline{\eta}$ %
No Mg	450-690 (3X)-450	577	27.2	71.5	11.3
with Mg	450-690 (3X)-450	580	28.1	72.1	11.8

Table III : Results for Al-metallization to which Mg has been added

The results listed in table III clearly show that the addition of Mg does not degrade the cell performance. Since Mg does eliminate the problem of bubble formation and does not introduce thermal mismatching it seems to be the best material to be added to the aluminium pastes.

5. The use of printed antireflection coatings

The aim is to develop a true integral screen printing process in which screen printing is used not only for junction formation and metallization, as described in the previous sections, but also for the anti reflection coating. This section describes the first experiments performed to evaluate the potential of this approach.

5.1. Description of the technology

1. Cells : 1.8 x 1.8 cm^2, scribed out of 2 inch wafers;
   (111) orientation, resistivity = 1.5 to 3 ohm-cm
2. Cell preparation : etching, cleaning, junction formation by screen printing, etching and cleaning by the standard techniques described
3. Printing of the ARC
   - pastes : Engelhard E422 and A4125. Both are Ta_2O_5 pastes and must be diluted with Terpinol or/and ethylene-glycol to obtain the exact thickness of the coating.
   - a mesh of 165 wires per inch has been used : this allows to print thicker layers than with e.g. a mesh with 325 wires and to dilute the pastes.
   - after printing the wafers are kept at room temperature during 10 minutes to level off the pastes and to eliminate, as far as possible, the mesh pattern in the paste. Then, they are dried in air during 10 minutes at a temperature of 100°C.
4. Then the paste is baked at a temperature of 300 to 500°C in air to dry off most of the remaining organics.
5. Immediately after printing of the ARC the front contact grid is printed using the method described in earlier sections, and the silver paste is fired in the conveyor furnace (820°C effective firing temperature during 7 minutes).
6. Next treatment : printing of the aluminium at the backside, firing, scribing and testing (see previous sections).

One of the problems encountered with the total integral process is that a trade-off must be made between a rather small shunt resistance and a rather high series resistance. The reason for the existance of this small shunt resistance is not well understood; it can be considerably smaller than for a spin-on ARC.
Apparently both shunt and series resistance are a strong function of the baking temperature during the treatment before the Ag printing.

5.2. Recent results

The next table (V) shows a comparison between the results obtained with several types of antireflection coatings, used for our fired-through standard process.

	V_{oc}(mV)	J_{sc}(mA/cm^2)	FF(%)	η(%)
evaporated Ta_2O_5	573	28	67.1	10.8
screenprinted Ta_2O_5	568	27.5	71.5	11.2
spin-on TiO_2	569	28.7	72.0	11.8
No coating	552	17.5	68.7	6.6

Table V : Cell parameters for different antireflection coatings (average of 8 cells, 3.24 cm^2, AM1, 28°C)

The data in Table V indicate that the spin-on TiO_2 coating results in an average efficiency gain of a factor 1.78 over the cells without any antireflection coating. This is a result of the higher short circuit current caused not only by the lower reflectance, but also by the longer diffusion length (resulting from the creation of a barrier against the fast diffusers) and of the better fill factor (lower contact resistance and better 'second' diode quality). These points have extensively been discussed in the previous sections.

It is also interesting to note that the firing through an evaporated ARC does not yield the same large FF; it has not been determinated yet whether this is a result from a higher series resistance or from a poorer 'second diode' quality.

The initial results using a screen printed Ta_2O_5 coating are encouraging as can be concluded from table V. The best cell with a screen printed ARC (Ta_2O_5) had an efficiency of 12.3%, a V_{oc} = 580 mV, a J_{sc} = 28.9 mA/cm^2, a FF = 73.4% (AMI, 3.24 cm^2 cell).

However the main problems that remain were the lack of reproducibility and the fact that the average efficiency for large batches is considerably smaller than for cells with a spin-on ARC. We hope that using the new screen printing equipment, that has just been installed, to increase the repeatability of the process.

6. Implementation of the integral screen printing process on potentially low cost industrially available silicon

A comparison of 3 potentially low-cost materials for use with the integral screen printing process has begun. Our aim is to utilize baseline processing established for single crystal crystal silicon through cell characterization and material evaluation, optimize the printing process for non-conventional materials. The materials being studied included;
1. Silso Blanks, 10 x 10cm, average thickness 450μm, 5 ohm-cm; p-type, supplied by Wacker;
2. HEM : Heat Exchanger Method grown large grain polycrystalline silicon, 10 x 10cm, average thickness 400μm, 1 ohm-cm, p-type, supplied by Crystal Systems;
3. MHO : Dislocated single crystal silicon, CZ grown <111>, 94 cm^2 cut semi-square from a 4.5 inch crystal, average thickness 460μm, 2 ohm-cm, p-type, 5000 and 50000 dislocations/cm^2, supplied by Metallurgie Hoboken Overpelt.

The best results obtained on these materials with the integral screen-printing technique are shown in Table VI.

Material	Area (cm^2)	Voc (mV)	Jsc (mA/cm^2)	FF (%)	Efficiency (%)
Wacker	3.2	545	26.1	70.7	10.1
	100	554	22.6	69.3	8.8
HEM	3.2	554	26.4	72.5	10.6
	100	564	23.1	67.4	8.9
MHO	3.2	561	27.8	75.7	11.8
(50000)	86.3	551	25.4	70.1	9.9
(5000)	86.3	569	26.6	73.8	11.2
Monocrystal	78.5	595	29.2	72.5	12.7

Table VI : Best results on low cost materials, based on total cell area, 28°C, AMI (Spin-on ARC)

Reasonable efficiencies on all materials have been achieved in small batch quantities. Efficiency distributions are similar to the single crystal cells but with a lower efficiency. However, the process parameters and ink compositions have to be modified since their optimal values differ from those on single crystalline silicon. The initial surface preparation and the firing procedure are more critical for polycrystalline silicon materials than for single crystal material.

7. Implementation of the integral screen printing process on thin film silicon

With thin film silicon solar cells it is possible to achieve reasonable efficiencies because more than 70% of the incident light will be absorbed in the first 10 microns. We have used upgraded metallurgical (UGM) silicon as substrates; the UGM substrates have been obtained starting from conventional 98% pure MG silicon that has been purified by acid leaching to a level of 99.9%. This purified MG Si is then CZ pulled; during this operation further upgrading occurs due to segregation. Due to the low segregation coefficient of B and the high initial Al concentrations, UMG Si will be strongly p^+. We tried several different cell structures that are listed below.
1. First we tried to make cells directly into the UGM silicon
2. The second structure is a front surface (FSF) cell obtained by growing an n-type epi-film onto the p^+ substrate, followed by the diffusion of a shallow n^+ layer yielding an $n^+ n p^+$ structure. This structure will only yield reasonable efficiencies if the minority carrier diffusion length is larger than the epitaxial layer thickness.
3. The third structure is a back surface field cell (BSF) obtained by growing a p-type epi-film onto the p^+ substrate, followed by the same shallow n^+ diffusion as described above.

Results, obtained using the integral screen printing process to fabricate these structures, are given in Table VII.

Substrate	Voc mV	Jsc mA/cm^2	FF %	η %
UMG, no epi	257	9.7	49.9	1.2
UMG, n type epi	565	23.9	66.5	9.0
UMG, p type epi	516	24.0	66.1	8.2

Table VII : Screen printed solar cells on epi-grown films
Epi-thickness is 15 μm, cell area : 3.24 cm^2

The following conclusions can be drawn from table VII :
1. Making cells directly into the UMG substrates yields poor results
2. Rather surprisingly the front surface field structure with an n-type epi yields good results; this indicates that the junction properties are acceptable although the junction is formed at the epi-MG substrate interface. It also indicates that the diffusion length indeed exceeds the epi thickness.
3. Our p-type epi process is not optimized yet.

8. Conclusions and future work

Solar cells made on single crystal silicon in small batch quantities (400 wafers) using the printing technique for junction and electrode deposition have mean efficiencies over 11% with good yields and reproducibility.

Our initial work indicates that the glass frit is very critical for the optimization of the front electrode metallization as well as for the formation of a BSF. Printing the ARC, followed by a firing through operation is possible but more critical than the spin-on technique, especially with the rather poor screenprinter that we used.

Cell processing that is optimized for single crystal silicon is not directly transferable to polycrystalline silicon materials. The initial surface preparation and the silver ink firing temperature are more critical for polycrystalline silicon materials. Our initial results indicate that there is a reduction in cell efficiency of about 2 to 3 absolute percent when implementing the screen printing fabrication technique on polycrystalline materials. This loss factor is similar to that when other more convential fabrication technologies are implemented on the same materials.

Initial results on epitaxially deposited thin films on upgraded metalurgically grade silicon obtained with the screenprinting technology are encouraging since efficiencies of over 10% seem possible (9% has been reached).

Future works will include :

1. The achievement of a better understanding of the front electrode contact formation.
2. Continue using screen printing technology with thin film silicon such as thin epitaxial layers on upgraded MG.
3. Small production batches on large area 100 cm^2 cells on polycrystalline silicon and on defected single crystal will begin.
4. Continuation of grain boundary studies by implementing a hydrogen passivation technique for polycrystalline silicon cells.

REFERENCES

1 A.D. Haigh, Int. Conf. on Solar electricity, Toulouse, 1976, p. 183.
2 K. Firor and S. Hogan, Solar Cells, 5(1981-1982), p. 87.

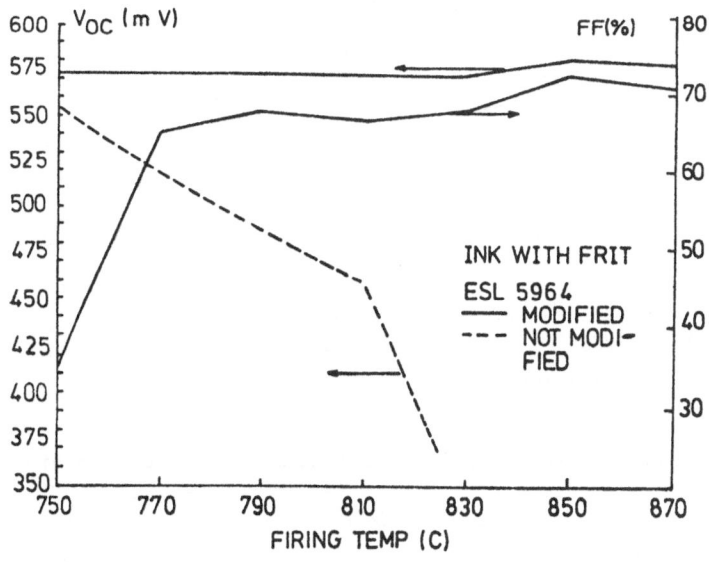

Fig. 1 :
Voc and FF for
inks with frit:

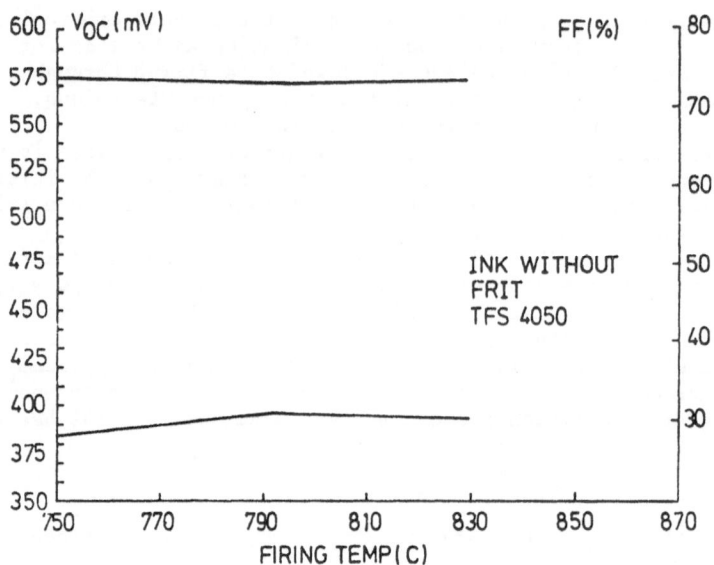

Fig. 2 : Voc and FF for a fritless ink.

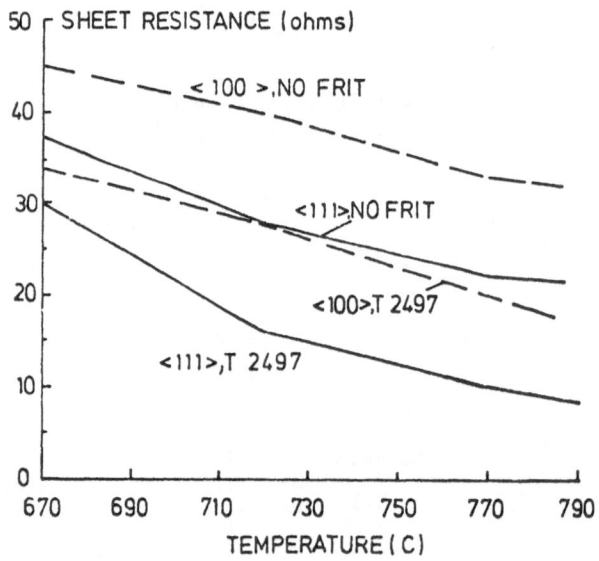

Fig. 3 : sheet resistance of the p$^+$ BSF layer versus firing temperature

STUDY OF MONO OR POLYCRYSTALLINE SOLAR CELL PROCESS

USING SCREEN PRINTING TECHNOLOGY

Authors   : J. DONON, H.LAUVRAY, P. LOUBLY, P. AUBRIL

Contract number   : ESC-R-019-F

Duration   : 18 months   1 July 1980 - 31 December 1981

Head of project   : J. DONON (PHOTOWATT International S.A.)
                       D. DIGUET (RTC)

Contractor   : PHOTOWATT International S.A.

Address   : 125, rue du Président Wilson
                       F92303 LEVALLOIS PERRET

Summary

The objectives of this contract were to develop a cost effective process
for solar cell manufacturing. Dry etching technologies and general use
of screen printing have been investigated. Plasma etching has been used
for surface sawing damage removal, edge leakage current reduction (ring
etching), selective oxide etching. Diffusion from screen printed doping
layers has been used for single step or two step structure.
The premilinary application of the above technologies to semicrystalline
material was done. Economical analysis showed that some of the above
process can be used now by the solar cell manufacturers.

1. Introduction

L'objectif de cette étude consiste à étudier un procédé de fabrication économique de réalisation des cellules photovoltaïques par l'intermédiaire de procédés secs et par la généralisation de la sérigraphie.

L'étude a permis d'étudier d'une part les décapages plasma en surface pour éliminer la couche perturbée dûe au sciage et la couche d'oxyde de diffusion, sur les bords pour éliminer la jonction et supprimer ainsi la fuite de bord.

D'autre part la réalisation de la jonction par l'intermédiaire de la diffusion du phosphore contenu dans une encre déposée par sérigraphie.

Ces techniques ont été utilisées pour les matériaux mono-cristallins et polycristallins.

2. Etude des décapages plasma

Le décapage plasma peut être utilisé pour trois opérations différentes :
. Elimination de la zone perturbée dûe au sciage,
. Ouverture de la jonction située sur la tranche de la plaque,
. Décapage sélectif de la couche d'oxyde.

Les gaz utilisés classiquement pour la gravure plasma sont le tétrafluorométhane, l'hexafluorure de soufre ou du tétrafluorométhane plus oxygène : $4 CF_4 + Si \rightarrow 2 C_2 F_6 + Si F_4$

En réalité plusieurs espèces réactives se forment telles que F, CF_3 mais celles-ci sont absorbées par les surfaces de silicium avec lequel elles réagissent chimiquement à nouveau : $CF_3 + F \rightarrow CF_4$ et $2 CF_3 \rightarrow C_2 F_6$

La concentration en $C_2 F_6$ est le principal facteur de vitesse d'attaque en tant que générateur de l'espèce fluor $2 CF_4 \rightarrow C_2 F_6 + 2F$.

L'oxygène introduit dans le CF_4 réduit le taux de recombinaison des espèces actives et permet une augmentation de la vitesse de gravure.

2.1 Elimination de la zone perturbée dûe au sciage

La première opération qui doit être faite lors de la fabrication des cellules, consiste en l'élimination de la zone perturbée dûe au sciage du lingot en plaques.

Pour le sciage classique à l'aide d'une scie à lame diamantée, il faut éliminer environ 15 um par face, pour un sciage futur à l'aide d'une scie à fil 5 um par face est suffisant.

Le tableau I suivant donne les vitesse de décapage des différents procédés :

TABLEAU I		$\langle 100 \rangle$	$\langle 111 \rangle$	POLYCRISTAL
Voie humide	Acide	13 μm/mn	13 μm/mn	13 μm/mn
	Base	1 μm/mn	3 μm/mn	3 μm/mn
Plasma $CF_4 + O_2$	$CF_4 + O_2$	0,4 - 0,5 μm/mn	0,3 - 0,5 μm/mn	0,3 - 0,4 μm/mn

A partir de ces premiers résultats les paramètres plasma ont été étudiés afin d'optimiser les vitesses d'attaque : nature du gaz, puissance RF, débit de gaz, temps de décapage ont été étudiés et font l'objet du Tableau II :

TABLEAU II		150 cc/min			200 cc/min			250 cc/min		
	15 min	x 0,58 µm/mn			0,3	0,56	0,36	0,33	0,4	0,43
CF_4 + 10% O_2	25 min		x		x 0,38	x 0,53	x 0,35	x 0,34	x 0,4	x 0,43
	35 min		x		x 0,34	x 0,4	x 0,35	x 0,33	x 0,4	x 0,43
	15 min				x * 0,36	*		x	* 0,28	x
CF_4 + 4% O_2	25 min				x * 0,32	*		x	* 0,28	x
	35 min				x * 0,34	*		x	* 0,44	x
x Si mono ⟨100⟩ * Si Poly		500 W	650 W	750 W	500 W	650 W	750 W	500 W	650 W	750 W

Les résultats montrent qu'il n'y a que peu d'influence du débit sur la vitesse d'attaque à faible puissance. Pour la puissance à débit donné la vitesse d'attaque augmente avec la puissance puis diminue, car le débit est alors trop faible pour cette puissance.

Le temps est un paramètre important, de même que la quantité d'oxygène qui doit être élevée.

Des cellules ont été réalisées les plaques étant soit en position verticale, soit en position horizontale dans l'appareil de décapage ; le tableau III donne les résultats obtenus en valeur moyenne.

TABLEAU III	Isc mA	Voc mV	Im mA	FF %	η %	IR - 400 mV mA	IF + 400 mV mA
Mono ∅ 100	2095	575	1950 à 0,455 V	73,5	11,3	14	24
Semi ∅ 100	2250	535	2000 à 0,43 V	70	8,4	10	30

2.2 Ouverture de la jonction sur la tranche

Les premiers essais ont été effectués sur des structures N+PP+ obtenue sur le silicium monocristallin du plan ⟨111⟩ ; les tranches de silicium sont empilées les unes sur les autres.

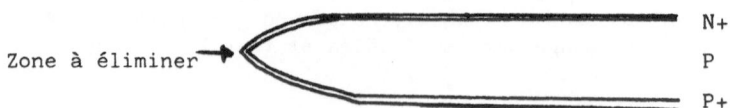

Zone à éliminer ➤

N+

P

P+

L'appareil plasma utilisé est un appareil de type Barrel de marque
IPC 220 T et les paramètres utilisés sont les suivants : $CF_4 + 4\% O_2$

P = 150 W
Pression 1 tour

Les caractéristiques électriques des cellules obtenues sont sur les
Figures 1 à 4.

Les essais suivants ont permis de vérifier la possibilité de décaper
150 à 200 plaques à la fois et la reproductibilité du procédé.

Les plaques de silicium utilisées ont été des plaques <100> décapées
soude pour éviter l'effet lentille du décapage acide qui donne une légère
attaque sur la face avant. Un appareil plasma de marque différente a été
utilisé, qui permet le décapage d'empilement de 200 plaques. Les paramètres utilisés pour le décapage ont été optimisés sur de plus faibles
quantités : $CF_4 + 10\% O_2$

P = 850 W
Débit de gaz 200 cc/nm
Temps de décapage 15 mn

Les essais effectués ont montré qu'il n'y avait pas d'influence du
nombre de plaques mises pour le décapage et que l'uniformité du décapage
entre le haut et le bas de l'empilement était bonne.

La reproductibilité du décapage a ensuite été vérifiée sur plusieurs
lots de 200 tranches de silicium.

Enfin il a été vérifié que le décapage pouvait être effectué après
diffusion et avant pose des contacts ou bien après diffusion et après
pose des contacts.

2.3. Décapage de l'oxyde de four

Le décapage plasma pourrait remplacer le traitement à l'acide
fluorhydrique réalisé après la diffusion.

Les essais ont été effectués avec les paramètres suivants :
$$CF_4 + 10\% O_2 \quad P = 850 W$$
Temps de traitements 2 à 16 minutes.

Les résultats ont montre que seuls les temps de décapage courts permet-
taient de réaliser des cellules ; pour les temps plus longs, la jonction
est détruite par le traitement Tableau IV.

TABLEAU IV		Plaque entière	1/2 droite	1/2 gauche
	Icc mA	1460	530	930
Temps 2 mn	Vco mV	545	525	550
	FF %	33	28	36

Des essais avec des temps de décapage plus cours ont été effectués et ont
donné des résultats légèrement meilleurs que les précédents. (FF = 45 à 60%)

Néanmoins l'uniformité de l'attaque n'a jamais pu être obtenue entraînant la présence de zone où l'oxyde de four n'est pas encore totalement attaquée et des zones ou celui-ci est totalement éliminé et où la jonction elle-même est percée.

3. Diffusion à partir d'une source diffusante sérigraphiée

L'objectif consiste à réaliser des structures N/P à partir d'une source diffusante déposée par sérigraphie puis diffusée sous gaz oxygène :
. Simple jonction,
. Double jonction, dont une localisée et profonde sous les contacts (contacts Aluminium).

La source dopante utilisée consiste en un mélange du produit N250 Emulsitone avec des liants et solvants appropriés pour obtenir une viscosité adaptée :

Proportion en poids du véhicule organique : 90 % Terpineol
10 % ethyl cellulose

Le mélange final est composé de 50 % de véhicule organique et de 50 % de N250.

Le procédé de réalisation des cellules est le suivant :

Nettoyage - décapage - sérigraphie de l'encre diffusante - diffusion - sérigraphie des encres de contacts - traitements thermiques - décapage des bords (plasma) - dépôt de la couche anti-reflet (spray) - test.

Le dépôt de la source diffusante est réalisée par sérigraphie à travers un écran en acier de 200 mesh trame orientée à 45°, pour les simples jonctions recouvrement à 100 % de la surface. Pour les doubles jonctions : premier dépôt d'encre diffusante localisée sous forme d'un peigne aux dimensions légèrement supérieures au peigne qui formera le contact N - diffusion profonde $x_j = 2\ \mu m$ 1h 30 950°C $R_\square = 10/15\ \Omega/\square$ - deuxième dépôt sur 100 % de la surface - diffusion normale $x_j = 0,6\ \mu m$ 1h 30 850°C $R_\square = 30\ \Omega/\square$.

Les résultats obtenus sont les suivants en simple jonction.

	η%	11,8	12	IR mA - 400 mV	7	4,8
MONOCRISTAL Valeurs moyennes	FF%	72,7	73	IF + 400mv mA	11,2	9,3
	Tableau V	Acide	Soude		Acide	Soude

	η%	8,4	8,7	IR mA - 400 mV	10	5,5
SEMI CRISTAL	FF	72,3	74,5	IF mA + 400 mV	28	23
		N250	$POCl_3$		N250	$POCl_3$

Les essais de fiabilité réalisés en vapeur d'eau bouillant sur des cellules faites avec le N250 ou le $POCl_3$ ne montrent pas de différence entre les deux types de diffusion.

Les résultats obtenus en double jonction avec des contacts aluminium n'ont pas abouti car l'aluminium perce la jonction de 2 um, d'autre part le positionnement relatif de la grille sur la jonction profonde est difficile à réaliser. Les photographies 1 et 2 montrent la coupe d'une diffusion localisée profonde ainsi que la vue de la superposition des motifs.

La meilleure cellule obtenue avec cette technique a permis un rendement de 9 % avec un facteur de forme 65 %.

4. Etudes économiques préliminaires
4.1 Décapage des bords

Le tableau VI donne un comparatif des différentes techniques utilisables pour l'ouverture de la jonction sur les bords.

TABLEAU VI	Investissement pour 1 MN	Matière	MØ
Décapage plasma	150 KF	0,76 Fr/pl	0,05 Fr/pl
Silox et attaque chimique .	1000 KF	0,07 Fr/pl	0,7 Fr/pl
Masquage par sérigraphie protectrice et attaque chimique	200 KF	1,65 Fr/pl	0,5 Fr/pl

4.2 Décapage de surface

Les appareils existant à l'heure actuelle ne permettent pas de faire ce type de décapage de façon économique car les capacités sont trop faibles, une première approche est donnée dans le tableau VII suivant.

TABLEAU VII	Investissement pour 1 MW	Matière	MØ
Décapage plasma	600 KF	0,10 Fr/pl	0,20 Fr/pl
Acide	800 KF	2,00 Fr/pl	0,15 Fr/pl
Soude	200 KF	0,27 Fr/pl	0,15 Fr/pl

4.3 Sources dopantes

Du point de vue économique les sources dopantes de sérigraphie ne sont pas encore rentables : 1800 Fr/Kg avec objectif 900 Fr/Kg en grosse quantité soit 0,36 et 0,18 Fr/plaque ce qui est à comparer à 0,06 Fr/plaque pour la diffusion $POCl_3$ classique. D'autre part l'investissement pour une capacité de 1 MW sera double.

5. Conclusion

Cette étude a permis de démontrer la faisabilité concernant la réalisation de cellules à partir de procédés secs :

Décapage plasma des zones perturbées dues au sciage,

Ouverture des jonctions par plasma,

Dépots des dopants par sérigraphie.

Tous ces points sont à l'heure actuelle techniquement vérifiés mais pas toujours économiquement rentables.

Le décapage de l'oxyde de four n'a pu être réalisé correctement par plasma du fait de la sélectivité trop faible et d'une uniformité d'attaque insuffisante.

Des dépot dopants par sérigraphie réalisés en deux étapes et permettant de faire une jonction profonde sous le contact n'ont pas donné satisfaction avec des contacts aluminium par le fait d'une diffusionde l'aluminium trop importante et d'un alignement trop difficile à réaliser.

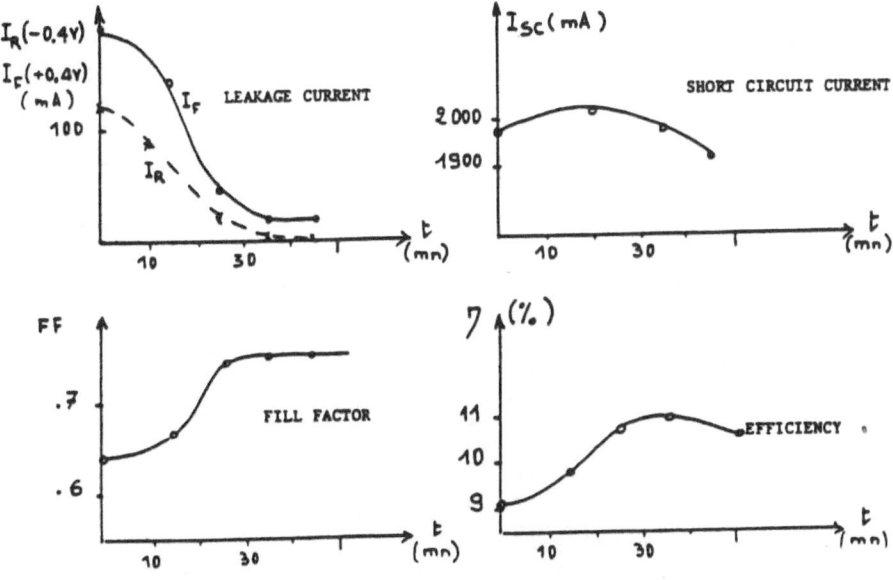

Figures 1 à 4

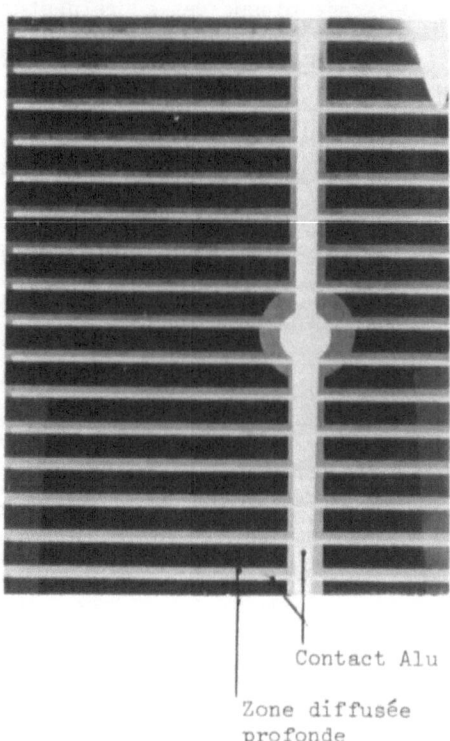

PHOTO 1

Contact Alu

Zone diffusée
profonde

PHOTO 2

Coupe d'une diffusion localisée profonde

TECHNOLOGY OF AMORPHOUS SILICON THIN FILMS FOR SOLAR CELLS
AND
APPLICATIONS TO POWER SYSTEMS

Authors   : Part I. D.Girginoudi, A.Thanailakis, P.Abarian and J.G.
                       Antonopoulos
                     Part II.G.Vachtsevanos, K.Kalaitzakis and E.Gribas

Contract number   : ESC-R-079-GR(B)

Duration   : 18 months   1 January 1982 - 30 June 1983

Total budget   : DRA 7.117.250   CEC Contribution: DRA 1.860.000

Heads of project   : Prof. Anthony Thanailakis
                     Prof. George Vachtsevanos

Contractor   : Democritus University of Thrace, School of Engineering

Address   : School of Engineering,
                     Laboratory of Electric Circuits Analysis
                     Democritus University of Thrace,
                     Xanthi - Greece.

Summary

Preliminary attempts have been made to grow non-hydrogenated a-Si films by
evaporation in an ordinary vacuum system (ultimate pressure 10^{-6} Toor) fitted with a six-hearth electron gun and a substrate heater. This divergence
from the original proposal was due to the fact that our RF sputtering system has not as yet been completed. The substrate temperature was varied
from room temperature up to 400 ^{O}C, and the deposition rate for the a-Si
films in the range of 5 to 40 Å/s. Studies of the film structure, using
transmission electron microscopy, showed the characteristic features of a-
morphous material. n-type, p-type or intrinsic a-Si films were obtained when
the evaporation source contained doped (n-type or p-type) or undoped polly-
crystalline silicon respectively.
    Regarding the second major research area of this program a system si-
mulating the voltage current output characteristics of a photovoltaic array
was designed and tested succesfully. The simulator has the advantages of
low implementation cost and circuit simplicity. It is usable in the kilowatt power range.
    Last, a maximum power transfer tracker was designed, constructed and
tested. This system ensures the best dynamic matching between a PV array
and the utility grid, making possible the maximum power transfer under all
conditions. It is making use of relatively simple electronic circuitry and
is implemented easily.

1. Introduction

There is a great interest in a-Si, because of its potential as a material suitable for the manufacture of low-cost photovoltaic solar cells [1].

Kilgore and Roberts [2] prepared, in 1963, a-Si thin films by resistive heating of a single crystalline silicon wire in a UHV system (pressure $\approx 5 \times 10^{-10}$ Torr), in order to study the chemical behaviour of the film surfaces. Kaplan et. al [3] have also prepared a-Si films, by vacuum evaporation (pressure better than 10^{-8} Torr and evaporation rate about 30 Å/s), which subsequently they hydrogenated, by heating in a hydrogen plasma, achieving electrical properties similar to those of a-Si:H prepared by glow-discharge decomposition of silane. Recently, Schubert et.al [4] prepared photovoltaic solar cells of n-i-p n-,....i-p and p-i-n p.....n type, by e-gun high vacuum (pressure about 10^{-6} Torr) evaporation of a-Si on stainless steel substrates and subsequent hydrogenation by heating at $400°C$ for 1 h in an atmosphere of molecular hydrogen.

There is no publishe work, as far as we know, on photovoltaic solar cells of Schottky diode type, using a-Si prepared by vacuum evaporation. Moreover, the relevant solar cell theory for a-Si and a-Si:H is at present far from being complete, whereas the maximum reported thus far cell efficiency does not exceed the value of 8% [5,6]. There is, therefore, a wide field of research in the area of photovoltaic solar cell technology using amorphous silicon films.

This paper is a contribution in this direction, presenting preliminary experimental results obtained from structural studies of a-Si thin films, prepared by vacuum deposition on different substrates and different substrate temperatures, using transmission electron microscopy techniques.

2. Experimental

The starting material, used for the preparation of a-Si films studied in this work, was polycrystalline silicon of 99.9999% purity or single crystalline silicon of n or p-type. Before use, the silicon was cleaned successively with trichloroethlylene, acetone and methanol. Then it was etched in HF, washed in deionised water and placed in one of the six cermet hearths of an electron gun, which was fitted in the chamber of a high vacuum system (pressure about 5×10^{-6} Torr). The substrates for the deposition of a-Si films were placed at different distances above the e-gun evaporation source, which was always operated under identical power supply conditions. Films of a-Si for thickness measurements and structural studies were prepared simultaneously with films of a-Si suitable for Schottky diodes. The thickness of a-Si films, used in structural studies, was not allowed (by means of a shutter) to exceed 0,4 μm, in order to be transparent to the electron beam of the transmission electron microscope. A substrate heater and a substrate temperature measuring device were constructed and fitted into the same vacuum chamber. The substrates used for the deposition of a-Si were stainless steel or aluminium films deposited (by vacuum evaporation) on glass. The substrate temperature was varied from room temperature up to 400 °C. The film thickness was measured using a multiple beam interferrometer, and the deposition rate was calculated from the measured thickness and the duration of the evaporation. The structural studies were carried out using the JEOL 120CX electron microscope by means of bright field, dark dield and diffraction techniques.

3. Results and Discussion.

The results of measurements for the callibration of the a-Si film de-position rate as a function of the source-substrate distance, at various substrate temperatures, are shown in figure 1. These curves were used in this work in order to prepare silicon films of predetermined thickness, in the range of 0.2 - 1.4μm.

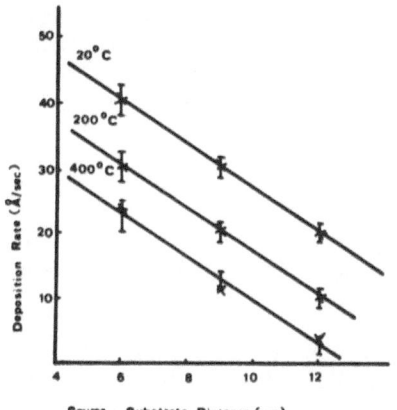

Figure 1. Silicon deposition rate ver-sus source-substrate distance, at different substrate tem-peratures.

The silicon films studied using transmission electron microscopy tec-hniques showed the characteristic diffraction pattern of amorphous mate-rial, irrespective of the nature and temperature of the substrate and of the deposition rate. Typical photographs of the grown silicon films are shown in figure 2. We can therefore, grow a-Si films on to which we can subsequently prepare Schottky diodes as solar cells.

(a)   (b)

Figure 2. Typical photographs of vacuum deposited silicon films
          on stainless steel substrate.
          (a) A low magnification bright field image showing the
              general appearance of the films.
          (b) The corresponding diffraction pattern, characteristic
              of an amorphous material.

The conductivity type of the above grown a-Si films was determined by means of the Hall effect and was found to be n-type, p-type or intrinsic, respectively, when the starting silicon (under evaporation) was single cry-stalline n-type or p-type silicon or pure (undoped) polycrystalline silicon.

The bulk of our work, up to now, which is presented in this paper, is concerned with the preparation of a-Si films and the study of their structure.

4. Conclusions

The silicon thin films prepared by vacuum deposition on stainless steel and aluminium-on-glass substrates were found to be amorphous, irrespective of the substrate temperature in the range of room temperature-400°C and of the deposition rate, in the range of 5-40 Å/s. The conductivity of these films was found to be n-type, p-type or intrinsic, respectively, when the starting material under evaporation was n-type silicon, p-type silicon or pure (undoped) polycrystalline silicon.

References

[1]. Hovel, H.J. "Photovoltaic materials and devices for terrestrial solar energy applications", Solar Energy Materials, Vol.2, No3, pp 277-312 (April-June 1980).

[2]. Kilgore, B.F., and R.W. Roberts "Preparation of evaporated silicon films", The Review of Scientific Instruments, Vol 34, No 1, pp 11-12 (January 1963).

[3]. Kaplan, D., N. Sol, and G.Velasco "Hydrogenation of evaporated amorphous silicon films by plasma treatment", Appl. Phys. Lett., Vol 33, No 5, pp 440-442 (September 1978).

[4]. Schubert, C.C., P.H.Fang and J.H.Kinnier "Electron-beam evaporated amorphous silicon multiple cascade solar cells", Japanese Journal of Applied Physics, Vol 20, No 6, pp 437-438 (June 1981).

[5]. Krühler, W, M.Möller, R.Plättner, H.Pfleiderer and B.Rauscher "Stability of amorphous silicon solar cells", Proc. Photovoltaic Solar Energy Conferrence, organised by the Commission of the European Communities, Stresa, Italy (May 10-14, 1982).

[6]. Tawada, Y, K. Tsuge, M.Kondo, H.Okamoto, and Y. Hamakawa, " 8% Efficiency a-SiC:H/a-Si:H heterojunction solar cells", Proc. Photovoltaic Solar Energy Conference, Organised by the Commission of the European Communities, Stresa, Italy (May 10-14), 1982).

PART II

Introduction

The research effort during the period of 1/1-30/6/1982 was concentrated upon the following issues: First, design, construction and testing of a photovoltaic simulator. A novel design of a PV array simulator was undertaken using a readily available dc source such as a battery bank or a dc power supply and appropriate electronic waveform conditioning apparatus. The system simulates the voltage-current characteristics of a photovoltaic array and was designed as a respond to the need for experimenting with photovoltaic systems capable of delivering electrical power in the kilowatt range with different voltage-current characteristics. The main reason that limits considerably the possibility of experimenting with actual devices is their relatively high cost.

The block diagrammatic form of the proposed PV simulator is shown in Fig. 1.

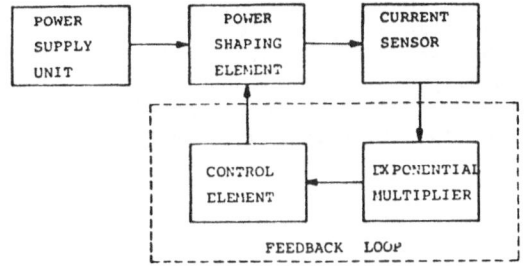

Figure 1. Block diagram of the photovoltaic
simulator.

The system consists of (i) a d.c. Power Supply Unit (PSU) (ii) a Power Shaping Element (iii) a Current Sensor (iv) an Exponential Multiplier and (v) a Control Element.

Secondly, initial design, construction, and testing of appropriate power conditioning apparatus for the inter connected operation of PV arrays with the utility grid were undertaken. As a result of this activity a conceptual design of a power conditioning system with a provision for maximum power tracking was developed and implemented.

The design was based upon a dual methodology - a theoritical system simulation study of the interconnected PV array-grid system and parallel experimental testing of actual prototypes.This dual approach has resulted in optimum energy transfer efficiencies with an increased quality and reliability of performance of the PV array-utility grid system. Figure 2 shows a conceptual block diagram of the proposed scheme.

For any solar intensity H, the array output feeds into the inverter via a PV interface circuit which assists in improving signal quality while maintaining an approximate "static" matching between the generator and load impedance characteristics.The feedback loap maintains a dynamic matching between the array and the grid.

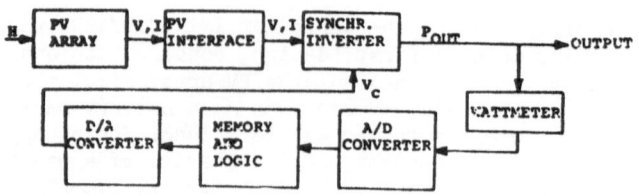

Figure 2.

Results and conclusions

Typical V-I characteristics obtained from an experimental simmulator model are shown in Figure 3(a). As it can be readily observed, there is a very good agreement between these characteristics and those of Figure 3(b) corresponding to an actual PV array as given by the manufacturer. The system may be scaled up to higher power levels by simply transforming the Power Shaping Element and the Current Sensor Circuit so that they can handle the increased voltage and power levels while the rest of the circuit remains essentially the same. The simplicity of the circuit design as well as the use of commercially available electronic components guarantee a low overall cost for the simulator; this is estimated to be of the order of $10 per Kilowatt for power capacities in the range of tens of Kilowatts.

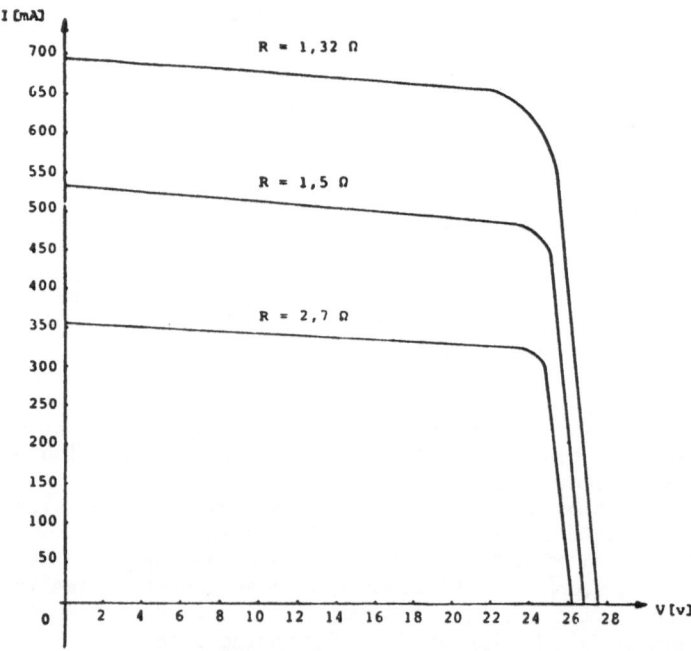

Figure 3(a). Experimental voltage - current characteristics of the model simulator.

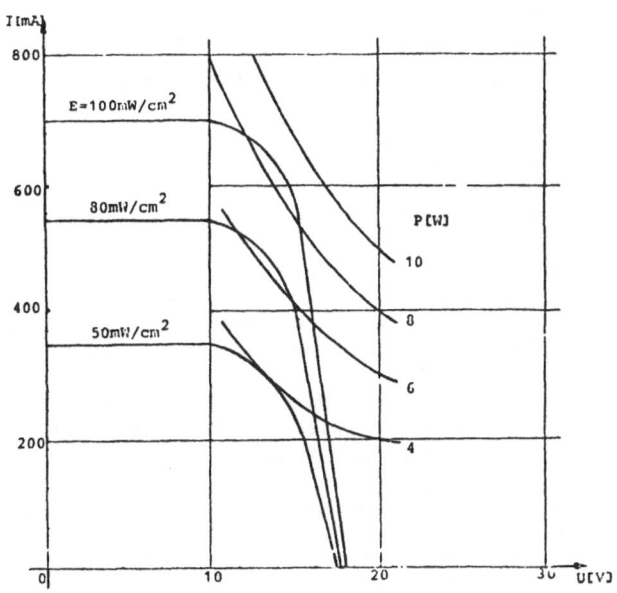

Figure 3(b). Voltage-current characteristics of a commercial
photovoltaic array.

Figure 4 shows a typical power output vs. solar radiation intensity
characteristic with and without the maximum power tracker in operation.

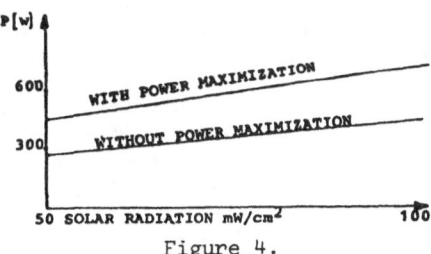

Figure 4.

- 35 -

ION IMPLANTATION WITHOUT MASS ANALYSIS AND LASER ANNEALING

FOR FABRICATION OF SILICON SOLAR CELLS

Authors   : J. COM-NOUGUÉ, J.C. MULLER

Contract number   : ESC-R-048 F

Duration   : 16 months - 1 July 1980 - 30 October 1981

Head of project   : Mr. J-P. DUMAS, assisted by Mr. J. COM-NOUGUÉ
                       Materials Department - Laboratoires de Marcoussis

Contractor   : Laboratoires de Marcoussis, Research Center of C.G.E.
                       in cooperation with CRN-STRASBOURG, PHASE Group,
                       (MM. SIFFERT and MULLER).

Address   : Laboratoires de Marcoussis - F-91460 MARCOUSSIS

                       CRN - Groupe PHASE, 23 rue du Loess
                       F-67037 STRASBOURG

The study is part of a general program supported in part by COMES, now AFME.

SUMMARY

The work outlined in the present report concerns the optimization
of an ion incrustation technique, that is, implantation without
mass separation, and pulsed YAG or continuous CO_2 laser annealing
for the production of junctions on silicon. The ion incrustation
process employed, developed by the Strasbourg CRN, is based on
gas discharge in a gas containing the dopant (PF_5). The machine
developed includes two work stations processing the wafers continuously. The first station performs doping, while the second can
perform activation annealing, e.g., by laser.
In the framework of the present contract, ion incrustation and
pulsed laser annealing were applied to the production of monocrystalline silicon solar cells. The photovoltaic efficiencies
obtained were up to 13.3 % AM1 on active areas of 10 cm^2. These
performances can be compared with those attained by cells manufacturing by ion implantation with mass separation and YAG laser
annealing. The process designed was subsequently applied to polycrystalline silicons : efficiencies better than 10 % AM1 were
achieved. The results show the advantages of ion incrustation,
which is a more simple and inexpensive technique than tradi-
tional ion implantation.
From a production point of view, pulsed annealing laser machines
are already available but their throughputs are not yet compatible
with a mass production. On the contrary, high power CO_2 lasers may
be considered to meet this goal. For this reason, continuous CO_2
laser annealing is presently studied and promising results have
been obtained on solar cells. The work in progress concerning
this technique is also reported.

1. INTRODUCTION

Ion implantation and surface annealing (laser, electron beam, incoherent light) are regarded as one means of bringing about a significant reduction in the cost of manufacturing solar cells (1). Several laboratories are working in this field. Most of them use low current implanters but different types of laser processing to activate the implanted dopant : Q-switched Nd: glass (2) or ruby laser (3) for example. More recently, R.F. WOOD et al. used a pulsed Excimer laser operating at .3 micron in order to produce implanted silicon solar cells (4). Polycrystalline silicon cells processed by ion implantation and laser annealing - pulsed ruby or CW argon - were reported to exhibit good efficiencies (5, 6).

CRN-Strasbourg developed a technique simpler than implantation with mass separation to form solar cell junctions. This process is based on the shallow implantation of ions produced by the D.C. glow discharge containing gas (ion incrustation). It must be noted that since the design of this process by CRN, other laboratories have developed similar techniques (4, 7, 8).

The work covered by the contract was the first phase of a project aimed at the optimization of the ion incrustation and laser annealing technique for the continuous production of junctions. This first phase was concerned with the achievement of mono. Si solar cells using the glow discharge technique and a Nd:YAG laser annealing. The objective was to produce 10 % AM1 cells. As a reference basis, cells were also processed by ion implantation with mass analysis.

Since the expiration of the contract, the research work has been relating to the elaboration of poly. Si solar cells and to a continuous CO_2 laser annealing technique. The results of this work in progress are also indicated in this report.

2. ION INCRUSTATION (implantation without mass analysis)

The technique employed is as follows. A gas discharge is created in a glass chamber in the presence of PF_5. Plasma is established at a voltage of 7 to 10 kV ; the ion beam thereby created is then accelerated to the substrates to be doped at a voltage of 10 kV. A prototype was developed at the CRN on the basis of this principle and a second unit was built in the Marcoussis Laboratories. The units consist basically of a doping module followed by a second module designed to receive a surface annealing device, e.g. a laser. These units can perform continuous processing of silicon wafers or ribbon which move under the ion beam coming from the source oscillating in a plane perpendicular to the silicon's movement. Figure 1 gives an overall view of the laboratory unit.

In order to plot the mass spectra of ions produced by the ion source, the source was mounted on the CRN laboratory ion accelerator with its own power supplies to ensure strictly the same discharge conditions. The spectrum for ions produced by discharge in PF_5 gas is shown in figure 2. The discharge current affects the temperature reached by the source, and in particular by the silicon anode. When the power to be dissipated is high (> 100 W), i.e. for discharge currents of around 14 mA, one observes a marked reddening of the anode and increased ionization which is more to the advantage of the lighter $^{31}P^+$ ions. On the other hand, for all discharge cases, the heavier molecular ions such as PF_4^+ and PF_5^+ are only slightly ionized.

3. LASER ANNEALING

3.1. Pulsed laser processing (liquid phase epitaxy)

The PHASE laboratory acquired (for the purposes of the present contract) a YAG laser (at present mounted with a frequency doubler $\lambda = 0,53$ µm) from the Company Quantronix (USA), Epitherm M 610 model. Concerning the tests reported hereunder, the process conditions were the following : pulse duration 100 nsec, spot diameter 100 microns, repetition rate 10 kHz. An adequate spot overlapping rate had to be determined to ensure an effective overlap of the energy density used in practice, which is usually double the fusion threshold energy. Figure 3 shows the evolution of the calculated melted thickness as a fonction of the laser energy density. In the case of the YAG laser, the spread of the melted zone depends slightly on the thickness of the amorphous layer, and the "threshold" energies are very similar for the two extreme cases : amorphous silicon - crystalline silicon - due to the high rate of light absorption at 0.53 µm. The curves obtained for the ruby laser (pulse duration 25 nsec) are given for purposes of comparison.

Residual damage subsists in the doped layer after implantation and laser annealing. Recent studies using DLTS techniques have shown that the observed defects are rather related to the ion implantation process. Practically, in the case of solar cells processed with the above techniques, a heat treatment at 600 °C allows a complete curing of the defects and brings about an improvement of the cells performances (9).

3.2. Continuous CO_2 laser processing (solid phase epitaxy)

At a 10.6 µm wavelength, the absorption mechanism in silicon essentially results from a free-carrier excitation that directly transforms the absorbed energy into heat (10). This absorption can be enhanced by a simultaneous illumination by photons of a convenient wavelength or by a moderate heating of the substrate during the laser treatment. In these conditions, a solid phase regrowth of the implanted silicon can take place by using a CO_2 laser operating in the continuous mode.

The experiments were performed with a CW CO_2 laser operating at a power of 15 W. The beam, 5 mm in diameter, was focused on the wafer holder which can be heated up to 500 °C. The scanning was achieved through X-Y displacements of the holder under the beam. The speed in the X direction was 1 mm/s and the displacements in the Y direction could be adjusted at the desired step. The power density was controlled by moving the focusing lens with respect to the sample. Figure 4 gives the scheme of the system.

We investigate the influence of the annealing parameters on the activation of implanted dopant, namely the beam power density, the substrate temperature and the beam overlapping. The annealed samples were then electrically characterized. Sheet resistance and local photoresponse measurements were used to investigate the irradiation effects (11).

The laser treatment induces a change of the surface colour. Local photoresponse indicate that this modification is connected to an activation of the dopant (figure 5). At energies lower than 4 kW/cm², solid phase recrystallization takes place on the whole width where a surface aspect change is induced. The size of this zone corresponds approximately to the spot diameter. When a melting of silicon occurs, the molten zone presents an important degradation of the collection efficiency, as can be seen in figure 5.

Sheet resistances (R/□) were measured on silicon wafers implanted with energies between 1 and 4 kW/cm² at various temperatures ranging from 250 to 450 °C. Figure 6 shows that i) R/□ decreases from values higher than 100 Ω/□ at 1.1 kW/cm² to 75 Ω/□ at 3.9 kW/cm², and ii) R/□ decreases from values of

700 $\Omega/\square$ at 250 °C to 100 $\Omega/\square$ or lower at 450 °C.

Concerning the overlapping rate, it was shown that an excessive value (~ 55 %) may cause a local degradation of the collection efficiency similar to the one observed in the case of a surface melting. More generally, the results show that the overlapping brings about no improvement of the collection efficiency. Consequently, the scanning can be performed with a very low overlapping rate.

4. APPLICATION TO SOLAR CELLS

4.1. Fabrication process

. Substrates

The substrates were wafers of diameter 40 mm, of type P monocrystalline silicon, resistivity 1 Ω.cm, oriented (111). The as-sawn wafers underwent chemical attack, and after junction preparation, a mesa attack representing a total area of 11 cm^2 and an active area outside the metallization area of 10 cm^2. Furthermore, polycrystalline silicon wafers were also processed.

. Implantation conditions
- PF$_5$: dose 5.10^{16}/cm^2 - energy : 10 KeV
- P$^+$: dose $2.5.10^{15}$ and 5.10^{15} ions/cm^2 - energy 10 and 20 KeV.

. Annealing
- YAG laser   : energy : 2-2.5 and 3 J/cm^2
          post-laser treatment : 500 to 600 °C
- CO$_2$ laser   : energy : 1.5 to 2.2 kW/cm^2
          temperature : 440 °C
          scanning rate : 1 mm/s
- heat treatment : 50 min at 800 or 850 °C with steps at 580 °C.

. Metallization and antireflective coating

Metallization of the back contact and the grid on the front surface was obtained by evaporation of a triple layer of Ti-Pd-Ag. An antireflective coating of N$_4$Si$_3$ was deposited on the front surface of the cells.

The cells so manufactured were subjected to sheet resistance, dark current and I(V) under AM1 illumination measurements.

4.2. Results

. Ion incrustation and YAG laser annealing :

Table I gives the various characteristics measured under AM1 illumination on monocrystalline silicon cells. The results show a great influence of the laser energy and of the post-laser heat treatment (9). The effects of these two parameters is more particularly observed on open circuit voltage Voc. Values higher than 580 mV were measured for a 2.5 J/cm^2 energy and a 600 °C treatment. In these conditions, short circuit currents Jsc greater than 30 mA/cm^2 (active area) and fill factor greater than 0.7 were usually achieved. The corresponding best efficiencies were 13.3 % AM1. Furthermore, an initial assessment has shown that the ion incrustation process used in this study was highly reproducible.

. Ion implantation :

Table II gives the main results obtained on cells processed by ion implantation of phosphorus ; the activation was obtained by different methods : heat treatment, pulsed annealing by YAG laser, and continuous annealing with a CO$_2$ laser. Concerning the heat treatment activation, the optimum results correspond to a dose of 5.10^{15} ions/cm^2 implanted at 20 KeV and were obtained after annealing at 850 °C. The performances of the best cells indicate a short-circuit current of 32.5 mA/cm^2 and an efficiency of 13.5% AM1.

Similar results were achieved in the case of a YAG laser annealing done with a 2.5 J/cm^2 energy : efficiencies of 13.2% (active area) were measured under AM1 illumination. Concerning CW CO_2 laser processing, table II shows that efficiencies higher than 12% AM1 can be achieved ; the cells were annealed at 440 °C with a laser energy between 1.5 and 2.2 kW/cm^2.

The various processes are now applied to polycrystalline silicon such as SILSO, HEM as well as silicon under development at Laboratories of Marcoussis. The results obtained up to now relate to cells processed in the same conditions of implantation or incrustation and pulsed or continuous laser annealing as for monocrystalline silicon. In the case of a YAG laser processing, the results show that efficiencies higher than 10% AM1 can be easily achieved with the different polycrystalline materials.

5. DISCUSSION - PROSPECTS

The most important point that emerges from the study concerns the ion incrustation technique : the results show that this process gives photovoltaic junctions as performant as junctions elaborated by ion implantation with mass analysis. This is an interesting point since incrustation has significant advantages by comparison with the latter process. The incrustation unit is of simpler design then the implanters employed in the semiconductor industry, the initial cost of an incrustation machine is lower and maintenance requirements are less onerous.

Another important point relates to pulsed and continuous laser processing. Comparison of tables I and II indicates that a pulsed annealing gives better performances and more precisely better Voc than continuous annealing. The difference must be related to the fact that a liquid phase epitaxial regrowth takes place during the short YAG-laser pulses - typically 100 nsec. This liquid phase epitaxy brings about an activation of the whole implanted dose and a perfect recovery of the damaged zone. In the case of the solid phase recrystallization induced by the CO_2 laser irradiation, Rutherford Backscattering analysis was done on As implanted (111) silicon annealed with an energy density between 2 and 4 kW/cm^2. The results indicate that 34 % of As is on the substitutional lattice sites and that the implanted zone still presents defects (χ_{min} (Si) $\simeq$ 10%).(100) oriented silicon would have probably shown a higher activation and a better recovery for the same conditions of implantation and annealing.

This point can probably be improved and it can be expected that implanted cells processed with a CW CO_2 laser will present performances identical to those obtained with a pulsed laser annealing. In particular, a higher activation level and a better recrystallization of the damaged zone could be achieved by using laser beams with high energy densities. Furthermore, high processing throughput rates could be so achieved. The calculation of the temperature rise during a CO_2 laser annealing of silicon has shown that temperatures higher than 1200 K can be reached in times shorter than 1 ms (12). Thus the use of high laser beam powers would allow throughput rates higher than 10 m^2/h. This type of annealing in conjunction with ion incrustation can also be regarded as a mean of achieving significant reductions in solar cell production costs.

REFERENCES

(1) H. GOLDMAN and WOLF - 14th IEEE Phot.Spec.Conf. SAN DIEGO (1980)p.923.

(2) J.S. KATZEFF, M. LOPEZ, R.H. JOSEPHS - 3rd EC Phot.Solar Energy Conf. CANNES (1980) p. 708.

(3) R.T. YOUNG, R.F. WOOD, G.E.JELLISON, W.H. CHRISTIE in ref.2, p. 703.

(4) R.F. WOOD, R.T. YOUNG, G.A. van der LEEDEN, R.D. WESTBROOK - 16th IEEE Phot. Spec. Conf. SAN DIEGO (1982).

(5) R.T. YOUNG, R.F. WOOD, J. NARAYAN and C.W. WHITE - Laser and electron beam processing of materials (1980) - Academic Press, p. 651.

(6) G.F. GIBBONS et al., 3rd EC Photovoltaic Solar Energy Conference, 27-31 Oct. 1980, CANNES.

(7) L.D. NIELSEN and P. BALSLEV in ref.1, p. 698.

(8) M.D. SIRKIS and D.L. SALTZMAN - 15th Photovoltaic Specialists Conf. (KISSIMMEE) 1981, p. 981.

(9) J.C. MULLER, A. MESLI, P. SIFFERT, J. COM-NOUGUÉ, C. TESSARI, J.P.DUMAS 4th EC Photovoltaic Sol. Energy Conf. STRESA (1982), p. 994.

(10) W. SPITZER and H.Y. FAN - Phys. Rev. 108, 167 (1959).

(11) J. COM-NOUGUÉ, C. TESSARI, E. AUGARDE, J.C. MULLER - 16th IEEE Phot. Spec. Conf. SAN DIEGO (1982).

(12) Results to be published.

TABLE I : ION INCRUSTATION - YAG LASER ANNEALING

LASER ENERGY (J/cm^2)	POST-LASER ANNEALING	Voc mV	Jsc mA/cm^2	FF	η % AM1
2	550 °C	564	32.4	.64	11.6
2.5	550	548	31.1	.56	9.5
	600	582	30.8	.71	12.7
3	550	576	30	.71	12.2
	600	585	32.6	.7	13.4

TABLE II : ION IMPLANTATION

PHOSPHORUS IMPLANTATION	ANNEAL	R/□ Ω/□	Voc mV	Jsc mA/cm^2	FF	η % AM1
1°) Thermal annealing						
$2.5.10^{15}/cm^2$ - 10 KeV	850°C	57	576	30.3	.71	12.4
$2.5.10^{15}/cm^2$ - 20 KeV	850	51	560	30.7	.73	12.5
$5.10^{15}/cm^2$ - 10 KeV	800	89	576	33.3	.56	10.6
	850	46	574	30.3	.67	11.7
$5.10^{15}/cm^2$ - 20 KeV	800	73	568	31.4	.58	12.2
	850	39	580	32.2	.72	13.5
2°) YAG Laser annealing 2.5 J/cm^2						
$5.10^{15}/cm^2$ - 20 KeV	-	-	575	32.7	.7	13.1
3°) CO_2 laser annealing (work in progress						
$5.10^{15}/cm^2$ - 20 KeV	-	90	550	32.1	.73	12.8

Fig.1 : Glow discharge implantation machine and schematic of the source.

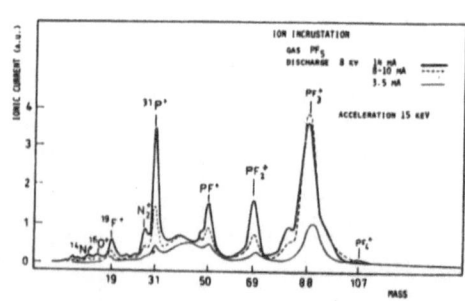

Fig.2 : PF5 ion mass distribution spectra for various discharge currents.

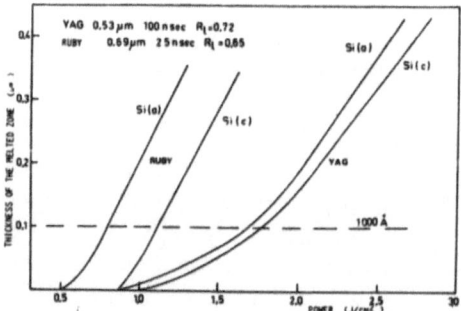

Fig.3 : Influence of the energy of the laser beam on the thickness of the melted layer (ruby and YAG lasers).

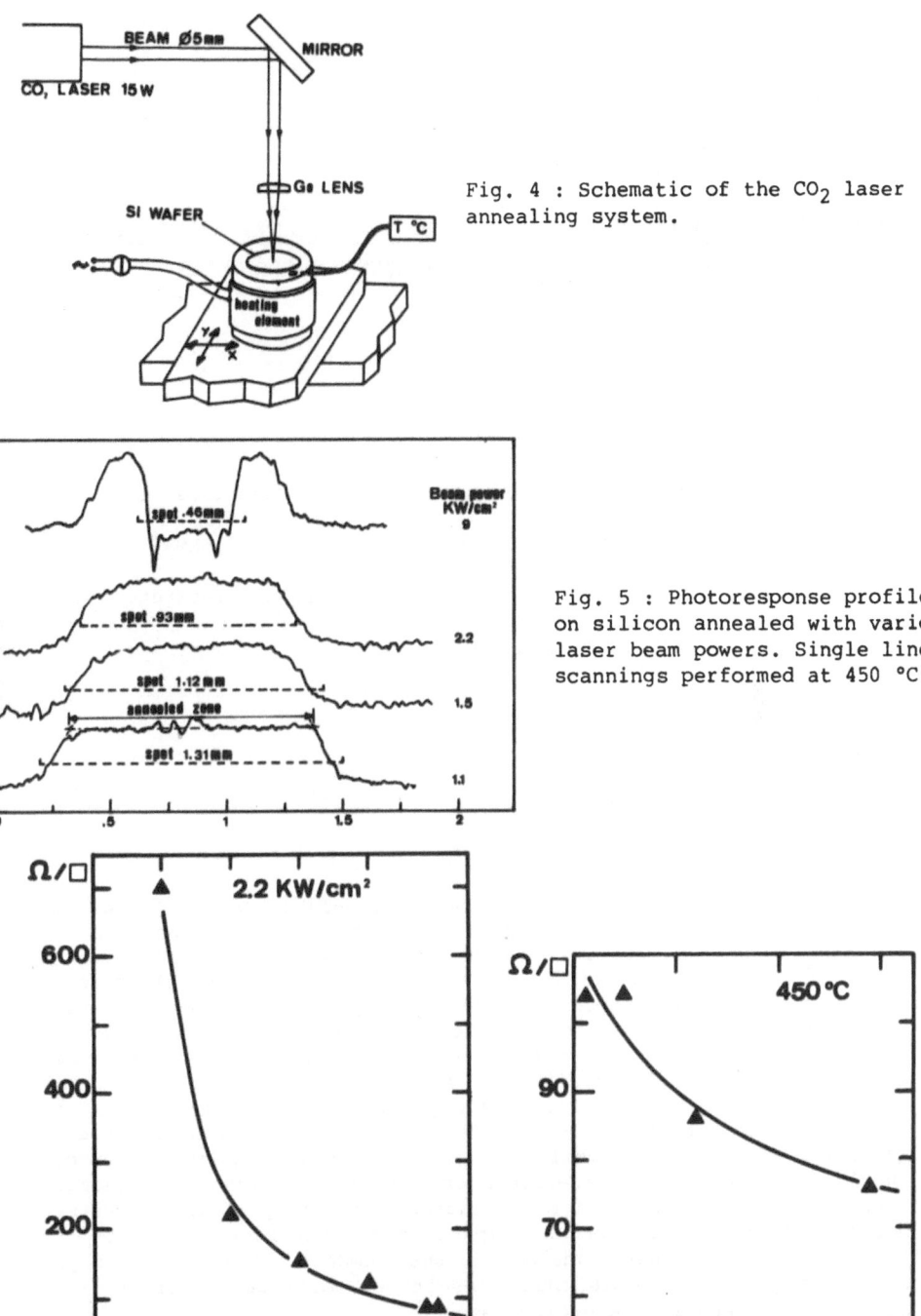

Fig. 4 : Schematic of the CO_2 laser annealing system.

Fig. 5 : Photoresponse profiles on silicon annealed with various laser beam powers. Single line scannings performed at 450 °C.

Fig.6 : Influence of the beam energy and the substrate temperature on the sheet resistance of the implanted silicon (implantation 5.10^{15} p+/cm2).

INVESTIGATION ON THE POTENTIALITY OFFERED BY ION IMPLANTATION AND ELECTRON
BEAM ANNEALING TO OBTAIN HIGH EFFICIENCY SOLAR CELLS

Authors: : R. GALLONI, P.G. MERLI

Contract number   : ESC-R-041-I (S)

Duration   : 30 months : 1 January 1981 - 30 June 1983

Head of project   : R. GALLONI, P.G. MERLI - CNR Istituto LAMEL

Contractor   : Consiglio Nazionale delle Ricerche - Istituto LAMEL

Address   : CNR Istituto LAMEL
                        Via Castagnoli, 1
                        40126 BOLOGNA (Italia)

Summary

The basic features of a highly automated apparatus especially designed
for solar cell production are discussed. Progress in production rates
and cost reduction depends both on the design of implantation equipment and on the choice of a proper radiation damage annealing technique. An electron gun especially designed for research purposes in the
field of semiconductor thermal treatments has been set up and its
capabilities are described here. The apparatus will provide information
on the possibility of annealing the radiation damage due to the implantation process through a fast solid phase process, able to keep pace
with the implantation rate of new developing ion accelerators especially designed for solar cell production. The fast annealing process
should also have positive effects on carrier life-time. A preliminary
investigation of life-time modification induced by thermal treatments
has also been carried out. A computer model has been set up to study
the effects of stress induced in the samples by fast solid phase
annealing processes. Preliminary results of solar cell characteristics
made by this technique are discussed.

1. STATE OF THE ART AND PERSPECTIVES

Today the available ion implantation equipment routinely used in the semiconductor industry, allows implantation of 7.62 cm diameter wafers for solar cell application (5-30 KeV energy, 2×10^{15} at/cm^2 dose) at a rate of about 400 wafers/hour. This output has been made possible by a high qualitative improvement in the equipment and particularly in the ion source design. Beam currents of 1-5 mA can now be used compared to a few μA of first generation ion implantation machines.

As is well known an annealing treatment must follow the implantation process to remove the radiation damage introduced. At the energy and doses needed for solar cell fabrication in fact, radiation damage is so severe that a surface layer about 0.1-0.2 μm thick, in which most of the ions come to rest, becomes completely amorphous. Thermal annealing in a furnace is the technique normally used to remove lattice damage and to restore electrical properities of ion implanted Si. However, new beam processing techniques, in which the heat treatment is induced by energetic beams of photons, electrons or ions, have been recently investigated. From a physical point of view they can be divided into two general categories:
a)   those which perform the annealing as a solid phase process;
b)   those which produce melting and risolidification of the thin implanted layer in a very fast process (10-100 nsec).
Annealing by furnace treatment, by continuous wave (argon, kripton or CO_2) laser irradiation, by continuous electron beam or by incoherent light belong to the first category of techniques; pulsed laser (Q-switched ruby, Nd:YAG, Nd:glass, excimer) and pulsed e-beam (PEBA) belong to the second group.

The general problem is to design an automatic plan for solar cell production which uses ion implantation doping followed by the annealing technique which can provide the highest cell efficiency together with lower cost and higher automation (1).

Let us consider the ion beam current (phosphorus or arsenic) as the cell production limiting factor; we find that $\sim$ 2 mA current, typical of today's most diffused ion implanters allows 7.62 cm diameter wafers to be implanted at 10-30 keV energy and 2×10^{15} cm^{-2} dose in about 9 sec. A 10 mA beam current which is already available in the most advanced ion implanters produces 1 wafer every 1.5 sec.; 100 mA current beam which is now considered a medium term final goal could allow implantation of 1 wafer every 0.15 sec.

By considering a cell efficiency of 14% (which is obtained by implanting only one side of the wafer (i.e. without back surface field) and a solar irradiation of 0.85 KW/m^2 (AM1.5), we would have an annual production of solar cells which could supply respectively 1.2 MWp, 7.2 MWp and 72 MWp of photovoltaic power. It must be pointed out that implantation at beam currents higher than 50 mA not only involves the problem of the ion source, but mainly of beam handling. In this regard studies concerning the possibility of using non mass analyzed beams are most important and are carried out in several laboratories (group PHASE-CRN Strasbourg, SPIRE Co., JPL, Motorola, Oak Ridge Nat. Lab. U.S.A.).

If we now try to sort out the most suitable annealing technique to be employed at different production rates, we can see that up to about 7.2 MWp/year production (10 mA beam current) a simple 15 min. furnace treatment after implantation at a temperature of about 700°C could be performed by using a standard type furnace 3 m long, with the wafers moving through at the speed of 0.3 cm/sec. Such a set up does not require any new technological advancement except some mechanical movement to transfer the samples from the accelerator to the furnace. The same process at the 72 MWp/year production (100 mA beam current) would require a 30 m furnace to anneal all the samples at the same rate at which they are implanted. Economical evaluations (2) show that under these conditions a pulsed annealing technique such as the PEBA (set up by SPIRE Co.), or a laser pulse (Nd: glass - JPL or XeCl excimer - Oak Ridge, Helionetics Co.) would be preferable. The total cost of a photovoltaic panel (calculated with the aid of SAMICS methodology) made with this technique would be 0.66 $ (1980)/Wp with an energy pay back time of 9 months (RTR silicon, Motorola technology) (3). The post annealing treatment (about 400°C, 1-2 min) that has been found to be necessary after most pulse annealing processes, could still be performed in a furnace 3 m long with a belt transfer speed of 3 cm/sec.

It has been shown (4,5) that if several thermal cycles are introduced instead of a single one the efficiency of the final device is considerably improved. The reason for this behaviour is mainly to be attributed to the increased ability of removing the radiation damage introduced by the implantation process and of preserving or improving bulk carrier life time. However multi-step furnace annealing is not a treatment that can possibly be introduced in a processing sequence as fast as the one needed for solar cell fabrication. For this reason, a research effort has been made at LAMEL to set up an electron gun especially designed for research in this field. It has been especially designed to produce solid phase regrowth of implanted layers in time intervals ranging from 10^{-4} sec. to several minutes. We chose an electron beam suitable for promoting solid phase epitaxy instead of liquid phase in order to avoid dopant diffusion and to preserve the possibility of programming the profile shape as discussed in more detail further on. Although very little is known on the kinetics of the process of solid phase annealing by electron irradiation, the first experimental results obtained show that it is possible to operate in a wide range of power densities and times.

In the following sections a brief overview of the results obtained at LAMEL in the research program undertaken under CNR-CEE contract n°ESC-R-041--I(S) will be given.

2. ELECTRON GUN (6)

A picture of the electron beam annealer set up at LAMEL is shown in Fig.1. It provides electron beams having a maximum power of about 10 KW ($E \simeq 50$ KeV , $I_{max} \simeq 200$ mA). A diagram of the electron optical column, is shown in the insert of Fig. 1. It consists of :
1) Thermoionic triode gun with adjustable interelectrode distance. The massive tungsten cathode is indirectly heated by electron bombardment from three filaments on the back side. In spite of its complexity the system provides a good thermal stability and regularity of emission.
2) Gate valve allowing the vacuum to be preserved in the region of the gun when the specimen chamber is opened to change the specimen.
3) Beam defining aperture.
4) Variable focal length magnetic lens (1cm minimum at 50 KeV).
5) Deflection system providing a scanning frequency of up to 3 KHz in X as well as in Y over a square area of 100 cm^2.
6) Wide specimen chamber containing wafer holders of different types and a device for the measurement of beam current and shape. Sample holders allow irradiation of samples of different shape and dimension under two different boundary conditions: i) thermal insulation; ii) good thermal contact to a substrate kept at room temperature.

This column has been designed to work in two different modes, namely "defocusing mode" and "scanning mode". The defocusing mode is obtained by focusing the e-beam high above the specimen in order to produce a large beam pattern at the target plane (6). This mode allows the production of isothermal treatments at temperatures up to the melting point with processing times ranging between 10^2 sec (which is the minimum reliable processing time of a conventional furnace) and less than 1 sec with a single electron pulse

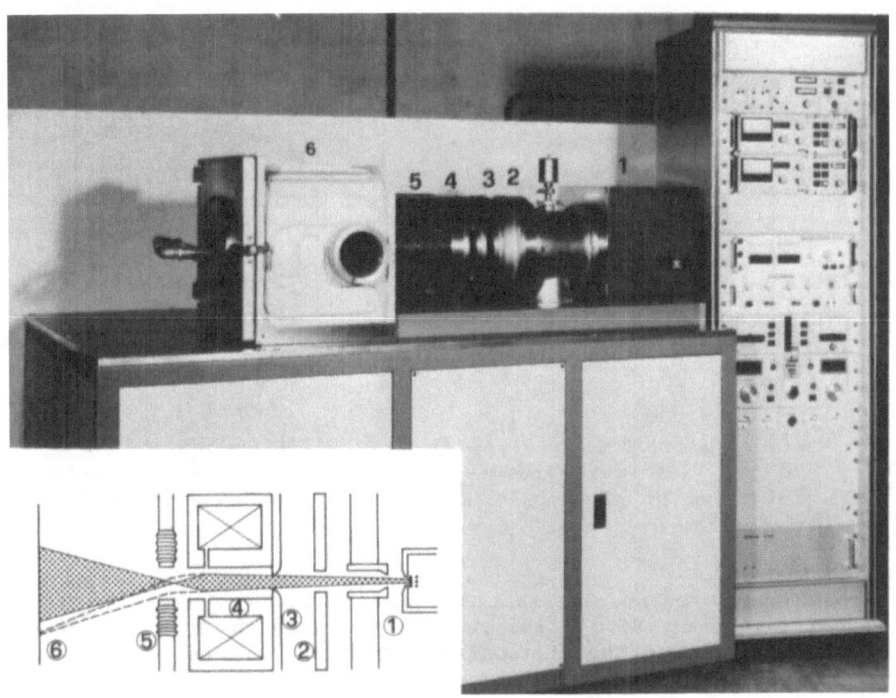

Fig.1. Electron beam annealer developed at LAMEL C.N.R.: (1) electron gun,
(2) gate valve, (3) beam defining aperture, (4) magnetic lens, (5) deflection system, (6) specimen chamber. A sketch of the electron trajectories
operating in two modes, defocousing and scanning (dashed line) is reported
in the insert.

over the whole area 10cm in diameter. In this case the electron beam irradiation provides a "fast furnace" facility. The power density required to
operate in this mode is of the order of 10 watt/cm^2. Pulses as short as 10^{-2}
sec can also be used to rise the surface sample temeprature up to the
melting point over smaller areas (few cm^2). The scanning mode of operation
is achieved by focusing the beam onto a spot with a radius of about 1 mm
size and scanning it over the sample surface at adjustable frequency . In
this way it is possible to reach power densities of the order of 10-100
KW/cm^2 which produce a non isothermal heat treatment with in depth thermal
gradient which depends on processing time and boundary conditions. Fig.2
shows the surface temperature vs. time as obtained by one dimensional
numerical simulation (which represents a good approximation for a beam
radius much larger than the specimen thickness) for two different values of
the power density. Also reported in the drawing is the calculated solid
phase epitaxial (SPE) regrowth thickness of amorphous Si P$^+$ implanted after
the same temperature history. The calculation has been made by extrapolating
to high temperatures data on the regrowth rate obtained in the range
475<T<525 (7). From these data it is possible to deduce that the minimum
dwelling time to regrow an amorphous layer 0.2 μm thick by SPE is of the
order of 10^{-4} sec. With this operating mode it is possible to anneal samples
at a very fast rate, working for instance at 30 KeV, I=100 mA, r=1 mm (P=90
KW/cm^2) and dwelling time of 10^{-4} sec (3 KHz scanning frequency), the time
to anneal a 7.62 cm diameter wafer is only 0.15 sec, which, as previously

seen, is the time needed for the implantation of the same sample by a 100 mA beam current accelerator.

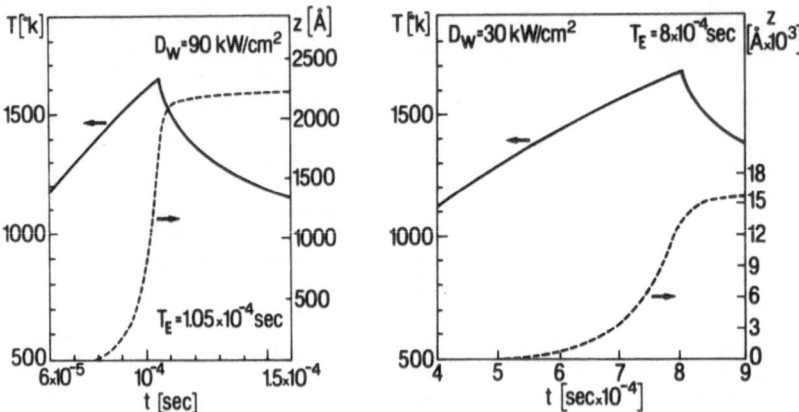

Fig.2. Surface temperature profile for two different values of the power density and two different exposure times. The dashed line represents the regrowth thickness of an amorphous layer P$^+$ doped according to data reported in ref. (7).

3. STRESS ANALYSIS

When energy pulses or fast scanning beams are employed to anneal implanted layers in Si, introduction of lattice defects can take place (8,9). An analysis of the thermoelastic effects induced by fast transient thermal treatments is then needed to determine the best annealing conditions to be used. Fig.3 shows the results obtained by a one dimensional computer model in case of irradiation of a whole single crystal Si wafer by short energy pulses (10^{-4} - 5×10^{-4} sec) (9). The amorphous layer considered in this case is 0.1 μm thick. The main result is that the elastic limit is generally exceeded on the front surface for pulse duration t $< 5 \times 10^{-4}$ sec if enough energy to performe the annealing is supplied; damage introduction can be avoided if the sample is pre-heated to a suitable temperature.

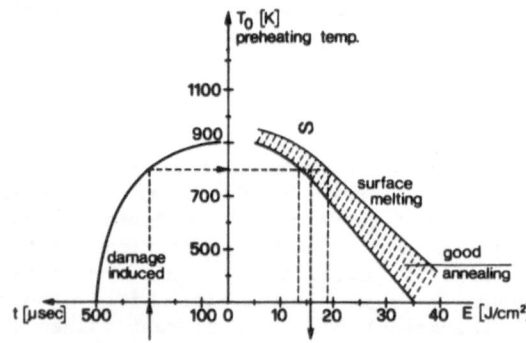

Fig.3. Relationship among pre-heating temperature, pulse duration and energy density for SPE regrowth of 0.1 um amorphous Si layer. By entering a pulse duration, the corresponding minimum pre-heating temperature and energy density interval to obtain annealing without introduction of lattice defects can be deduced by following the illustrative path (and inversely).

4. SOLAR CELLS

One of the most praised peculiarities of ion implantation in the semiconductor technology, is the possibility to program the profile shape with much greater freedom than by using thermal diffusion. In the case of emitters for solar cell utilization, the problem to be solved is particularly tricky because a series of competitive requirements must be fulfilled:

a) the emitter region must be thin so that most of the incident photons generate electron-hole pairs in the lightly doped bulk region;

b) emitter sheet resistivity must be of the order of ∼ 50 Ω/square or less to allow use of a simple grid contact pattern;

c) a concentration vs. depth gradient should be produced to improve carrier collection;

d) thermal treatments must be such as to preserve or improve the bulk- -carriers diffusion length;

e) possibly, residual defects should be outside the p/n junction deple- tion region where their effect is most deleterious.

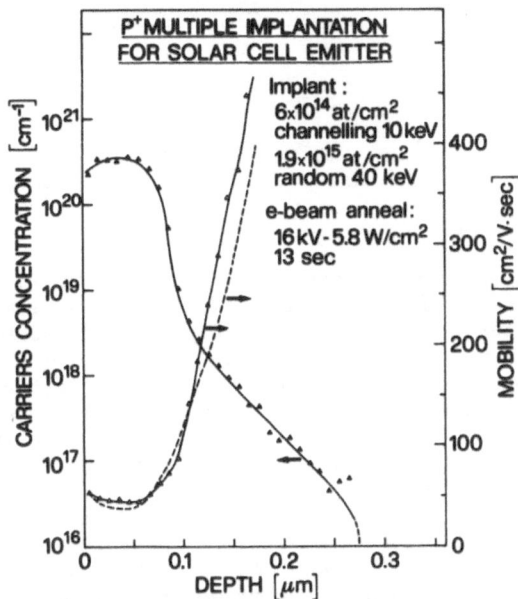

Fig.4. Carrier concentration and mobility profiles of P$^+$ tailored implantation. Dashed curve is the reference mobility vs. concentra- tion trend (12).

Several profile shapes have been analysed by using our computer model (10,11) and a detailed analysis of the results obtained will be reported elsewhere (12). Fig. 4 shows the doping profile experimentally measured by using the fully automated apparatus set up at LAMEL (13) relative to the best implantation conditions determined by using the computer model. The profile is the result of overlap between a 10 KeV implantation under chan- nelling conditions (obtained by simply setting the (100) oriented Si samples at right angles to the beam in our Lintott III accelerator, the critical angle at this energy being 5.2°) and a 40 KeV random oriented implantation (8° off axis) (14). The 10 KeV implantation produces a random concentration peak very near the sample surface (∼ 140 Å depth) (15) and a channelling peak at about 0.25 μm depth, while the 40 KeV random has a 490 Å deep projected range. The purpose of the 10 KeV implantation is twofold: a) to obtain a surface concentration value up to the solubility limit and b) to reduce the radiation damage in the junction depletion region (16,17); the 40 KeV implantation has the purpose of lowering the sheet resistivity of the emitter (the 10 KeV implantation alone would give ∼ 130 Ω /square sheet resistivity). Good annealing of the implanted layer, without dopant diffu- sion effects, has been obtained by e-beam irradiation, using the electron gun described before in the "defocusing mode": 16 KeV monochromatic energy, 5.8 W/cm^2 power density. The irradiation allowed the heating of the sample to ∼ 720° C for 13 sec. Under these conditions good bulk carriers life-time was obtained, while heating for the same time at 930° C caused a severe de- gradation of this important parameter. As will be reported in detail in the final contract report, bulk life time degradation in Si single crystals is strongly correlated to the content and distribution of O_2 and C in the virgin samples and it seems possible to improve the quality of devices by controlling these factors. It has also been verified that, under well controlled conditions, low annealing temperatures and short times are sufficient to obtain considerable life-time improvements (18).

Annealing	$I_{sc}^{(max)}$ (mA)	ARC	$V_{oc}^{(max)}$ (mV)	ARC	η_{max}%	ARC	$\overline{\eta}$%	$\dfrac{\Delta\overline{\eta}}{\overline{\eta}}$
e-beam (750°C,15sec)	94	131	555	564	9.5	13.4	9.3	2%
furnace (650°C,30min)	93	129	565	572	9.9	13.6	8.9	12%

TABLE 1 - (2x2) cm^2 Si solar cells; TiO_2, ARC.

Carrier mobility values experimentally detected are also shown in Fig. 4 and compared with carrier mobility values, corresponding to the same dopant concentration, that are available in literature (doping obtained by thermal diffusion). The fact that our mobility values are higher than the reference ones along the profile tail is probabily due to the high lattice perfection present in these region after annealing the sample implanted under channelling conditions. Preliminary results on a set of solar cells obtained by a simple technique (no BSF, Al back contact, TiAg front grid - 1 mm distance, ~ 8% coverage ratio - TiO_2 ARC) are shown in TAB. 1 and compared with the best results obtained by furnace annealing of a similarly implanted layer. Shown in Fig.5 are the quantum efficiency spectra and I-V characteristics of the best cells obtained by e-beam and furnace annealing. As can be seen a better short wavelength response is obtained in case of the e-beam but V_{oc} is lower mainly due to higher residual damage still present in the p/n junction region.

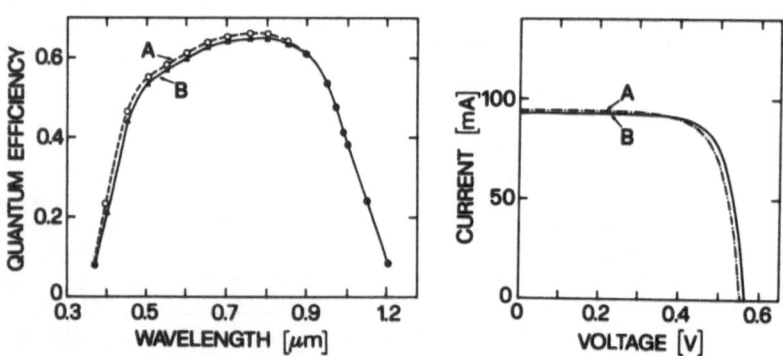

Fig.5. a) Quantum efficiency of (2x2) cm^2 Si solar cells by tailored ion implantation and (curve A) electron beam annealing; (curve B) furnace annealing (650°C, 1/2 h). b) I-V characteristics of the same cells.

The preliminary experiments performed up to now show that it is possible to anneal ion implanted layers by e-beam irradiation and that results similar to a much longer furnace annealing can be obtained. With the aid of

DLTS analysis and Double Crystal X-Rays Diffractometry it will be possible to have a better physical knowledge of the phenomena involved in the electrical activation process of implanted species by fast SPE. A good knowledge of these phenomena on Si single crystals is of primary importance to face the more complex problem of annealing treatments on solar grade Si with due capability.

AKNOWLEDGEMENTS

The authors wish to point out that the results reported in this paper have been obtained with the active contribution of different groups at LAMEL. More detailed results on the specific subjects, which emphasize these contributions, will be published elsewhere.

REFERENCES
1   R.Galloni, F.Zignani,"Photovoltaic cells: state of the art and perspectives for a highly automated high speed production line", (to be publ.).
2   SPIRE - Final Report DOE/JPL 954786-79/02
3   M.G.Coleman, MOTOROLA Final Report DOE/JPL 954847-80/8
4   A.C.Greenwald, R.P.Dolan, S.P.Tobin,"Laser and Electron Beam Solid Interactions and Materials Processing", J.F.Gibbons, L.D.Hess, T.W. Sigmon, North Holland, p.321 (1981).
5   J.A.Minnucci, K.W.Matthei, A.C.Greenwald, R.G.Little, A.R.Kirkpatrick, "Laser and Electron Beam Processing of Materials", C.W.White, P.S. Peercy Acad.Press, p.658 (1980).
6   G. Lulli, P.G. Merli, M. Vanzi, Proceedings of the ninth international Conference on Electron and Ion Beam Science and Technology, ed. by R. Bakish, p.626, The Electrochemichal Society, Inc., N.J., USA (1980).
     G. Lulli, P.G.Merli, Optik,$\underline{60}$,29(1981).
     G. Lulli, P.G.Merli, (to be publ.).
7   L.Csepregi, E.F.Kennedy, T.J.Gallagher, J.W.Mayer, T.W.Sigmon, J.Appl. Phys.,$\underline{48}$,4234(1977).
8   K.H.Heinig, K.Hohmuth, R.Klabes, M.Voelskow, H.Woittennek: Rad.Effects, $\underline{63}$,115(1982).
9   G.G.Bentini, L.Correra, J.Appl.Phys. (in press).
10   P.U.Calzolari, A.M.Mazzone, Proc. First Photoltaic Solar Energy Conf., Luxembourg (1977).
11   G.F.Cembali, R.Galloni, G.Lulli, A.Mazzone, P.G.Merli, R.Nipoti, F.Zignani, Fourth EC Photovoltaic Solar Energy Conf., Stresa (1982).
12   F.Cembali, L.Favero, R.Galloni, F.Zignani (to be pub.).
13   R.Galloni, A.Sardo (sent for publ. Rev.Sci.Instrum.).
14   R.Galloni, L.Favero, A.Carabelas, Rad.Effects Lett.,$\underline{68}$,39 (1982).
15   J.F.Gibbons,W.S.Johnson,S.W.Mylroie,"Projected Range Statistics", Halsted Press (1975).
16   F.Cembali, R.Galloni, F.Zignani, Rad.Effects, $\underline{26}$,161(1975).
17   R.Galloni, L.Pedulli,F.Zignani, Rad.Effects,$\underline{32}$,223(1977).
18   E.Susi, G.Passari, (to be publ.).

AN EXPERIMENTAL INVESTIGATION OF ION IMPLANTATION COMBINED WITH
LASER AND INCOHERENT-LIGHT ANNEALING, AND OF LASER-INDUCED
DIFFUSION FOR THE PRODUCTION OF SOLAR CELLS

Authors: A. NYLANDSTED LARSEN, G. SØRENSEN, F. NIELSEN,
                            L.D. NIELSEN, and V. BORISENKO

Contract number: ESC-R-020-DK(G)

Duration: 36 months   1 July 1980 - 30 June 1983

Total budget: DKR. 3,237,200

CEC contribution (50%):   DKR. 1,618,600

Contractors: Physics Laboratory III, Technical University
                           of Denmark, DK-2800 Lyngby
                           Laboratory for Semiconductor Technology, Technical
                           University of Denmark, DK-2800 Lyngby
                           Institute of Physics, University of Aarhus,
                           DK-8000 Aarhus C

Supervisory staff: Professor N.I. Meyer
                           Associate Professor O. Leistiko
                           Professor J.U. Andersen

Summary

    Ion implantation with and without mass separation and laser-induced
diffusion have been investigated as possible means of introducing dopants into silicon for making the p-n junction of silicon solar cells.
The ion implantations have been combined with pulsed-laser and incoherent-light annealing to remove the damage inherent to the ion-
implantation process. Results are presented for boron-fluoride implantations with and without mass separation, combined with pulsed-laser
and incoherent-light annealing, for As implanted in single and polycrystalline silicon combined with incoherent-light annealing and for
laser diffusion of As and Sb in silicon. The experimental results are
discussed with reference to solar cells.

1. Introduction

Although ion implantation is a well established technique within semi-conductor technology, its implementation with respect to production of photovoltaic solar cells has not yet been successful. However, there is a need for developing fast doping procedures in order to make solar cells of cheap polycrystalline silicon. Both ion implantation with and without mass separation and laser-induced diffusion processes have to be considered.

The present report summarizes the results obtained in a collaboration between the Technical University of Denmark and the University of Aarhus under contract ESC-R-020(DK). A more detailed discussion of the results may be found in the literature (1-9), including the analytical methods developed as a part of this project.

2. Experimental Details

Ion implantation without mass separation has been performed in the energy range 10-30 keV, using BF_3 as a feed gas to the ion source. Ion implantation with mass separation has been performed in the energy range 1-500 keV, using the 90-kV ion accelerator and the 600-kV heavy-ion accelerator at the University of Aarhus.

Pulsed high-energy lasers (Nd glass (Aarhus) and ruby (Strasbourg)) as well as a xenon-lamp system (1.6 kW) have been used either for annealing the damage caused by the ion bombardment or for doping.

Van de Graaff accelerators (2 and 4 MeV) have been applied in the profiling of the dopant and in lattice location by channeling.

The samples were n- or p-type single or polycrystalline silicon from Wacker Chemietronic.

3. Laser Annealing Following Ion Implantation

Pulsed high-energy lasers have been used to anneal structural damage caused by the ion bombardment. When implantations are performed without mass separation, using BF_3, ion species such as B^+, F^+, BF^+, BF_2^+, BF_3^+, SiF_2^+, and CF_2^+ are all present in the ion beam. Depending on the plasma conditions in the ion source, the surface layers will be doped with a varying concentration of boron and fluorine, which may influence the radiation damage and the absorbed laser energy.

For the study of the importance of surface fluorine, mass-separated B^+, F^+, BF^+, BF_2^+, and BF_3^+ were implanted into silicon and, in turn, laser annealed (Nd glass, 40-ns pulse length).

Fluorine depth profiling of laser-annealed BF_2^+-implanted silicon single crystals has been carried out using the 340-keV resonance in the cross section of the nuclear reaction $^{19}F(p,\alpha\gamma)^{16}O$ to study the redistribution of fluorine as a consequence of Nd-glass laser exposure. The depth resolution was estimated to be 175 Å. Diffusion of fluorine into the crystalline substrate was not observed, neither was any accumulation of fluorine, e.g., at the interface between the recrystallized amorphous layer and the crystalline substrate, as observed by Tsai et al. (10) in thermally annealed BF_2^+-implanted silicon single crystals. Diffusion to the surface, on the other hand, was observed at all laser-energy densities; however, about 50% of the implanted fluorine is still retained in the implanted surface layer after a laser exposure at $2J/cm^2$.

The recrystallization of the silicon single crystals and the electrical activity of the boron were studied for mass-separated B^+, BF^+, BF_2^+, and BF_3^+ implanted to the doses 5×10^{14} cm^{-2} and 5×10^{15} cm^{-2}. The implantation energies of the different ions were such chosen that the projected range of the boron dopants was identical for all implantations (400Å); hence only the fluorine concentration in the surface layers and the radia-

tion damage were different. The implanted crystals were studied by 2-MeV α
-particle channeling, transmission-electron-microscopy, and sheet-resistivity measurements. Figure 1, which shows the results from ion-channeling

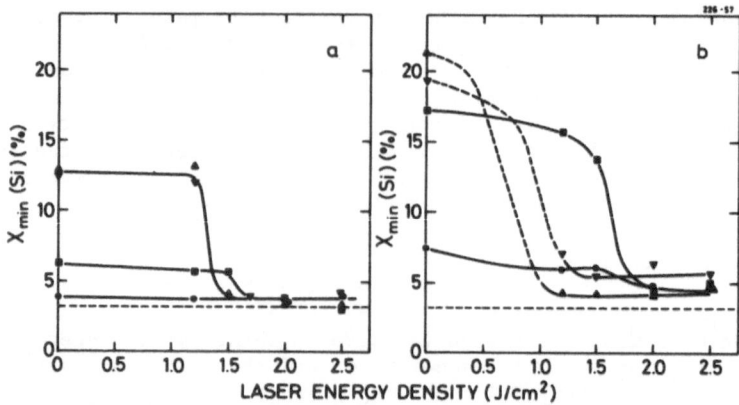

*Fig. 1. Results from 2-MeV channeling measurements along the <100>
axis in <100>Si single crystals for implanted doses of 5×10^{14} cm^{-2}
(a) and 5×10^{15} cm^{-2} (b). (o), B, 11 keV; (□), BF, 30 keV; (Δ), BF_2,
50 keV; (∇), BF_3, 68 keV. Dashed line gives the χ_{min} value for a
virgin <100> Si crystal.*

measurements, indicates a threshold laser-energy density at about 1.5 J/cm^2
for BF, BF_2, and BF_3 implanted at low dose and for boron, and BF implanted
at high dose. However, for the fluorine excess ions, BF_2 and BF_3, implanted
at high dose, the data show recrystallization already at 1.2 J/cm^2, which
is below the assumed threshold value for recrystallization. It should be
mentioned that the channeling spectra of the recrystallized samples showed
no indication of point-defect accumulation or growth of extended defects.
Sheet resistivities have been determined for all implanted samples, and the
results are displayed in Fig. 2. The most spectacular result is that almost
identical values are observed for BF, BF_2, and BF_3 implanted at high dose
despite the fact that the channeling analysis yielded quite different results in the three cases. Low sheet resistivities are obtained already at
a laser-energy density of 1.2 J/cm^2, and a constant value of about 50 $\Omega/\square$
is reached at 1.5 J/cm^2. The transmission electron microscopy of the 2.0-J/
cm^2 laser-annealed 5×10^{15} cm^{-2} implantations yielded results equivalent to
those obtained by channeling. The samples were defect-free for all four implanted species, and the diffraction patterns indicated complete recrystallization. The results for test cells made from these samples and from
samples implanted without mass separation are in agreement with these findings. The internal quantum efficiencies reach 50% at wavelengths between
400 and 425 mm and drops below 50% at wavelengths between 980 and 1000 mm.
Thus, the bulk minority-carrier lifetime has not been effected significantly, and there seems to be no surface dead-layer.

4.   Incoherent-Light Annealing Following Ion Implantation
     In relation to solar-cell production, it appears that annealing with
an incoherent-light source may be advantageous from an experimental point
of view and inexpensive from an economic point of view. In the present project, arsenic as well as compounds of boron and fluorine have been used as
dopants. In the case of arsenic implantations, the implantation energy and
dose, anneal temperature and anneal time have been varied. The implantation

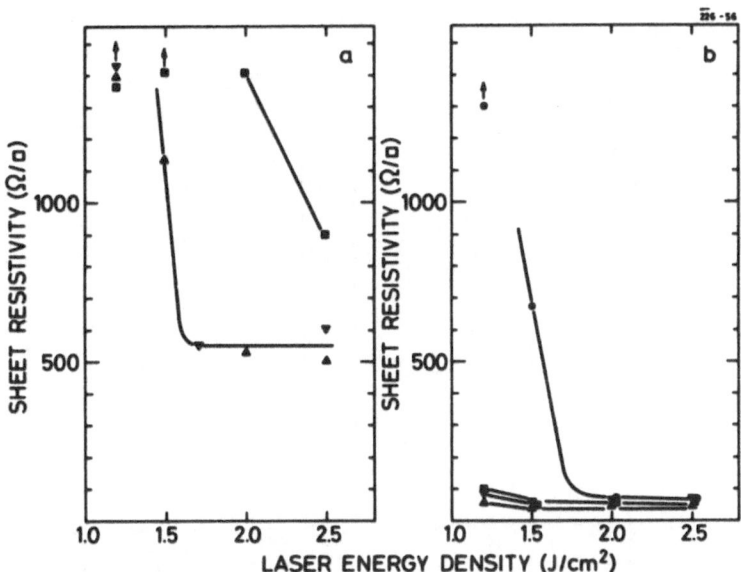

Fig. 2. Results from sheet-resistivity measurements for implanted doses of 5×10^{14} cm^{-2} (a) and 5×10^{15} cm^{-2} (b) in Si single crystals. (o), B, 11 keV; ($\square$), BF, 30 keV; ($\triangle$), BF$_2$, 50 keV; ($\triangledown$), BF$_3$, 68 keV.

conditions correspond to average depths of arsenic between 100 Å and 600 Å and peak concentrations between 0.2×10^{21} cm^{-3} and 5×10^{21} cm^{-3}. Figures 3 and 4 display measured sheet resistivities as a function of these parameters for single crystals. A distinct dependence on implanted dose and implantation energy is observed, the smallest sheet resistivity being obtained

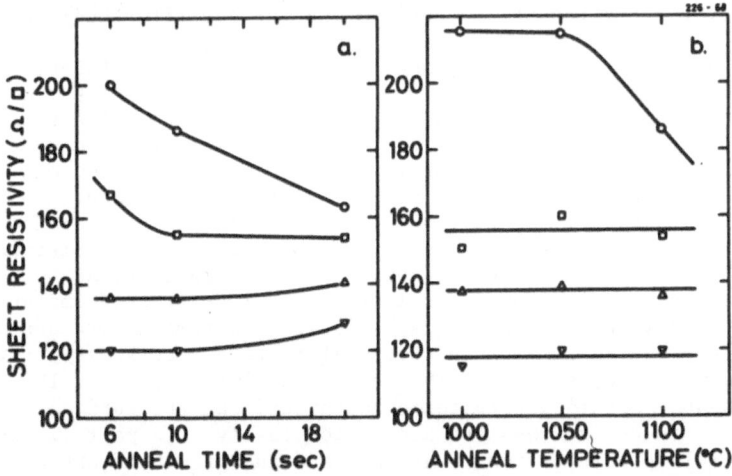

Fig. 3. Results from sheet-resistivity measurements for As implanted in <100> Si single crystals to a dose of 1×10^{15} cm^{-2}. (o), 10 keV; ($\square$), 30 keV; ($\triangle$), 50 keV; ($\triangledown$), 70 keV. (a) Anneal temperature 1100°C; (b) anneal time 10 sec.

for the highest implantation energy and dose (as-implanted average depth

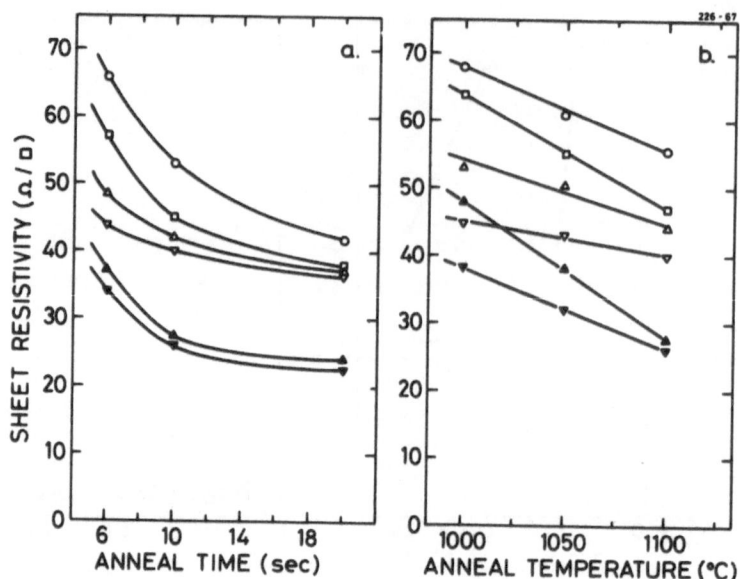

Fig. 4. Results from sheet-resistivity measurements for As implanted in <100> Si single crystals to doses of 5×10^{15} cm^{-2} (open symbols) and 1×10^{16} cm^{-2} (closed symbols). Otherwise as Fig. 6.

and concentration: 600 Å and 2×10^{21} cm^{-3}). As regards the annealing conditions, low-sheet resistivities are obtained for anneal temperatures above ∿1050°C and for anneal times longer than ∿10 sec. Typical experimental values are for a dose of 1×10^{16} cm^{-2}: 25 Ω/□, 5×10^{15} cm^{-2}: 40 Ω/□, and 1×10^{15} cm^{-2}: 150 Ω/□. These samples have also been studied with channeling to follow the recrystallization of the silicon and the substitutionality of the arsenic. Figure 5 shows χ_{min} values for silicon and arsenic measured with 2-MeV α particles aligned along the <100> direction. Except for the 950°C anneal and the 20-sec anneal, all the χ_{min} values for silicon are small, scattering around 3.2%, which is the value found for a virgin crystal. A temperature of 950°C in 10 sec is not sufficient for complete recrystallization as the damage peak in the channeling spectrum was found to be high, indicating the existence of point defects in the surface layer. The channeling spectrum of the 1100°C, 20-sec annealed sample showed a perfect silicon surface layer; however, the dechanneling at greater depths was high, indicating a high concentration of extended defects. As to arsenic, low χ_{min} values are also found for most of the investigated implantation and anneal conditions. Exceptions are (i) the low-temperature annealing at 950°C probably due to incomplete recrystallization (see above), (ii) the long-time annealings, and (iii) the low-energy implantations. The latter two are probably due to precipitation of arsenic at the surface; however, more detailed investigations are needed to clarify this part. It should be noted that this precipitation of arsenic and the growth of extended defects, as observed at long-time anneals, are not reflected in the sheet-resistivity results. Redistribution of the implanted arsenic impurities was observed for high-dose implantations, but detailed depth profiling has not yet been carried out for these samples. The substitutional or equivalent fractions, defined as $E = (1-\chi_{min}(As))/(1-\chi_{min}(Si))$, are between 94% and 97% for the well annealed samples, depending upon the implantation conditions. Combining the channeling and sheet-resistivity results, it appears that a

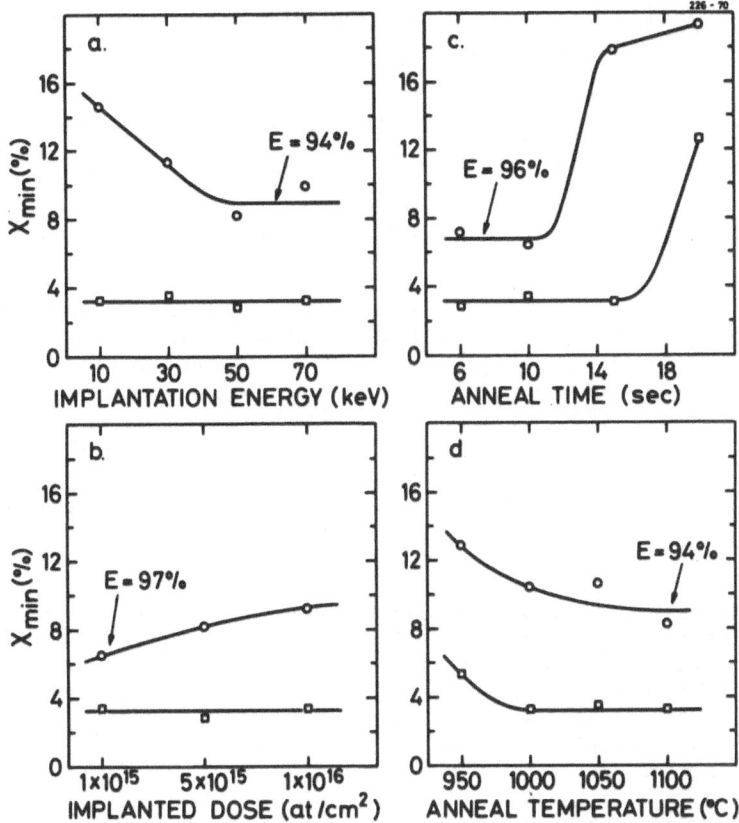

Fig. 5. Results from 2-MeV He⁺-channeling measurements along the <100> axis in As-implanted <100> Si single crystals. (o), $X_{min}(As)$, (□), $X_{min}(Si)$. (a) Impl. dose: $5×10^{15}$ cm⁻², anneal temp. and time: 1100°C, 10 sec. (b) Impl. energy: 50 keV, anneal temp. and time: 1100°C, 10 sec. (c) Impl. energy and dose: 50 keV, $1×10^{15}$ cm⁻², anneal temp.: 1100°C. (d) Impl. energy and dose: 50 keV, $5×10^{15}$ cm⁻². Anneal time: 10 sec.

10-sec anneal at 1050°C is optimum. For a 50-keV, $5×10^{15}$ cm⁻² implantation, these annealing conditions result in a sheet resistivity of 40 Ω/□ and a substitutional fraction of 94%.

Polycrystalline silicon has been implanted with 15- and 30-keV arsenic to doses of $1×10^{15}$ cm⁻² and $5×10^{15}$ cm⁻² in the average depths of 150 Å and 260 Å and peak concentrations between $6×10^{20}$ cm⁻² and $3×10^{21}$ cm⁻². The samples have been annealed with incoherent light, keeping the anneal time constant at 10 sec but varying the anneal temperature between 900°C and 1100°C. These measured sheet resistivities are displayed in Fig. 6. On the lines with arsenic implanted into single-crystalline silicon, a temperature of 1050°C seems to be sufficient to anneal the damage. It should be noted that the sheet resistivities obtained for polycrystalline silicon are almost identical to those for single-crystalline silicon for similar implantation and annealing conditions.

Boron-fluoride compounds implanted without mass separation into sili-

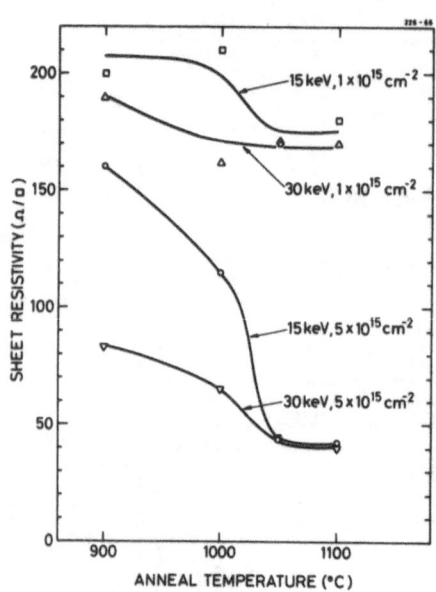

con single crystals and mass-separated BF$_2$ implanted into silicon single crystals have been studied at different annealing conditions. Also in these cases, anneal temperatures and times of about 1050°C and 10 sec seem to be optimal; however, none of the samples studied was perfectly annealed. The ion-channeling results can be summarized as follows: A surface layer ranging to about the interface between the damaged layer and the crystalline substrate was found to be almost perfect; near the interface, a damage peak was situated, and below this damage peak, the dechanneling was high, indicating a high concentration of extended defects. Similar ion-channeling results were obtained for mass-separated fluorine implantations and for furnace-

Fig. 6. Results from sheet-resistivity measurements for As implanted into poly-crystalline Si. Anneal time: 10 sec.

annealed samples; however, the furnace-annealed samples had in general more residual damage than the incoherent-light-annealed samples. The sheet resistivities of the implanted layers were found to be low, in agreement with the damage-free surface layers observed by channeling. A typical value for an implantation without mass separation at 30 keV to a dose of 2x10^{15} cm^{-2} was 25 Ω/□. In agreement with these findings are the results from test cells made without mass separation. The short-wavelength response is very good, with an internal quantum efficiency of more than 50% at a wavelength of 400 nm, however, the internal quantum efficiency drops below 50% already at a wavelength of 750 nm, indicating that the bulk minority-carrier lifetime has been affected, probably as a result of the fluorine content.

5. Laser-Induced Diffusion

The study of laser-induced diffusion reported here is based on spin-on-deposited films containing the dopant. Two types of dopant solutions have been tried: (i) A commercial arsenic-containing doping emulsion (Emulsitone) and (ii) SbCl$_3$ dissolved in absolute alcohol.

The commercial doping emulsion is a liquid formulation, which, after spinning and drying, forms an arsenic-doped SiO$_2$ film. Laser-induced diffusion in <111>-oriented silicon single crystals from these films were done with 40 ns laser pulses from a Q-switched Nd-glass laser at laser-energy densities between 1.5 and 4.0 J/cm^2 and for film thicknesses between 1500 and 5000 Å. The spin-on-deposited films were analyzed by Rutherford backscattering prior to laser exposure to estimate the film thickness and stoichiometry. After laser exposure, the samples were rinsed in HF to remove any residual film from the surface and subsequently analyzed by Rutherford backscattering. Only low doping levels were found, the maximum being ≤4x10^{14} cm^{-2}, obtained from a 300-Å thick film (∿10^{17} As/cm^2) and a laser energy density of 3.2 J/cm^2. This corresponds to a utilization of only about 0.4% of the arsenic in the film. It is speculated that this low utilization factor could be correlated to the higher melting point of SiO$_2$ than that of silicon. If the silicon substrate melts before the SiO$_2$ film, a high vapour pressure might build up between the silicon substrate and the film, which could result in a peeling-off of the film before it can react with the silicon (11).

The deposition of SbCl$_3$ by spin-on technique results in a dopant film with an antimony-chlorine-oxygen ratio of approximately 2:1:2, as revealed by Rutherford-backscattering analysis. The ratio is probably due to the formation of oxochlorines during the spinning process since no special precautions were taken to avoid atmospheric moisture. The melting point of this film is expected to be considerably smaller than that of silicon. A detailed investigation of laser-induced diffusion from these films in silicon single crystals has been performed by varying the film thickness, laser-energy density, laser wavelength (Nd-glass, λ = 1.06 μm, ruby, λ = 0.69 μm), and silicon-substrate orientation (<111> and <100>). All samples have been analyzed by Rutherford-backscattering technique after laser exposure to estimate the depth profiles and concentration of the antimony impurities, and selected samples have been analyzed by ion channeling to estimate the substitutionality. Figure 7 displays measured surface concentrations after exposure to light from the Nd-glass laser. The surface concentration of antimony atoms is strongly dependent on the laser-energy density, the thickness of the dopant film, and on the crystalline orientation of the silicon substrate, the highest surface concentration being obtained in <100>-oriented silicon for the thinnest film exposed to the highest laser-energy density. The utilization

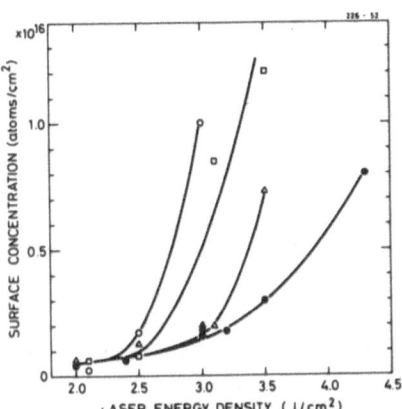

Fig. 7. Surface concentrations in Si single crystals after Nd-glass-laser exposure of spin-on-deposited SbCl$_3$. Open symbols: <100> Si; closed symbols: <111> Si. Film thickness: (o), 0.8×10^{17} Sb/cm^2 (∿340 Å), (□), 1.1×10^{17} Sb/cm^2 (∿550 Å), (△), 1.5×10^{17} Sb/cm^2 (∿750 Å), (●) 0.8×10^{17}Sb/ cm^2 (∿340 Å), (▲), 1.4×10^{17} Sb/cm^2 (∿600 Å), (■), 10×10^{17} Sb/cm^2.

factor is about 30% under these conditions. The difference between <111>- and <100>-oriented silicon might indicate that solid-state diffusion plays a role. Diffusion with the aid of light pulses from a Ruby laser is different from that of the Nd-glass laser, as shown in Figs. 8 and 9: (i) The antimony-surface concentration peaks at laser-energy densities of about 1.6 J/cm^2 in <100> silicon but not in <111< silicon. This behaviour was observed for all film thicknesses investigated (400-2000 Å). (ii) The antimony-surface concentration peaks at film thicknesses of about 1400 Å for both <100> and <111> silicon for all laser-energy densities investigated (1.2-2.0 J/cm^2). (iii) Diffusion seems to be favoured in <111>-oriented silicon. (iv) Utilization factors are smaller than for Nd-glass laser, the highest measured being about 5%.

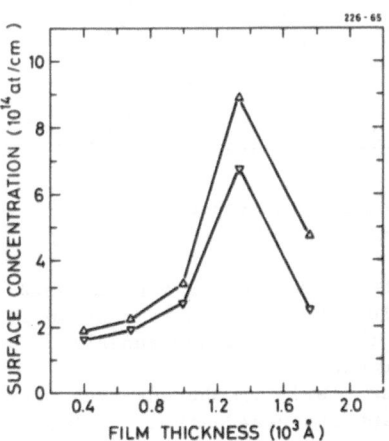

Fig. 8. Sb-surface concentrations in <100> Si single crystals after ruby-laser exposure of spin-on-deposited SbCl$_3$. (▽), 1.4 J/cm^2, (△), 1.8 J/cm^2.

The ion-channeling measurements gave in general the same results for all samples investigated, showing that in a surface layer of less than 400 Å, the antimony impurities are mainly in nonsubstitutional sites; below the layer, the substitutionality is very high, i.e., close to 100%.

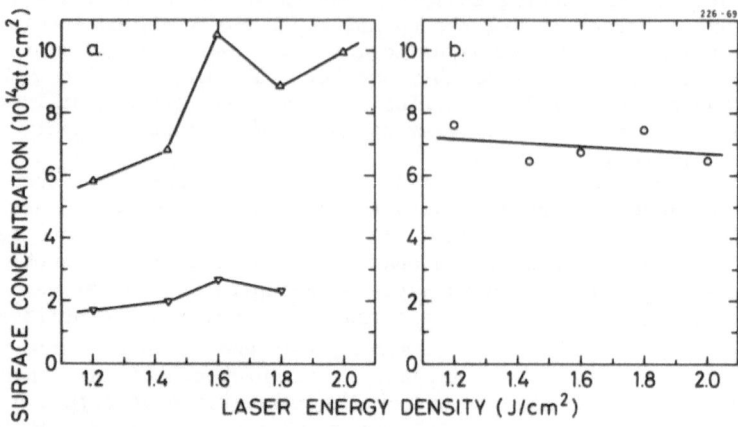

Fig. 9. Sb-surface concentrations in Si single crystals after ruby -laser exposure of spin-on-deposited $SbCl_3$ on <100> Si (a) and on <111> Si (b). Film thickness: (∇), 1.5×10^{17} Sb/cm^2 (∿675 Å), (Δ), 3×10^{17} Sb/cm^2 (∿1350 Å), (o), 1.4×10^{17} Sb/cm^2 (∿575 Å).

The sheet resistivities were in general smaller for the samples having been exposed to light from the Nd-glass laser than that from the ruby laser, in agreement with the observed surface concentrations. Typical values were between 30 and 60 $\Omega/\square$ for the Nd-glass laser and between 60 and 100 $\Omega/\square$ for the ruby laser. Results of test solar cells (0.5 cm^2) from ruby-laser-induced diffusion of antimony in silicon single crystals are collected in Table I. Good solar-cell parameters are obtained, and there seems to be no significant difference between the two experimental conditions.

TABLE I. Test solar cells from ruby-laser-induced diffusion of Sb in Si single crystals without antireflection coating.

Si-substr. orient.	Film thickness [Å]	Laser-energy density [J/cm^2]	Sb-surf. concentr. [at/cm^2]	Sheet resist. [$\Omega/\square$]	V_{oc} [mV]	I_{sc} [mA/cm^2]	FF	η [%]
<100>	1325	2.0	1.0×10^{15}	60	545	23.3	0.68	8.6
<111>	575	1.8	0.75×10^{15}	103	542	23.7	0.72	9.3

6. Conclusion

As a result of the analytical part of the present project, it has been possible to draw some conclusions with respect to the further development of low-cost solar cells.

Much effort has been devoted to doping procedures, using ion implantation without mass separation, in which case fluorine is often an unavoidable contaminant. From the present study, it seems clear that laser annealing with pulsed lasers is an adequate annealing procedure.

Due to the recent success of the much less costly incoherent-light annealing discussed briefly in this report, it appears necessary to include this possibility in future proposals for a continuation of the photovoltaic programme. From the findings of the present study, fluorine contamination in the near-surface region of the solar cell has to be avoided.

With respect to laser in-diffusion, this doping procedure has an interesting perspective because it is simple and needs no vacuum operations. The results in single-crystal silicon have been encouraging, but further evaluation is necessary.

A very promising production process seems to be ion implantation without mass separation, in combination with incoherent-light annealing, but more fundamental research is needed, especially in the case of polycrystalline silicon.

References
 (1) L.D. Nielsen and P. Balslev, Proc.3rd Photovoltaic Solar Energy Conf., Cannes, France, 1980, ed. W. Palz (Reidel, Dordrecht, 1981) p. 698
 (2) L.D. Nielsen, Phys.Scripta 24 (1981) 390
 (3) J.L. Borowitz, F. Nielsen, and G. Sørensen, in Ref. (1), p. 635
 (4) F. Nielsen, A. Nylandsted Larsen, and G. Sørensen, Proc.10th Nordic Semiconductor Meeting, Elsinore, Denmark, 1982, p. F1:1
 (5) A. Nylandsted Larsen, F. Nielsen, and G. Sørensen, Proc. 4th Photovoltaic Solar Energy Conf., Stresa, Italy, 1982, eds. W.H. Bloss and G. Grassi (Reidel, Dordrecht, 1982) p. 901
 (6) A. Nylandsted Larsen and R.A. Jarjis, Appl.Phys.Lett. 41 (1982) 366
 (7) G. Sørensen, Nucl.Instrum.Methods 186 (1981) 189
 (8) M. Oron and G. Sørensen, Appl.Phys.Lett. 35 (1979) 782
 (9) S. Eskildsen, in: Ion Implantation into Metals, eds. V. Ashworth, W.A. Grant, and R.P.M. Proctor (Pergamon Press, 1982) p. 315
(10) M.Y. Tsai, D.S. Deng, and B.G. Streetman, J.Appl.Phys. 50 (1979) 188
(11) A.K. Jain, V.N. Kulkani, and D.K. Sood, Appl.Phys. 25 (1981) 127

OPTIMIZATION OF POLYCRYSTALLINE SILICON SOLAR CELLS
PRODUCED BY ION-IMPLANTATION AND PULSED LASER ANNEALING

Authors   : W. SINKE, W. VAN SARK, S. DOORN, F.W. SARIS

Contract number : ESC-R-021-NL

Duration   : 36 months   1 July 1980 - 30 June 1983

Total budget   : f 1 650 590   CEC contribution: f 772 441

Head of project : Prof.dr. F.W. Saris, FOM-Institute

Contractor   : Stichting voor Fundamenteel Onderzoek der Materie

Address   : FOM-Institute for Atomic and Molecular Physics
                            Kruislaan 407
                            1098 SJ Amsterdam

Summary

Silicon solar cells have been made by shallow, mass-analyzed ion-beam implantation followed by Q-switched ruby laser annealing. The influence of surface texture of the 3" single-crystalline (100) substrates and of a thermal treatment at $400^{\circ}C$ during 40 min. either prior to, during or after pulsed annealing on the performance of the cells has been studied. Structure and composition of implanted and annealed silicon have been investigated by RBS and channeling. Solar cell performance was characterized by measuring the I-V curve. In general, cutting the 45 x 45 mm^2 center part out of the 3" wafers gave improved results. Both the FF and V_{oc} are superior in case of pulsed laser annealing of heated substrates. In contrast to earlier results on polished wafers, the etched wafers showed metallization problems which are associated with the combination of screen printing and sintering on the rather shallow p-n junctions made here. Best results, 12.7% AM1 efficiency, were obtained on wafers with pyramide-shape surface texture.

1.1 Introduction

The objective of this research project is to define the optimum fabrication mode for 100 cm^2 poly-Si solar cells with 10% efficiency. Technical and economical requirements for producing cheap silicon solar cells might be fulfilled using ion implantation and pulsed laser annealing for: 1) this is a low temperature process applicable to solar grade poly Si, 2) majority carrier profile in front layer is well controlled resulting in high blue response, hence high conversion efficiency, 3) only dry steps are involved which are suitable for automation and mass production.

Our approach is to define the optimum parameters first on single crystalline silicon and then to try and optimize further on poly-Si. In previous reports we have shown that ion-implantation and pulsed laser annealing may give results which are comparable to classical diffused cells in the same substrate material (1)(2). Those studies, however, were typical laboratory experiments making use of small (1x1 or 2x2 cm^2) polished silicon samples and evaporated contacts, whereas the solar cell manufactoring industry makes use of 3 inch diameter (or sometimes larger) wafers which are etched only and for metallization screenprinting is preferred. Thus we have extended our earlier work to such conditions.

Moreover, we have found that heating the substrate to 450°C for 10 min. after pulsed laser annealing. which was necessary to sinter evaporated front and/or backcontacts, had a favourable effect on cell performance. Thermal treatments at moderate temperatures before or during pulsed laser annealing have also been claimed by various authors to yield substantial improvements. Therefore, in the present work we compare the influence of thermal heating at 400°C during 40 min. either prior to, during or after pulsed laser annealing.

1.2 Experimental

The substrates used were (100)-oriented, single crystal p-type silicon wafers of 3" diameter. Before implantation, wafers were treated in NaOH which resulted in either a slightly roughened surface, or a rather pronounced structure, or a pyramide shaped surface texture. For comparison a set of wafers was subjected to an acid etch. The wafers were ion-implanted with a mass-analyzed 10 keV As beam up to a dose of 3x10^{15} cm^{-2}. Laser annealing was carried out with a single stage 5/8" diameter Q-switched ruby laser with maximum energy output of 4J and a pulse length of 20-25 ns. The laser-light was shone through a converging lense and a glass plate diffuser. The 3" wafers were annealed with overlapping pulses in a triangular pattern, with steps of 6 mm in one direction and 5 mm in the other. Energy output of the laser was adjusted such that each spot was at least once irradiated with an energy density in the range 1.0-1.4 J/cm^2 in the case of non-texturized wafers, and in the range of 0.8-1.2 J/cm^2 in the case of texturized wafers. For each type of surface texture 4 sets of 2 samples were made, each under different annealing conditions as follows: pulsed-laser annealing without any thermal treatment or with thermal annealing at 400°C for 40 min. before, during or after the laser. This annealing procedure was carried out in open air. The implanted and annealed wafers were processed further to provide a front- and backcontact by screenprinting and sintering. For comparison also evaporated contacts were applied in some cases. In total over 100 cells have been fabricated and tested in this program.

1.3 Results

In most cases cutting the 45x45 mm^2 center part out of the 3" wafers gave improved results. Therefore most of the data on the cell characteristics given in table I are shown for 45x45 mm^2 cells, except when indicated otherwise. The first set of tabulated data shows that texturzing the front surface improves I_{sc} and thereby η significantly. It should be noted, however, that the results have been given for the best cells as frequently rather poor fill factors were obtained due to the metal sintering step, see below. In the second set of tabulated data a comparison is made for screenprinted or evaporated contacts on wafers with a pronounced NaOH etched surface structure. The fill factors are clearly much lower in the case of screenprinted contacts due to the relatively high sintering temperature. Figure 1 shows an optical microscope picture of the texturized surface with small pyramide shaped surface facets. Figure 2 shows the rather pronounced square block structure. Clearly in the latter case the surface has areas which are parallel to the direction the ion beam during implantation and parallel to the laser beam during annealing. Hence these parallel surface areas will contain hardly any p-n junction and during the metallization step the metal will enter the cell there, thus destroying its fill factor. The last set of tabulated data shows that the results for acid etch are not as good as for NaOH etch and again fill factors are better for evaporated contacts.

Concerning thermal treatments one may conclude from table I that V_{oc} is higher when pulsed laser annealing is performed on heated substrates. We have observed that for these annealing conditions the p-n junction depth is about twice as deep (0.3 μ) as when laser annealing is done at room temperature. In addition one may expect that fewer defects are quenched-in if pulsed annealing is done on heated substrates. Consequently indiffusion of metal and shunting becomes much less of a problem as we have verified by measuring diode characteristics as function of post-evaporation annealing temperatures. Unfortunately deeper junctions also lead to a lower I_{sc} in most cells, probably due to enhanced photon absorption the effect of which may be reduced by using P^+ implantation instead of As^+.

2. Conclusions

The rather shallow p-n junctions, obtained after ion-implantation and pulsed laser annealing, are observed to be prone to shunting, high recombination currents and poor fill factors on silicon substrates which have not been polished. Especially when etching has resulted in pronounced square block structures with surfaces parallel to the ion and laser beam direction, will sintering of the front metalcontact lead to indiffusion of metal into the cell. This can be avoided by applying texturizing leading to pyramide-shaped surface facets. Comparison of thermal treatments prior to, during or after pulsed laser annealing indicates that pulsed laser annealing on heated substrates give best results: η = 12.7% AM1 efficiency measured on the 45x45 mm^2 center part cut out of a 3" wafer.

3. Acknowledgement

The authors wish to thank J. Donon of Photowatt Int. S.A. for his collaboration in preparing and characterizing the above cells.

4. References

(1) D. Hoonhout, F.W. Saris, J. Michel, C. Fages, E. Fabre, Proc. 15th IEEE Photovoltaic Specialists Conf. Orlando, 1981 p.253

(2) W. Sinke, D. Hoonhout, F.W. Saris, 4th E.C. Photovoltaic Solar Energy Conference, Stresa 1982, p. 1029

Figure 1. Surface of a texture-etched wafer   25μm

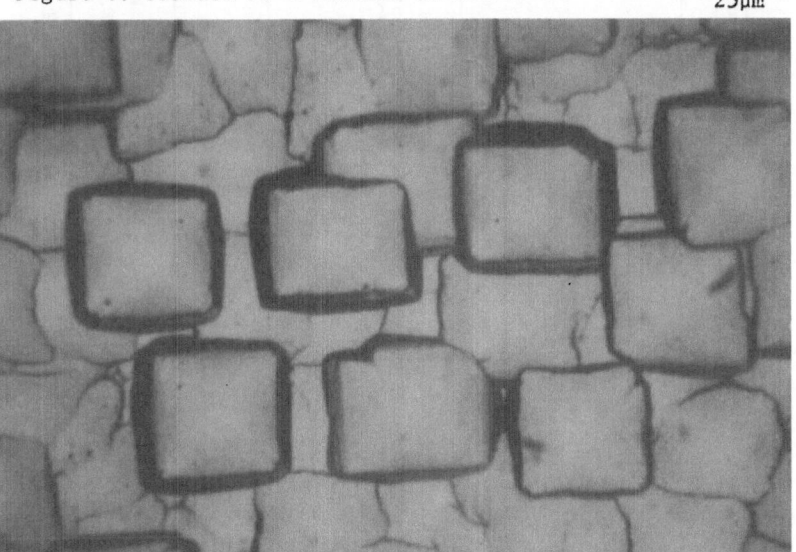

Figure 2. Surface of a NaOH(2)-etched wafer   100μm

TABLE 1. Comparison of screenprinted and evaporated contacts and of various surfaces Measurements on 45x45 mm^2 cells with A.R.C., best cells.

THERMAL TREATMENT	screenprinted, BSF NaOH(1) etch (little structure)				screenprinted, BSF texture etch			
	V_{oc} (mV)	J_{sc} (mA/cm^2)	FF	η(%)	V_{oc} (mV)	J_{sc} (mA/cm^2)	FF	η(%)
R.T.	550	30.6	0.57	9.7	540	32.1	0.51	8.8
40' 400°C before p.l.a.	545	31.6	0.61	10.6	510	31.1	0.50	7.9
p.l.a. at 400°C	565	29.6	0.67	11.3	560	34.1	0.66	12.7
40' 400°C after p.l.a.	540	30.1	0.58	9.4	535	35.1	0.60	11.2

THERMAL TREATMENT	screenprinted, BSF NaOH(2) etch (pronounced struct.)				evaporated, no BSF NaOH(2) etch (50x50 mm^2)			
	V_{oc} (mV)	J_{sc} (mA/cm^2)	FF	η(%)	V_{oc} (mV)	J_{sc} (mA/cm^2)	FF	η(%)
R.T.	555	25.7	0.56	8.0	550	27.6	0.60	9.1
40' 400°C before p.l.a.	555	26.2	0.52	7.6	555	27.4	0.67	10.2
p.l.a. at 400°C	555	25.9	0.52	7.6	560	25.8	0.72	10.4
40' 400°C after p.l.a.								

THERMAL TREATMENT	screenprinted, BSF acid etch (irregular structure)				evaporated, no BSF acid etch			
	V_{oc} (mV)	J_{sc} (mA/cm^2)	FF	η(%)	V_{oc} (mV)	J_{sc} (mA/cm^2)	FF	η(%)
R.T.	560	26.9	0.48	7.2	550	26.4	0.71	10.3
40' 400°C before p.l.a.	565	26.9	0.64	9.7	550	26.5	0.68	9.9
p.l.a. at 400°C	565	24.7	0.33	4.6	560	27.2	0.71	10.8
40' 400°C after p.l.a.								

LOW COST IMPLANTATION INTO SILICON

Author : J.C. MULLER

Contrat number : ESC-R-080-F

Duration : 18 months   1 January 1982 - 30 June 1983.

Head of : P. SIFFERT, Groupe de Physique et
project   Applications des Semiconducteurs (PHASE)

Contractor : Institut National de Physique Nucléaire et de
             Physique des Particules (IN2P3)

Adress : CENTRE DE RECHERCHES NUCLEAIRES
         Groupe de Physique et Applications des
         Semiconducteurs
         23, rue du Loess
         67037 STRASBOURG CEDEX (FRANCE)

Summary

The techniques of ion implantation and surface annealing
are one of the possible means of bringing a substantial reduction in the cost of manufacturing solar cells. The work of the
present report concern a large area, high current ion doping
procedure called "ion incrsutation" technique, that is,
implantation without mass separation. The annealing of the
ions induced damage is performed by using pulsed techniques,
especially high power laser. A full characterization of the
layers has been undertaken including ion distribution, damage
(macroscopic and microscopic) like residual defects) subsis-
ting after the liquid phase regrowth, post thermal treatment
use to remove this point defects. Ion "incrustation" with high
repetition rate YAG laser annealing where applied to the
production of monocrystalline or polycrystalline silicon solar
cells of 11 cm^2 area. The optimum photovoltaïc efficiencies
obtained were up to 13,4% AM1[+] for mono with open-circuit
voltage of 585 mV. Up to now, the first results on WACKER or
CGE polysilicon material give efficiencies over 10 %.

+ The solar cells results have been obtained during the collaboration with Laboratoire CGE Marcoussis contract ESC-R-048F
1.7.80-30.10.81.

1.1 Introduction

Sophisticated ion implantation machines have become available in a recent past. The popularity of this surface doping procedure is related to a series of advantages when compared to the conventional thermal diffusion. Today, the tendency is to increase the beam current density in order to achieve higher production rates over larger wafers ($\phi \simeq 4"$). However, this evolution complicates strongly the beam hangling and results in much higher cost of equipment.

Here, we propose a fully different method to achieve high current density ion doping by using a very simple equipment, which is of low cost, since no magnet neither highly stabilized power supplies are needed. Furthermore, the dopant profile, which is not gaussian, is advantageous for some devices like solar cells. We called this technique ion incrustation or AMI (atomic and molecular incrustation).

In this paper, we will review the research we did to develop the experimental set-up (1) and to characterize the doped layers as well as the devices (solar cells) realized by this procedure.

2.1 Description of the equipment

In principle, our technique of surface doping by ion bombardment is quite simple (figure 1) :
- from the plasma of an ion source, both atomic and molecular ions are extracted and accelerated directly towards the silicon target, without any mass analysis.
- the severe radiation damage induced by these low speed heavy ions is annealed by a pulsed laser or an electron flash of high energy, put on line in the same machine.

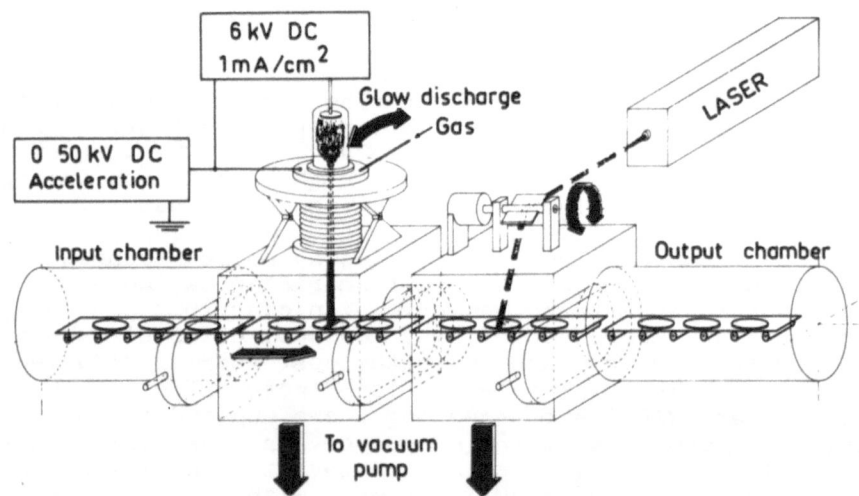

Figure 1 : Principle of the implantation and annealing machine with possibility of continuous operation of both processes.

2.2 Ion implantation machine
- with a single beam source (figure 1)

The ion source is a glass chamber, in which a glow discharge is established in presence of a gas containing the dopant, generally B_2H_6, BF_3 or BCl_3 for B and PH_3 or PF_5 for P, at a pressure of 10^{-2} to 10^{-3} torr. The plasma is adjusted through a 0-8 kV d.c. voltage, the ions being extracted and accelerated directly towards the samples, without any mass separation at voltages up to 50 kV. Current densities up to 1 mA/cm^2 can be obtained, the ions consist of both atomic and molecular species having not all the same velocity, by contrast to conventional ion implantation techniques.

The whole source can be tilted perpendicularily to the displacement of the samples, in order to cover large areas in a continuous production line. Besides the small size of the implant set-up (20 cm in length approximately), no stabilized power supplies are requested, due to the absence of any magnet A strong cost reduction is, therefore, achieved.

By contrast to the situation in conventional ion implantation, here the beam consists of a mixture of atomic and molecular ions which have not necessarily all the same velocity. For example with PF_5, only about 30 % of the ions are $^{31}P^+$, the others being PF^+, PF_2^+...(figure 2). When these molecular ions hit the silicon surface, they split into the components, each one having its proper energy, which is in the ration of atomic masses : for example for 15 keV PF_3^+, the p projectile will have 5.3 keV, whereas F has 3,2 keV. It should be mentionned that molecular ion implantation has been employed before, but in conventional implanters (2), the situation here is fully different, the real profile of phosphorus ions will be the addition of each contribution coming from the split of the impinging molecules as we have reported (figure 3) for $\sum_n PF_n$ (n = 0 at 5) at a total energy of 15 keV.

Two remarks can be made :
- we have a higher doping level at the surface ;
- the junction will be at the same position as classical $^{31}P^+$ implantation.

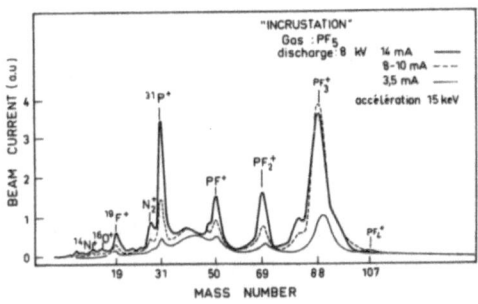

Figure 2 : Mass analysis of the ions extracted of the glow discharge ion source with PF_5 gas.

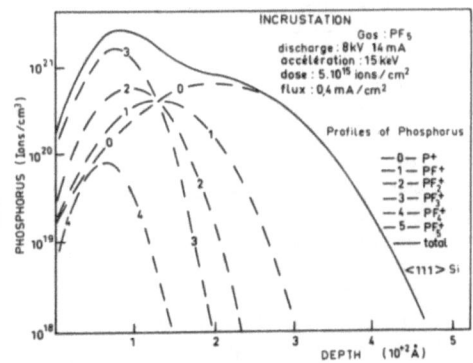

Figure 3 : Profile of phosphorus ions coming from the split of the molecules of PF_5.

- with a multiple beams ion source :

A more sophisticated source, which can be employed on the same implantation machine, has been employed. This source, with a hot filament is based on a multi-grid extraction (see figure 4) with more than 300 holes, which are small elementary ion guns. The total ion current is 30-40 mA for a beam surface of 40 cm^2. As a result the production rate will be increased by a factor in excess of 30.

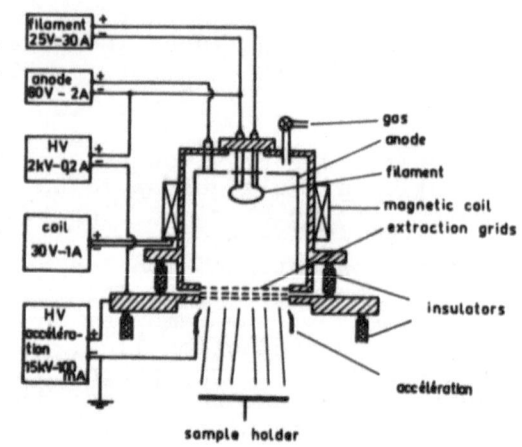

Figure 4 : Chematic of the multiple beam ion source.

2.3 Pulsed beam annealing

After the high dose rate incrustation, the silicon surface becomes fully amorphized. In order to restore the original cristallinity, the damaged layer is molten down to the virgin silicon interface by means of short duration pulses (lasers or electron gun).

- with pulsed laser

A high repetitive (up to 10 KHz) YAG laser operating at 0.53 μm giving by focusing a small spot diameter (0.1 mm) of 100 nsec. duration in the monomode regime at energy densities up to 3 J/cm^2. Large areas are covered by a microprocessor controlled set-up which determines the overlapping.

- with pulsed electron-beam

A pulsed electron beam (3) which has the particularity to have two storage capacities, a first one to initiate the discharge before the mean capacitor is entloaded. The energy of the electrons lies between 10-20 keV and the pulse duration is 0.3 μs, the beam diameter being limited to 13 mm.

The exact mechanism of regrowth is still under discussion (4,5), but most of the consequences can be described by considering liquid phase epitaxy. A computer model has been developped (6). The main result, obtained for the YAG laser at 0,53 μm are given on figure 5. together with those published earlier for a ruby laser(6).

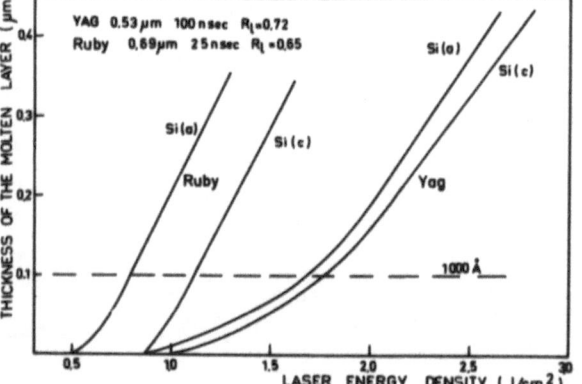

Figure 5 : Melt depth as a function of energy density.

3.1 Properties of the implanted layers
3.1.1 Crystal damage
Rutherford backscattering (RBS) of $^4He^+$ ions of 1 MeV energy under channelling conditions, has been used to investigate the amount and distribution of the displaced silicon atoms during the strong bombardment. A comparison of an analysis performed by our procedure and conventional ion implantation is reported on figure 6 for BF_3 glow discharge and $^{11}B^+$ well as BF_3^+ mass separated ions. As expected, much more damage are created by glow discharge. The defects extend to a depth corresponding to the range of the lightest ions (B^+). The surface is amorphized about 1000 Å in depth.

Previously published work (8) has shown that, by conventional thermal annealing of the glow discharge implantation, it is not possible to fully restore the original cristallinity even at high temperature and long times. The results are definitively better after laser annealing (9) : as seen on figure 7, only a 200 Å thick layer remains in close surface vivinity. The origin of this residual damage is may due to the presence of precipitation of fluorine (10) as has been seen by TEM.

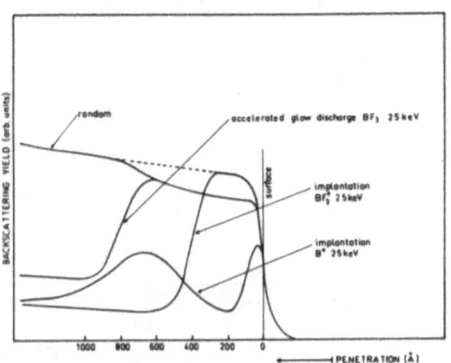
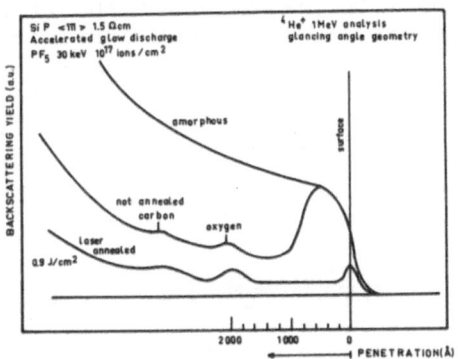

Figure 6 : Backscattering spectra of 1 MeV $^4He^+$ ions under channeling conditions for a classical ion implantation of boron and BF_3^+ and for an accelerated glow discharge of BF_3 at the same energy.

Figure 7 : RBS spectra under channeling conditions of a damaged layer by PF_5 molecular ions annealed by a ruby laser pulse of 0.9 J/cm².

3.2 Dopant profile
The dopant profiles have been recorded by secondary ion mass spectrometry (SIMS). It appears that the distribution of the dopants extends deeper into the silicon by increasing the laser energy density, the surface concentration being reduced. By scanning the laser beam, as previously indicated (10) the dopant spread is even more pronounced, due to the fact that each surface has been illuminated more than one time.

3.3 Residual defects after pulsed annealing
However, despite the good cristalline perfection after

pulsed annealing, in practice, no structure has been realized
up to now which has performance in excess of conventional
thermal diffused cells, due to the presence of residual
microscopic defects. Most of them are minority carrier traps,
which introduce compensation near the junction (11). As we
have previously shown, for ruby laser annealing on incrusted
layer as well as on classical implanted structures (12) that
the residual damage results mainly from damage generated in
the tail deeply penetrating of the dopant distribution (1 μm)
and which is not reached by the laser annealing.
        For YAG laser annealing at a equivalent energy (figure 6)
the number and the concentration of residual defects diminishes significantly (13). This can be explained by considering that :
- the pulse duration of the YAG (100 ns) is much longer than
that of the ruby (20 ns), giving rise to a smaller recristallization speed and a deeper penetration of the heat ;
- the high repetitive YAG laser, with overlapping pulses can
increases the mean temperature of the surface zone and melt
deeper-in.
        The results we observed with electron pulses (13) confirm
our interpretation of the origin of the levels observed with
lasers, namely a non sufficient depth of melting with regards
to the ion tailing effect. The electron annealed samples show
defect level due to the fast anneal process and not to implantation. Such defects are also visible if virgin silicon is
laser annealed.
        We found that all these levels disappear after a further
thermal annealing at rather low temperature (350-600°C) for a
short time (10-20 min).

4.1 Application to the preparation of solar cells
        Solar cells have ben preapred on both single and poly-
cristalline silicon by using conventional implantation or glow
discharge followed in all cases by laser annealing and a short
thermal heating, which is found necesary by all authors (14-
15). This post treatment can be done during the sintering
of the back contact in a industrial process.

4.2 Role of the implantation parameter and post annealing
        On figure 8, we have reported the variation of V_{oc} as a
function of the post laser annealing temperature for both kind
of pulsed laser annealing (ruby and YAG) for a very low
(2 keV) and (10 keV) incident acceleration of the ions in the
incrustation technique. It appears that the best annealing
temperature lies between 600 and 660°C for ruby laser treatment and around 550-600°C for the YAG laser. Furthermore, the
open circuit voltage V_{oc} increases for a given anneal temperature when the energy of the projectiles is diminished ; this
is true also for classical ion implantation (10) and may be
due to the reduction of the deeply penetrating "tail", we
already discussed (11-13).
        V_{oc} is systematically lower for ruby when compared to YAG
laser processing, for a same implantation energy. This may be
explained by the fact that we have a lower number of defects
as seen by DLTS after high repetition rate overlapping pulse

annealing.

The characteristics of the junction (in particular V_{oc}) are correlated at the quality of the recristallization during the pulsed process. We have shown that this regrowth start in the "tail" region of the bombardment doping procedure and the quality of the recristallization depends on the number of residual defects seen in this region, so that we found a direct correlation of the point defects in the depletion re - gion (starting from the melt depth) by DLTS and the quality of regrowth (seen on the dark I-V and C-V characteristics (10) and V_{oc} value).

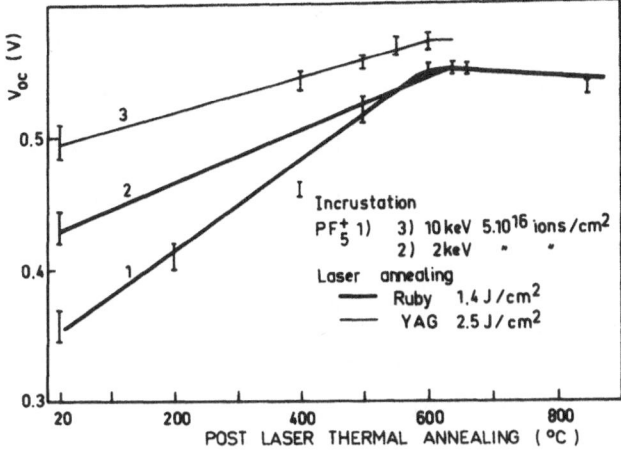

Figure 8 : Evolution of V_{oc} as a function of the post laser annealing temperature for both kind of pulsed laser annealing (ruby and YAG) for a very low (2 keV) and 10 keV) incident accelera tion of the incrusted ions.

4.3 Electrical characteristics

The cells have been tested on a similator at 100 mW/cm^2 (AM1). The performance have been measured at 28°C. The results obtained for "Incrustation" (AMI) with different annealing procedures (essentially Nd-YAG laser and electron beam) are given on the Table. The results obtained with the ruby laser have been previously published (10) as well as comparisons with conventionally implanted cells having been thermally or laser annealed (17).

The performance are given for cells having 10 cm^2 active area and the efficiency is taken after coating with Si_3N_4. Preliminary result obtained for pulsed electron beam annealing are given for 1 cm^2 cells (active surface) with SiO_x coating.

By considering the efficiencies achieved, it appears that incrustation give the same results that those obtained with classical ion implantation procedure. In practice, the effi- ciencies reported on the real active surface : 13-14 % on sin- gle crystalline silicon and 10-11 % on polycrystalline mate- rial are comparable to that of conventional commercial diffu- sed cells.

5.1 Conclusion

Our goal was, essentially, to demonstrated that ion "incrustation" is able to be used for solar cell manufacturing

exactly like classical ion implantation. By considering the
above results, it appears that this goal has been reached. The
two ion doping procedures can enter on automatic production
line, the absence of chemical treatments constitutes a further
advantage over thermal diffusion. The cost of a non mass analyzed ion doping technique is much lower than the cost of high
current ion implantation machine (18).

REFERENCES
1.   J.C. MULLER, J.P. PONPON, J. KUREK and P. SIFFERT, ANVAR
PATENT N° 77 08 138 18.3.77.
2.   H. MÜLLER, H. RYSSEL, I. RUGE, Ion Implantation in Semiconductors Springer (1971) 85.
3.   J. GEERK and F. RATZEL, Kernforschungszentrum Karlsruhe
(W.G).
4.   W.L. BROWN, Laser and Electron Beam Processing of Materials Academic Press (1980) 21.
5.   J.A. VAN VECHTEN, R. TSU, F.W. SARIS, D. HOONHOUT, Phys.
Lett. 74a (1979) 417 .
6.   R.O. BELL, M. TOULEMONDE and P. SIFFERT, Appl. Phys. 19
(1979) 313.
7.   M. TOULEMONDE, unpublished.
8.   J.C. MULLER, A. and J.J. GROB, R. STUCK, J.P. PONPON and
P. SIFFERT, 13th IEEE Photovoltaïc Specialists Conf. 5-8/6/78
Washington (USA) (1978) 711.
9.   J.C. MULLER, A. GROB, J.J. GROB, R. STUCK and P. SIFFERT
Appl. Phys. Lett. 33 (1978) 287.
10. J.C. MULLER and P. SIFFERT, Intern. Workshop on Ion Implantation Laser Treatment and Ion Beam Analysis of Materials,
Bombay (India) Feb. (1981) Radiation Effects 63 (1982) 81.

11. A. MESLI, A. GOLZENE, J.C. MULLER, B. MEYER, C. SCHWAB
and P. SIFFERT, Material Research Society Meeting Laser and
Electron Beam Interaction with Solids Boston (USA) (1981) to
be ed. by B.R. Appleton and G.K. Celler.
12. A. MESLI, J.C. MULLER, D. SALLES and P. SIFFERT, Appl.
Phys. Lett. 39 (1981) 159.
13. A. MESLI, J.C. MULLER and P. SIFFERT, proposed at Appl.
Phys. Lett.
14. F. ZIGNANI, R. GALLONI, L. PEDULLI, G.G. BENTINI,
M. SERVIDORI, F. CEMBALI and A. DESALVO, European Photovoltaïc Solar Energy Conf. BERLIN (W.Germany) (1979) 145,
D. Reidel Publ. Co., Holland.
15. J.A. MINNUCCI, K.W. MATTHEI, A.C. GREENWALD, R.G. LITTLE
and A.R. KIRKPATRICK, Laser and Electron Beam Processing of
Materials, Material Research Society Symposium, Cambridge USA
(1979) Ed. J.W. White and P.S. Peercy (1979) 658.
16. J.C. MULLER, P. SIFFERT, J. MICHEL and E. FABRE Conf.
Ref. 15, p. 278.
17. J.C. MULLER, A. MESLI, P. SIFFERT, J. COM NOUGUE and
J.P. DUMAS , 4th EC Photovoltaïc Solar Energy Conf. STRESA
(Italy) (1982) p. 994.
18. H. GOLDMAN and WOLF, 14th IEEE Photovoltaïc Specialist,
Conf. SAN DIEGO (USA) (1980) p. 923.

TABLE

	Impl. Cond. P-Si ⟨111⟩	Annealing Cond. Pulsed	Post treat.	R/□ Ω	V_{oc} mV	I_{SC} mA/cm²	FF	η % coating No	η % coating Yes
P-Si ⟨111⟩ — SILICON MONOCRYSTALLINE	AMI PF$_5^+$ 10 keV 5 10^{16} ions/cm²	Ruby 1.6 J/cm²	660°C 30 min	20	540	25.3	0.76	10.4$^+$	15.0$^+$ Si O$_x$
		Elec. gun 1.5 J/cm²	660°C 30 min	16 16	533 552	24.2 33.4	0.6 0.6	7.8$^+$	11.1$^+$ SiO$_x$
		Nd-YAG 2.0	550	30	564	32.4	0.64		Si$_3$N$_4$ 11.6
		2.5	600	23	586	29.7	0.72		12.5
					582	30.8	0.71		12.7
					571	33.3	0.70		13.3
		3.0	600	21	585	32.6	0.70		13.4
	31P$^+$ 20 keV 5 10^{15}	Nd-YAG 2.5	600	25	575	32.7	0.70		13.2
		Classical Thermal 850°C			580	32,6	0.72		13.6
POLYCRISTAL CGE — Wacker	PF$_5^+$ 10 keV 5 10^{16}	Nd-YAG 2,5 J/cm²	600		539	28,1	0,71		10,7
POLYCRISTAL CGE	31P$^+$ 10 keV 2,5 10^{15}	Nd-YAG 2,5 J/cm²	600		531	25,3	0,71		9,5

Cells manufacturing and testing by CGE Marcoussis
(AMI 100 mW/cm², active area 10 cm²)
+ small cells active area 1 cm²

DESIGN, CONSTRUCTION AND OPTIMIZATION ON THE INDUSTRIAL PROTOTYPE SCALE

OF A FURNACE ABLE TO PRODUCE POLYCRYSTALLINE SILICON INGOTS

AS MATERIAL FOR SOLAR CELLS

Author   : J. FALLY

Contract number   : ESC-R-042 F

Duration   : 24 months - 1 july 1980 - 30 June 1982

Head of project   : Mr. J.P. DUMAS, assisted by Mr. J. FALLY
                           Materials Department - Laboratoires de
                           Marcoussis

Contractor   : Laboratoires de Marcoussis, Research Center of
                          C.G.E.

Address   : Laboratoires de Marcoussis - F - 91460 -
                          MARCOUSSIS

SUMMARY

The feasibility of the elaboration of semicrystalline silicon ingots
by the process of unidirectional crystallization in the crucible
has been demonstrated on the scale of a laboratory furnace in a
previous step of our study.
The aim of the present work concerns i) the scale-up : design and
construction of a crystallization furnace to the industrial prototype level for the elaboration of 25 kg maximum unitary weight
ingots, and ii) the optimization of the properties of the semicrystalline material thus obtained, starting from electronic grade
silicon charges.
The main results obtained during this study are :
. basaltic semicrystalline silicon ingots 20 kg in weight were
   currently produced by means of the scaled-up furnace.
. photovoltaic efficiency of the elaborated semicrystalline cells
   (100 cm^2) using this material reach the 10% range.

1. INTRODUCTION

In order to sufficiently lower the production cost of silicon solar cells, we have developed a process intended to obtain silicon semicrystalline blocks to be wafered afterwards. The first part of our program has consisted in a feasibility study of the elaboration of such ingots by a unidirectional crystallization method using a fast variant of the so-called Bridgman process. The study led to a material quality sufficient to make solar cells of reasonably good photovoltaic efficiency ($\eta_{AM1} \simeq 8\%$) ; the weight of the silicon blocks was about 1 kg (1).

The main limitations of this material were related to crystallographic imperfections, combined with the relatively high bulk-grain impurity content coming from the crucible walls, both limiting the short circuit current density.

Beside the necessary correction of these quality parameters, and in order to thoroughly work out the economic potential of the chosen process, it was needed to increase the capacity of the equipment. So the work here presented concerns the design and construction of an industrial prototype furnace, able to permit the adaptation of the directional solidification process to the elaboration of semicrystalline silicon 20 kg ingots and to optimize the characterisatics of basaltic, electronic grade semicrystalline silicon blocks.

The use of a heating mode different of that used in the laboratory furnace (i.e. induction vs. resistance), combined with the adaptation to the important increase in weight of the ingots led us to modify consistently the crystallization method.

The objectives of the contract were :
- the construction and operation of a 20 kg-capacity furnace,
- the optimization of the elaboration parameters of 15-20 kg semicrystalline ingots in order to obtain 100 x 100 mm cells of an AM1 efficiency greater than 8%, and smaller cells in the 10% range.

2. ELABORATION OF SEMICRYSTALLINE INGOTS

A view of the scaled-up crystallization furnace is given in Fig. 1. The main features of this installation appears on Table I, which gives the current specifications.

As said above, we have introduced a modification of the crystallization method, which is described in Fig. 2. This figure indicates that the seedless crystallization step is operated in a crucible stationary with respect to the working coil (direct induction in the crucible walls).

During the whole solidification step the heating power is held constant. Heat extraction occuring by radiation directly from the bare bottom of the graphite crucible is sufficiently active to control the heat balance at the freezing front.

In comparison to the straight Bridgman process in which the crucible moves down the thermal gradient, the directional crystallization using the scheme described in Fig. 2 leds to a steadier solidification rate, and the shape of the solid/liquid interface is more regular, i.e. nearly flat over a substantial extent of its surface.

These two characteristics greatly improve the crystallographic properties (2) of the basaltic semicrystalline material (grain dimensions, EPD, lineages concentration). On the technological point of view, this evolution of the solidification technique permits a simplification of the furnace construction.

The original procedure we have experienced in order to extract the ingot out of the crucible during the first part of the program (1), was reconducted with the 20 kg furnace.

Some economic aspects of this crystallization process are indicated in Fig.3. The direct-cost of wafers includes : raw-materials, consumable goods, direct labor, amortization of investment, energy. The direct-cost shows a great sensibility to the ingot weight, especially under 30 kg. The production furnace capacity specification will be 35 kg.

The number of crucible uses is also of a great influence, the specification being 25. We have experienced over 15 uses of the same crucible without difficulty. The cycle duration is of a lower influence on the direct-cost of wafers, the chosen specification is the 10 hrs production cycle.

The general conditions concerning the block-sizing and wafering are somewhat optimistic, but are considered accessible in the mid-term future.

The current status of the CGE ingots growth method is summarized in Table II. The main attributes of the basaltic ingots elaboration process are : high throughput rate, reusable crucible, moderate furnace investment, and possible use of Solar Grade Silicon.

3. PROPERTIES OF THE SILICON INGOTS

The current material quality criteria of the obtained silicon ingots or wafers are summarized on Table II. We call "finished" ingot the state of the material just prior to the I.D. wafering, after the block-sizing operation.

4. SOLAR CELLS PERFORMANCES

Solar cells are realized using square 100 x 100 mm wafers, obtained by I.D. slicing industrial-sized ingots (15-20 kg unitary weight). The standard process used by PHOTOWATT S.A. includes : NaOH etching, $POCl_3$ diffusion silver screen-printing metallization and TiO_2 A.R. coating. Table II shows the photovoltaic results obtained on wafers cut parallel to the growth direction ("vertically sliced").

The statistic shown is related to about 70 cells taken in one ingot. The dispersion of results, as indicated by σ_{N-1} is relatively narrow, showing a good homogeneity of properties in planes parallel to the growth direction.

In order to study the variation of properties in the growth direction, we have laser-scribed the square 100 cm^2 cells in four 25 cm^2 cells, distributed vertically in four sections extending from the bottom to the top-regions of the ingot. The best results obtained on a "bottom" representative 25 cm^2 cell are given on Table II , AM1 efficiency of 10.25% was reached, and 10.1% on 50 cm^2 area solar cells. Photovoltaic efficiencies are related to the whole exposed area of the cell, including the metallization.

The best photovoltaic efficiency of 100 cm^2 cells is 9,45% (Table II, Fig.4).

The study of the vertical distribution of the photovoltaic parameters in this 15 kg ingot led to the results summarized in Fig. 5. We have observed an almost general trend on our basaltic material of decreasing photovoltaic characteristics, namely JSC, from the bottom to the top of ingots. This is ascribed to impurity segregation effect due to the directional crystallization (see also (3)).

This segregation of impurities on the top of the ingot is clearly shown in Fig. 6, the observed profiles on Fe, Zr, Mo and Ni constitutes a sufficient cause to the generally observed JSC lowering in the upper parts of the basaltic ingots.

The above exposed photovoltaic and analytical results were obtained with an ingot having the resistivity profile indicated in Fig. 7.

In Fig. 8, we show an experimental relationship between JSC and L_D (diffusion length of minority carriers), these results being obtained on basaltic ingots. L_D measurements were derived from the spectral response curves of solar cells (SPV method), the values of the absorption coefficient of silicon used are those published in (4).

The study in progress of the influence of slicing direction, i.e. "vertical" vs."horizontal" cells, in connection with the orientation of metallization pattern, was initiated following the hypothesis of a possible improvement of the photocurrent extraction, deduced from (5) , if the metallization fine lines are normal to the greatest length of the grains. The statistical comparison between the mean values of efficiency obtained with "vertical" and "horizontal" comparable cells populations, shows a significant superiority of 1.5% efficiency of the vertical solar cells. The generalization of this result, now established on one ingot, is under study.

The main limitation of the basaltic materials deduced from the above exposed results appears to be the impurity incorporation in the silicon during melting, coming from the crucible walls.

In order to avoid this mechanism we have studied, on the laboratory furnace scale ($\simeq 1$ kg ingots),a technique of "liquid encapsulation". This technique uses a melted composition which forms a continuous isolating film between crucible and silicon during the high temperature steps of the operation. The status of photovoltaic results using the liquid encapsulation in comparison with non-encapsulated materials is shown in Fig.9. This figure indicates that the liquid-encapsulated materials gives higher maximum values of JSC, the mean-values being located on the non-encapsulated ingots trend, with the exceptions of 0.11 and 0.7 Ω.cm resistivity ingots. No particular behaviour is to report concerning VOC.

The scale-up to 20 kg ingots of the liquid-encapsulation method is now under study.

5. <u>CONCLUSION</u>

The main results obtained during this study are :
- basaltic semicrystalline ingots 20 kg unitary weight were produced.
- AM1 photovoltaic efficiency of the elaborated 100 cm^2 semicrystalline cells, made from these ingots, reach the 10% range.

<u>ACKNOWLEDGEMENT</u>

This work is part of a program supported by AFME, CEC, and Société Nationale ELF Aquitaine.

REFERENCES

(1)   J. FALLY, C. GUENEL : "Study of elaboration of semicrystalline silicon
       ingots" - Proc. 3rd E.C. Photovoltaic Solar Energy Conf.
       Cannes, 27 - 31 Oct. 1980 - p. 598.

(2)   B. CHALMERS : "Principles of Solidification" John Wiley (1964) p. 306

(3)   J. OUALID et al., "Grain boundaries and intragrain defects dependence
       of local and global electronic and photovoltaic properties of CGE
       polysilicon"
       4th E.C. Photovoltaic Solar Energy Conf. Stresa, 10 - 14 May, 1982
       p. 421.

(4)   K.A. DUMAS, R.T. SWIMM "Minority carrier diffusion length and absorp-
       tion coefficients in silicon sheet material" SPIE vol. 248 -
       Role of Electro-Optics in Photovoltaic Energy Conversion - July 31,
       Aug. 1, 1980 - San Diego - p. 16.

(5)   H. MATARE "Defect Electronics in Semiconductors" Wiley Interscience
       (1971) - p. 282.

Fig. 1 - Crystallization Furnace.

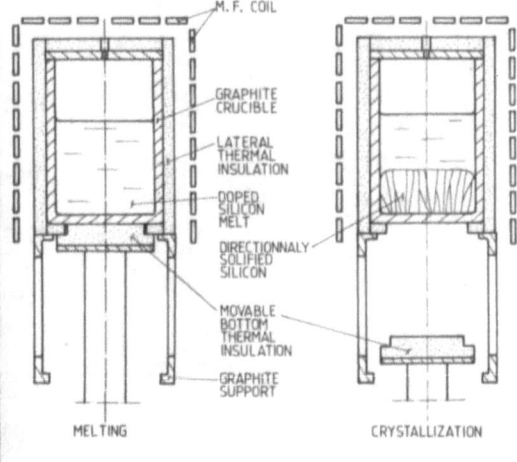

Fig. 2 - Crystallization method.

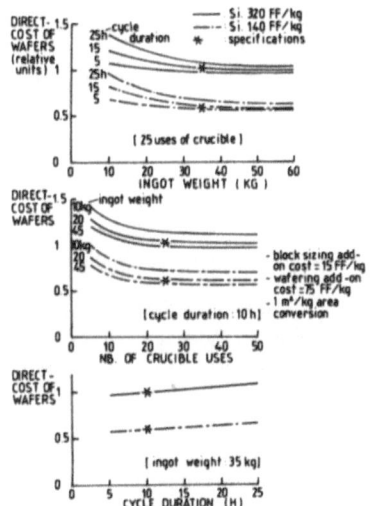

Fig. 3 - Economic aspects of the
process.

TABLE I

FEATURES.

FURNACE :

. DESIGNED CURRENTLY FOR A 25 KG SILICON CAPACITY,
POSSIBLE SCALE-UP TO 60 KG.

. HEATING IS PROVIDED BY MEDIUM FREQUENCY CURRENTS
(∼ 10 kHz).

. THE SQUARE SHAPED COIL SURROUNDS LATERALLY THE
CRUCIBLE, A PANKAKE COIL HEATS THE TOP.

. POWER ON COIL DURING MELTING IS 45 KW.

. WORKING ATMOSPHERE IS VACUUM.

CRUCIBLE :

. SQUARE SECTION REUSABLE CRUCIBLE, INTERNAL
DIMENSIONS : 230x230x360 mm
(MAX. CAPACITY = 25 KG SILICON ; TYPICAL CAPACITY
= 20 KG).

. CRUCIBLE IS ESSENTIALLY MADE OF SPECIAL GRADE
HIGH DENSITY GRAPHITE.

. DURING CRYSTALLIZATION, HEAT FLOW IS CONTROLLED
BY MEANS OF :

  - LATERAL AND TOP GRAPHITE FELT THERMAL INSULA-
  TION.

  - HEAT EXTRACTION BY RADIATION FROM THE BOTTOM
  OF THE CRUCIBLE.

TABLE II

STATUS

MATERIAL :
. 200x 200x160 mm FINISHED SEMICRYSTALLINE BASALTIC
  BLOCK
. SOLIDIFICATION RATE = 6 kg/HR
. TYPE P, RESISTIVITY : 0.5 - 1 Ω.cm
. TYPICAL DIFFUSION LENGHT LD = 35 μm
. TYPICAL EPD : 20% WAFER AREA $\geqslant$ 10^5 EPD.cm^2
. TYPICAL GRAIN WIDTH : 2 to 10 mm
. TYPICAL GRAIN LENGTH : 20 to 80 mm
. ELECTRONIC GRADE SILICON (SEG)

PHOTOVOLTAIC RESULTS :

(15 kg FINISHED INGOT - STANDARD PHOTOWATT S.A
PROCESS)

• ON 100 cm^2 "VERTICALLY SLICED" CELLS :

PARAMETER	ηAM1(%)	JSC(mA.cm^{-2})	VOC (mV)	FF(%)
MEAN VALUE	8.83	21.76	573	71.25
STD. DEVIAT.(σN-1)	0.30	0.40	2.90	1.42
MAX.VALUE *	9.45	22.45	577	73
MIN.VALUE *	8.3	20.85	567	70.2

• ON A 24.75 cm^2 "VERTICAL"CELL, BEST RESULTS :

PARAMETER	ηAM1(%)	JSC (mA.cm^{-2})	VOC (mV)	FF(%)
MAX.VALUE *	10.25	24.2	580	73

* simultaneously obtained on a cell.

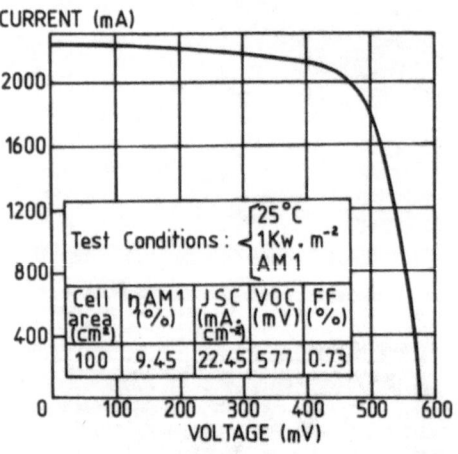

Cell area (cm^2)	ηAM1 (%)	JSC (mA. cm^{-2})	VOC (mV)	FF (%)
100	9.45	22.45	577	0.73

Fig. 4 - Electrical performance of
a 100 cm^2 basaltic cell.

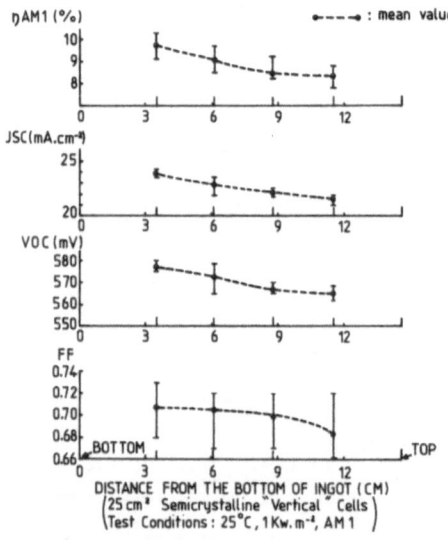

Fig. 5 - Vertical distribution
of the photovoltaic
parameters in a 15 kg-
ingot.

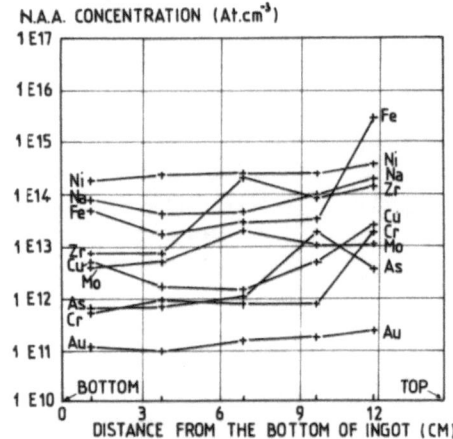

Fig. 6 - Vertical distribution along
the height of a 15 kg -
Semicrystalline ingot.

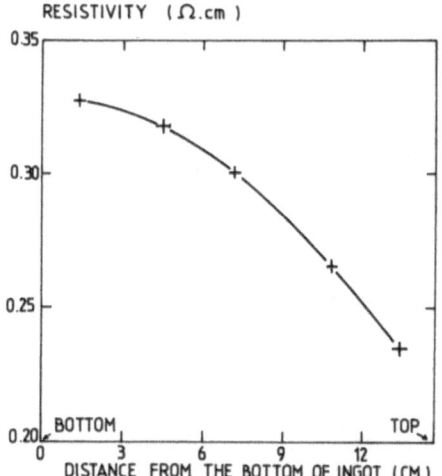

RESISTIVITY (Ω.cm)

Fig.7 - Variation of resistivity
vs. the height of a 15 kg
semicrystalline ingot.

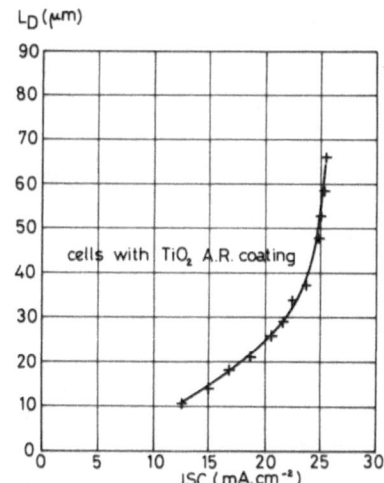

Fig.8 - Observed relationship
between diffusion length
LD and short-circuit
current density.

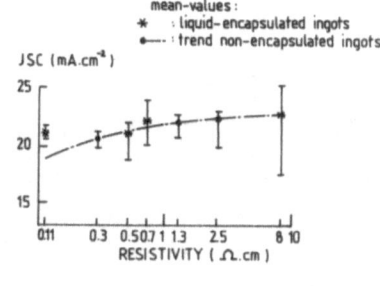

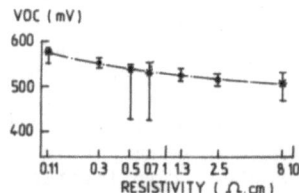

Fig.9 - Status of liquid-encapsulated semicrystalline
silicon materials
(~ 1 kg ingots, horizontal
wafers).

CLASSIFICATION OF CRYSTAL DEFECTS IN SOLAR BASE
MATERIAL WITH DIAMOND LATTICE

Authors   :   E. Sirtl, O. Spindler

Contract Nr.   :   652-78-10 ESD

Head of project   :   Prof.E.Sirtl, Heliotronic, Burghausen

Contractor   :   Consortium f. elektrochem. Industrie

Address   :   Zielstattstr. 20

                       D-8000 Muenchen

Summary

Starting from trigonometric evaluations of interstitial sites
in the most common structures of binary semiconductor compounds
their "ideal" radius has been correlated with the difference or
ratio of both tetrahedral covalent radii in a rigid sphere system.
In order to correlate theoretical results with experimental data
this work has been extended into two different directions:
a) Determination of the most realistic values of effective charges
(and hence effective atomic radius) and b) correlation of stability
ranges of the wurtzite lattice with the ratio of corresponding
effective radii assuming Frenkel defect mechanisms. It could be
finally demonstrated that the degree of polarity exceeds the
influence of steric factors (nearest neighbors) in a synergistic
system that has to select an optimum atomic structure of still
predominant covalent character. Investigations will be continued
with the revision of data for foreign point defects in silicon
based on corrections of their effective charge situation.

1. Introduction

After having categorized intrinsic defects in the diamond lattice during the earlier phase of this project, special emphasis has been put on the crystallographic analysis of interstitial sites in A1, B1, B3, and B4 lattices. As interstitial movement and occupation plays a major role in the kinetics of foreign atoms that are known to influence charge carrier lifetime, such studies are of direct importance to photovoltaics. The special geometry of space filling in a rigid sphere structural model had been largely reported already at the Sorrento Contractors Meeting.

2. The nature of interstitial sites

For the better understanding of our investigations described below, the most relevant results of our trigonometric studies, supplemented by the rock salt case, are brought again: a) In every case combinations of two simple interstitial sites (fcc or hcp) are the only existing types of interstitial grouping in the structural systems tested (see Fig. 1 a-d). b) In the wurtzite lattice the centers in each paired configuration are not identical. c) By variation of the ratio of (or difference between) the atomic radii in the two sublattices the size of the atomic radii changes. Whereas rocksalt, wurtzite, and diamond lattice exhibit maximum values of an ideal interstitial radius at $r_1 = r_2$ (see Fig. 2a and b), the sphalerite structure behaves quite differently here - due to the take-over between the nearest and next-nearest neighbor grouping.

3. Effective atomic radii

In order to test whether our fundamental studies could be correlated with the question of stability of the different phases, several diagrams have been put together. Fig. 3 indicates preferred positions of certain structures when the ratio of tetrahedral covalent radii is plotted versus the ratio of ionic radii. It becomes obvious that the different regions of coordination numbers (KZ) as postulated by the Goldschmidt rule are hardly in agreement with many of the compounds considered here. The different numbers indicating certain points are a part of the listing in the thesis of O. Spindler.[1] Generally, no strong point can be made in terms of the theoretically calculated special r_1/r_2 values.

It has been our intension, to revise the tabular radii given by Pauling[2] towards more realistic effective radii in many of the semiconductors. After thorough searches it was felt that the method to calculate effective atomic charges developed by Suchet[3] is the most appropriate approach available. Fig. 4 demonstrates the correlation of Suchet values versus literature data, mainly from Kimmel[4]. These values have to be converted then through semiempirical diagrams (q vs. r) into the corresponding effective radii.

4. Polarity aspects among tetrahedrally-bonded covalent lattices

It can be shown in Fig. 5 that better separation between the different structural categories is possible when such effective radii are plotted against $\Delta \chi$ standing for polarity. The major effect to show the dividing line, however, is due to the aspects of charge distribution in the lattice. Fig. 6 clearly indicates that the stability of wurtzite most strongly depends upon the amount of effective charge causing the c-axis to shrink and, thus, deform the electron cloud off the assumed (simplified) spherical shape An understanding of such trend can be derived from Fig. 7, where the eclipsed form of combining two layers of highest atomic density in tetrahedral covalent lattices (as in wurtzite) represents the more favored structure.

5. Outlook

The results of such studies will find their application in a better understanding of the behavior of foreign atoms in the silicon lattice in particular. Main targets are complex formation and segregation phenomena.

6. Literature references

1) O. Spindler, Doctoral Thesis, Ludwig-Maximilian University of Munich, in print

2) L. Pauling, "The Nature of the Chemical Bond", Cornell University Press, Ithaca, N.Y., 1973

3) J.P. Suchet, J. Electrochem. Soc. 124, 31C (1977)

4) H. Kimmel, Z. Naturforsch. 20a, 3 (1965)

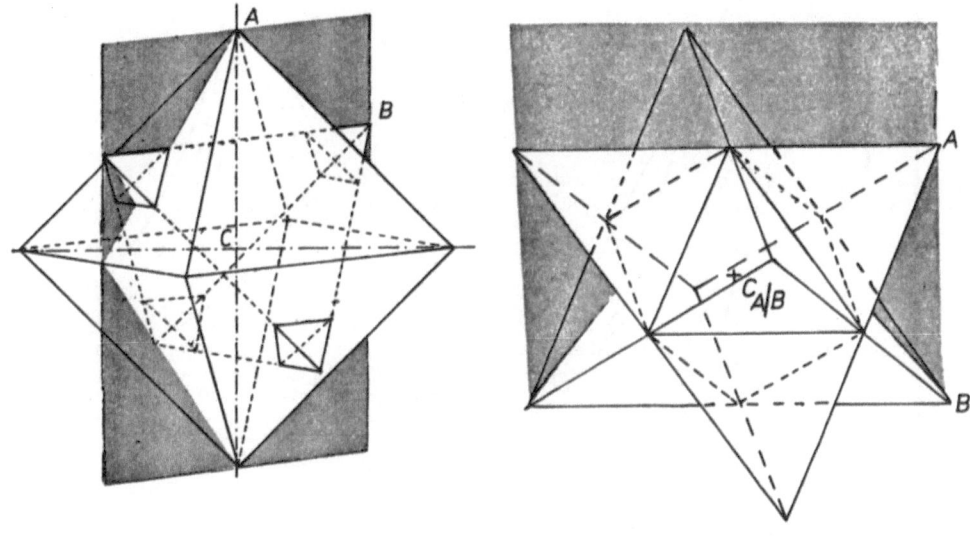

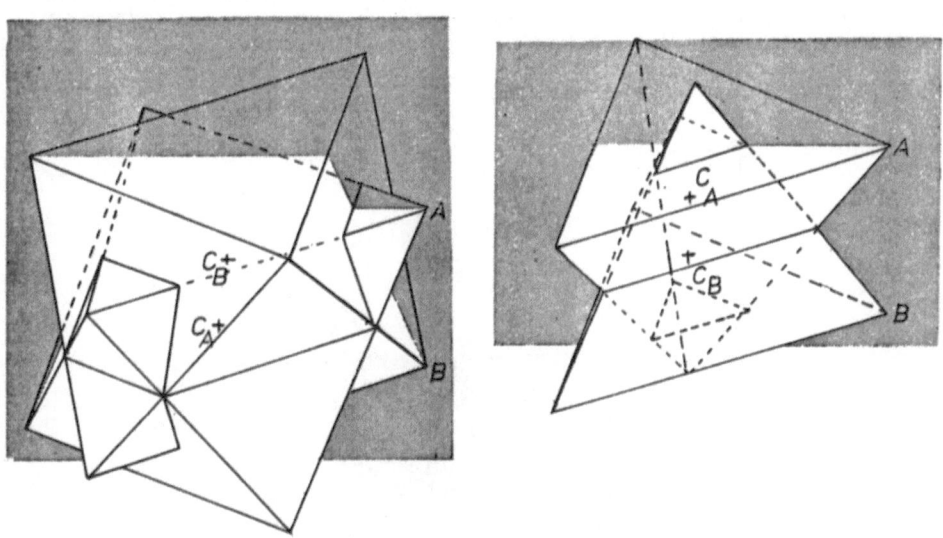

Fig. 1 - Crystallographic description of different types of interstitial
sites (A, B centers of sublattice atoms; C center of interstitial position)
in
a) sphalerite (TIS/OIS); b) rock salt (TIS/TIS); c) wurtzite (OIS/OIS);
d) wurtzite (TIS/TIS).

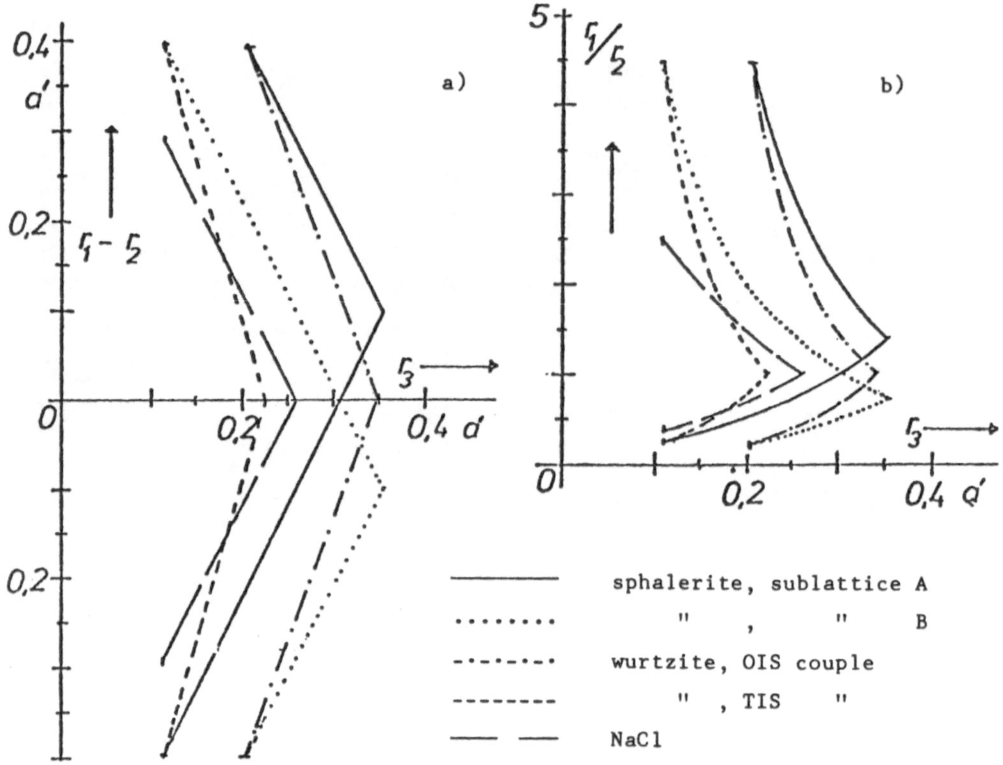

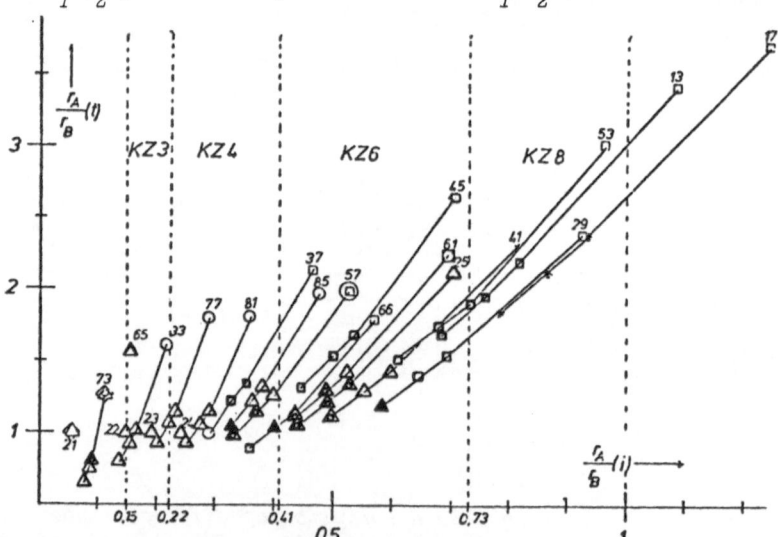

Fig. 2 – *Variation of geometrical confinement of interstitial positions (based on a rigid sphere model) relative to a' (edge of tetrahedron or octahedron, respectively). Radius of interstitial site (r_3) vs. a) difference in atomic radii ($r_1 - r_2$); b) ratio of atomic radii (r_1/r_2).*

Legend:
- ———— sphalerite, sublattice A
- ·········· " , " B
- —·—·—· wurtzite, OIS couple
- — — — " , TIS "
- — — — NaCl

Fig. 3 – *Ratio of ionic radii vs. ratio of tetrahedral covalent radii (after Pauling) –* △ *diamond or sphalerite;* □ *rock salt;* ○ *wurtzite;* ◇ *layer structure.*
Homologous groups of elements are linearly connected.

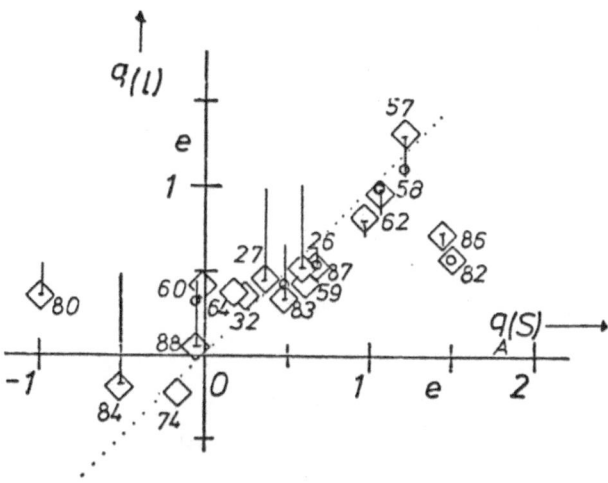

Fig. 4 - *Comparison of effective charges* $q_A(S)$ *calculated for the "cationic" partner after Suchet with data from literature sources* (q_1)

    ○ *measured by Kimmel*
    ◇ *calculated by Kimmel*
    | *other authors (height : limit of accuracy)*

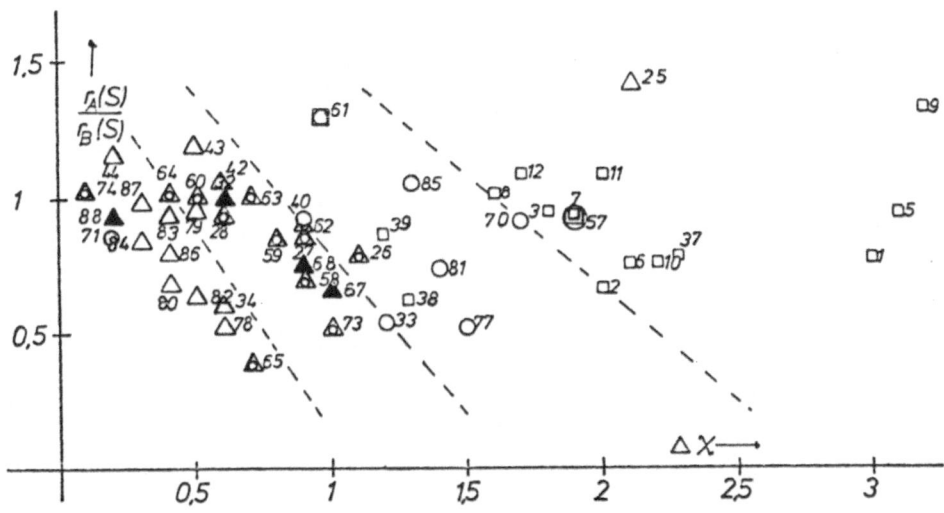

Fig. 5 - *Difference in electronegativity* Δχ *(after Pauling) vs. ratio of effective radii* $r_A(S)/r_B(S)$ *showing zones of preferred stability (symbols see Fig. 3)*

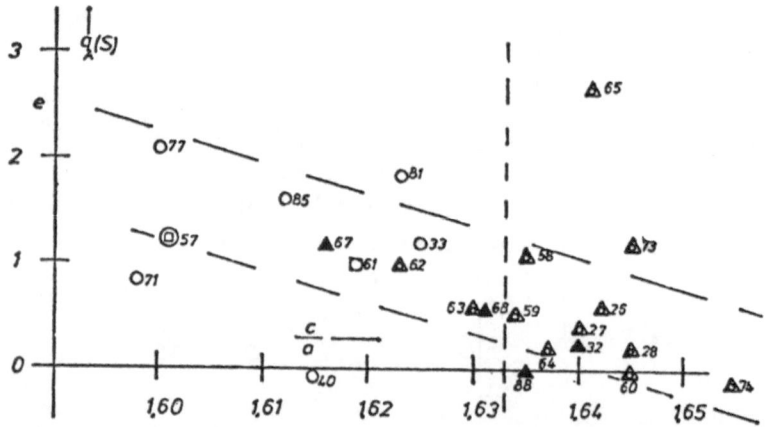

Fig. 6 – "Separation" of stable wurtzite phases due to polarity phenomena
Axial ratio (hexagonal) c/a vs. eff. charge $q_A(S)$.
Vertical (dashed) line : Ideal c/a value in a non-polar rigid sphere system.

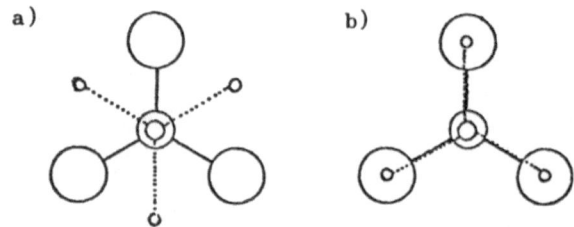

Fig. 7 – Combination of basic structural groupings :
a) sphalerite (staggered); b) wurtzite (eclipsed).

IMPLEMENTATION OF LOW COST SEMI CRYSTALLINE SILICON SOLAR CELLS

Authors and Head of Project : A.M. RICAUD

Contract number   : ESC R 0 7 3 F

Duration   : 24 months - 1 July 81 - 30 June 83

Contractor   : FRANCE PHOTON

Address   : Zone Industrielle LES AGRIERS
                             16015 ANGOULEME CEDEX

C.E.C. Contribution   : 642.500 FrF.

Summary

This project is related to the introduction of semicrystalline
(polycrystal with basaltic structure) silicon wafers into the low
cost process of manufacturing photovoltaïc cells implemented by
France-Photon since 1980 at Angoulême.
The aim of the research is to define which modifications are to be
brought to the standard process in order to maximize the photovoltaïc efficiency of polycristalline cells at a minimum cost.
Our investigations have already concerned, for différent materials:

1) Correlations between crystal defects (dislocations, punctual
   defects, precipitates) and photovoltaïc performances.
2) Slicingwafers from parallelipedic ingots.
3) Chemical etching optimization
4) Diffusion profiles
5) Front metallization with electroless Nickel plating
6) Transfert, conveyance, handling and mechanical stability
7) Tolerance to the hot spot problem
8) Economic considerations

It appears clearly that semicrystalline material is economically
ready to replace single crystal in 1983 although some fondamental
problems are not well understood within the grains and grain
boundaries and therefore are not well controlled at the manufacturing level, leading to a larger statistical dispersion on wafers
and cells than for single crystalline materials.
Best cells are obtained with the new HEM material of Crystal System
followed by Semix of Solarex and then Silso of Wacker.
Concerning the dispersion, Silso appears as the more industrial
product, followed by Semix and then by HEM which is probably
produced in small quantities.

Introduction :
        No doubt that cast semi crystalline material has many advantages
compared to the single crystalline ingots pulled by the Czockralski
or float zone method from molten electronic grade polysilicon.

- it is less expensive (reduced power consumption)
- it has a larger throughput
- cell unit area can be much larger
- square shape of the wafers gives a better fill coefficient at
  the panel level (very close to unity) allowing an unchanged
  electrical power per unit area with lower photovoltaîc efficiency.

For example, to reach 100 Wc/m2 we need only 10,5% efficiency with
a square shape against 14% with circular shape.

- it is less sensitive to the impurities, for impurities segregation takes place within the grain boundaries during solidification and does not disturb the bulk material properties.

- cells are much more capable of dissipating reverse power by
  grain boundaries leading to a better tolerance to the hot spot
  problem.

However a lot of problems had to be solved before introducing that
new material into the manufacturing.

- an insufficient knowledge of correlations between photovoltaîc
  performances (diffusion length and shunting effects) and :
      . the dislocations
      . punctual defects
      . incoherent grain boundaries
      . precipitates and impurity segregations.

- the selective chemical etching velocities depending upon grains
  orientations.

- optimization of the diffusion profiles
- optimization of front metallization and compatibility with the
  highly texturized surface.
- transfert, conveyance and handling of large square cells.

1/Correlations between crystal defects and photovoltaîcs performances

We had to analyse crystal defects such as dislocations, punctual
defects, twins, precipitates and impurities segregations.

We have followed an empirical approach which could give appreciable
results from the manufacturing point of view.

Three sorts of macroscopic defects have been correlated with poor
I/V curves on the Semix material. They are :

- sub-grains slighty desoriented in a large grain
- grit (or mush). Very small grains squashed together
- dendritic configurations associated with twins

We have studied the regions of sub-grains with electron microscope
and have correlated the poor diffusion length with high dislocation
density regions. By STEM we have also to verify if these regions
contain those precipitates observed by other searchers and try to
decide who is responsible of diffusion length degradation :
dislocations or precipitates ?

Fig. 1

Photograph of a Semix cell with the three
types of macroscopic defects.

Fig. 2

470	480	460	430	430	430	470	470	460	440
470	470	410	290	270	120	350	390	400	470
470	430	385	200	26	3	5	27	460	450
430	430	15	1,5	3	3	19	360	370	475
400	370	10	1	0	0	160	300	340	470
200	3%	2,2	2	1,5	4	40	190	230	450
410	390	340	1	0,8	2	110	150	280	430
420	400	370	19	1	7	30	140	400	410
440	415	390	40	9	24	190	380	370	
440	420	410	440	400	24	380			

Table of Voc (mV) under
20 mW/cm2 illumination
showing correlation with
macroscopic defects.

Fig. 3

5	9,5	11	10	10	7	10	9,5	7,5	10
5	9,5	6,5	6	5,5	2,8	5	5	5	9
5	8	5,5	4,5	2,4	0,5	0,9	2	5	9
5	7	1,5	0,2	0,7	0,7	2	5	4	9
4	5	1,5	0,2	0	0,3	0,7	9	4	4
3,3	6	0,5	0,3	0,35	1	5	8	18	7
5	6	3,5	0,2	0	0,6	3	4	11	0,5
5	7	5,7	1,8	0,4	1,5	2	5,5	28	13
5	28	20	2,3	1,4	4	8	11	20	
5	28	27	4,4	4	4	5,5	9		

Table of ISC (mA) under
100 mW/cm2 illumination
showing correlation with
macroscopic defects.

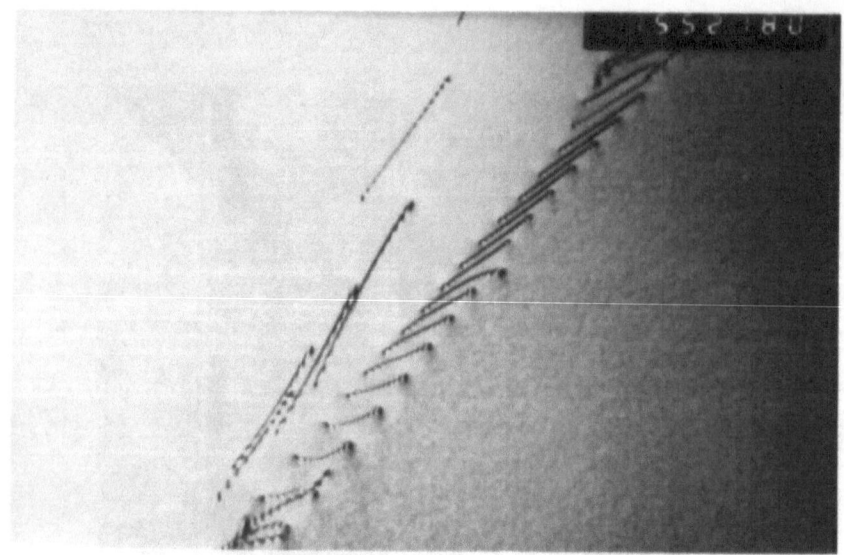

Fig. 4

A dislocation wall in a sub-grain. It is a regular arrangement of
dislocations belonging to the same family.

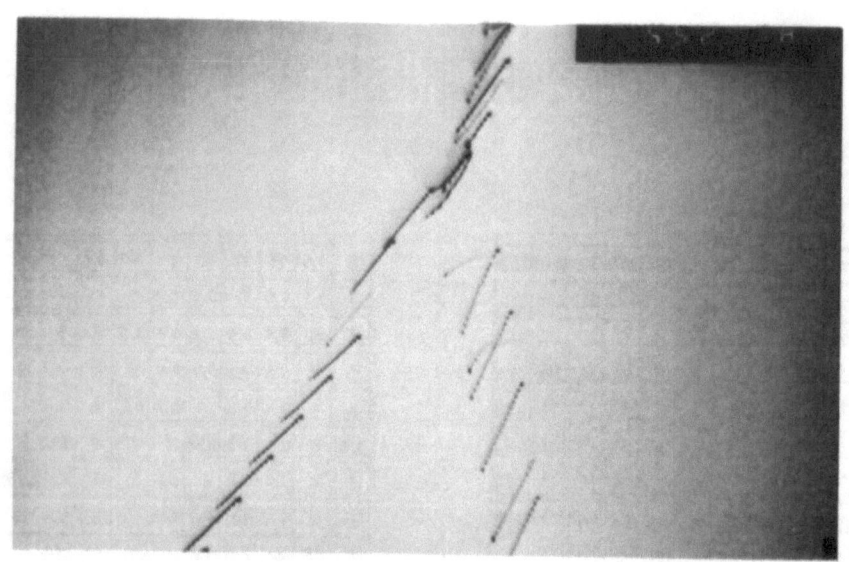

Fig. 5

Cross point of two dislocation walls.

In order to fill the gap in range of magnitude between photovoltaîc local measurements and TEM, we will continue the study of sub-grains by topography X (Begg-Barret)and EBIC. We shall try also to reduce the dimensions of elementary squared cell with a better dicing saw in order to reach the average size of a grain in Silso material.

2/ Sawing :

Comparison of a MBS VARIAN Model 7176 and IDS Mayer and Burger TS 23 over a few ingots of Semix material 10x10x12 cm.

	MBS	IDS
Wafer thickness (micron)	350	400
Kerf loss (micron)	250	350
Number of wafers/ingots	200	160
Bow and taper (micron)	20 to 50	10
Roughness	bad visual aspect	excellent aspect
Sawing damages (micron)	8	20
Wafers per hour	16	48
Amortization (FF)	0,45	0,26
Labour (FF)	1,15	0,31
Raw material (FF)	1,20	0,33
Silicon loss (FF)	1,40	1,60
Sawing cost per wafer (FF)	4,20	2,50

TABLE 6

MBS gives better result from the sawing damages point of view but to get the same throughput, a decrease in abrasive grain size must be compensated by an increase in the forth and back motion without enhancing the risks of breakage or flatness defects. So in practice IDS are frequently chosen for throughput and cost price reasons.

3/ Chemical etching :
3-1/ Optimization of sodium etch on SILSO material

Chemical etching is necessary to remove sawing damages and also to reduce the reflexion coefficient of silicon by multiple reflexions over a pyramidal texturization.

Table 7 summarizes geometrical and electrical parameters for different etching time with boiling Na OH. Those 150 wafers were cut with an ID saw.

Etching time (s)	30	45	60	75	90	105
removed thickness(μ)	17	20	21	25	40	50
texture aspect	saw marks every where	slight tracks	no saw mark	slight shift between certain grains	strong diffe-rences in orien-tation	deep steps between grains
R/□ (Ω)	34	33	32	32	29	31
Voc (mV)	515	531	535	541	545	550
Isc (mA)	2280	2330	2290	2330	2400	2200
PM (mW)	845	880	865	875	885	760
FF	0,720	0,710	0,707	0,693	0,676	0,630

TABLE 7

Increasing time increases Voc and decreases FF
optimum of Pmax at 90 s.
Etching time does not correlate well with Isc

3-2/ Comparison of alkaline and acid etch on Semix, Silso and HEM
(150 wafers). Dispersion is indicated in parenthesis.
In both cases, removed thickness was 40 microns. Diffusion not
optimized.

	SEMIX		SILSO		HEM	
	Alkaline	Acid	Alkaline	Acid	Alkaline	Acid
R/□ (Ω)	34,7 (1)	34,4 (3)	36,6 (1)	40 (2)	35,7 (2)	36,6 (2)
Voc (mV)	562 (2)	565 (8)	552 (2)	553(5)	559 (7)	566 (9)
Isc (mA)	2202 (41)	2348(47)	2096(17)	2337 (118)	2276(122)	2663 (127)
PM (mW)	859 (17)	887 (79)	797 (22)	935(50)	905 (50)	1033 (103)
FF	0,694	0,668	0,689	0,723	0,712	0,685

TABLE 8

4/ Diffusion

4-1/Optimization of diffusion :
Optimized process on 500 Semix Wafers compared to longer time and
higher temperature on 150 wafers from SILSO, SEMIX and HEM (sodium
etch optimized)

	SEMIX			SILSO		HEM	
	Optimized	15mn 910°C	27mn 910°C	15mn 910°C	27mm 910°C	15mn 910°C	27mn 910°C
$R/_\square$ (Ω)	45 (3)	34,7 (1)	29,1 (2)	36,6 (1)	29,1 (3)	35,7 (2)	32,9 (2)
Voc (mV)	556 (14)	562 (2)	561 (4)	552 (2)	544 (6)	559 (7)	547 (9)
Isc (mA)	2500 (201)	2202 (41)	2275 (42)	2096 (17)	2076 (89)	2276 (122)	2098 (125)
PM (mW)	910 (110)	859 (17)	761 (53)	797 (22)	581 (81)	905 (50)	729 (87)
FF	0,662	0,694	0,596	0,689	0,515	0,712	0,636

The more uneven sheet resistivity on a single cell has been measured on SILSO and the better uniformity on SEMIX (10% in average).

On a same SEMIX wafer, total oxide thickness variations was found to be less than 8% for the wide range of crystallographic orientation. While the sheet resistance variability (10%) is higher than for single crystal wafers (4%), this is not expected to cause significant variation in the cell characteristics.

Longer exposure at high temperature does not affect too much the value of Isc but degrade Vco and FF, probably because of a poor knee (non - Ideal disde coefficient) and also a slightly lower sh unt resistance.

4-2/ Gettering effect : 150 Silso wafers

	Standard 15mn 910°C	1 gettering with 1 sodium etch	2 gettering with 2 sodium etch
$R/_\square$ (Ω)	36,6	33,2	31,3
Voc (mV)	552	535	525
Isc (mA)	2096	2045	1960
PM (mW)	797	775	665
RSH ( )	12	35	5

One gettering degrades Vco and Isc but improves drastically RSH. The fill factor improvement is not sufficient to get better efficiencies but indicates that a part of the shunting effect is due to impurities.

Same effect has not been verified upon Semix material.

5/ Front metallization : electroless Nickel plating :

Alkaline bath of electroless Nickel have a tendancy to etch silicon before starting the autocatalytic deposition. It has a bad effect on uneven shallow diffusions since the junction can be shorted here and there in the thinnest regions.

This shunting effect is very wide spread on the grain boundaries of polycrystals, especially for the samples which have suffered a long alkaline etch exposure.

RSH varies from 35 Ω to 3 Ω when sodium etching time varies from 30 to 130s. On SILSO material we have noticed that this phenomenon increases with decreasing grain sizes.

6/ Transfert, conveyance, handling, and mechanical stability

The sawing process needs a minimum thichness of 300 microns to keep the yield over 90 %.

However, a thickness of 350 microns is the minimum acceptable value at the cells manufacturing level to maintain the breakage rate below 6 %.

Breakage increases with the grain size and is also depending upon the dislocation density. So, SILSO presents the lower rate of breakage then HEM and SEMIX arrives last probably because of high density dislocations in the twins vicinity.

Breakage increases with the grain thickness differences, hence with the chemical etching time.

Breakage decreases slightly when the corners are cut. We have not yet significant numerical values concerning the breakage rate ratio between poly and single crystalline silicon.

7/ Tolerance to the hot-spot problem

Some manufacturers have maximized the peak power of their cells by increasing the shunt resistance to maintain a high efficiency at low-level illumination. Correspondingly they have flat reverse curves and when shading occurs, the shaded cell moves up quickly into a high reverse voltage inducing serious failure if not protected, or in the best case an important power loss at the module level.

It is obvious that, the larger the cell area, the lower the shunt resistance effect compared to the one of series resistance.

For a semicrystalline cell (Semix 100x100 mm) the maximum power loss is less than 5 % when shunt resistance varies from 30 to 6 Ω .

Furthermore, polycristalline material has the additional advantage that, when reverse biased, heat is dissipated over a large percentage of the cell area. It has been demonstrated that the light emission patterns on a semicrystalline cell undergoing extreme reverse bias, are identical to the cell heat generation patterns with dissipation of heat over junction boundaries which are many times the area of a single crystal cell perimeter.

For RSH, an optimum value of 14 Ω has been induced, taking into account :

- the loss of efficiency at low level illumination
- the available direct power on a 36 cells module when a cell
  suffers complete occultation.
- the compromise between short-circuit current and fill factor
due to selective etching velocities.
- the mechanical stability

8/ Economic considerations :

Fig. 10 shows the recorded prices in constant FF 1980 for single crystal wafers ∅ 100 mm and for the share they represent in the final peak Watt.

These assumptions are based upon a purchase in large quantities, a breakage yield, a photovoltaïc efficiency at the cell level and a rabe of inflation of 10% to 11% per year.

The same recorded prices for semicrystalline wafers (100x100mm) are also represented.

It is obvious that the crossing point for the final peak Watt will appear during the course of 1983.

CONCLUSION :

Concerning their crystal structures, semicrystalline materials are not well controled from the dislocation point of view. Twins are rarely clean and the dendritic structures have a deleterious effect from the double point of view of diffusion length and cell manufacturing process.

Slicing is economically well done with IDS, but sawing damages have to be reduced at the MBS level.

Chemical etching is economically well done with NaOH but induces serious variations in grain thickness. Acid etch cannot be seriously considered for cost and security reasons.

Diffusion can be optimized in close relation with electroless nickel plating since maximizing the short-circuit current minimizes the shunt resistance.

In line breakage rgte is generally higher than for single materiel, whereas tolerance to the hot spot problem is much better.

Because of the constant increase of single material cost, semicrystal appears to be the only industrial alternative for the next few years.

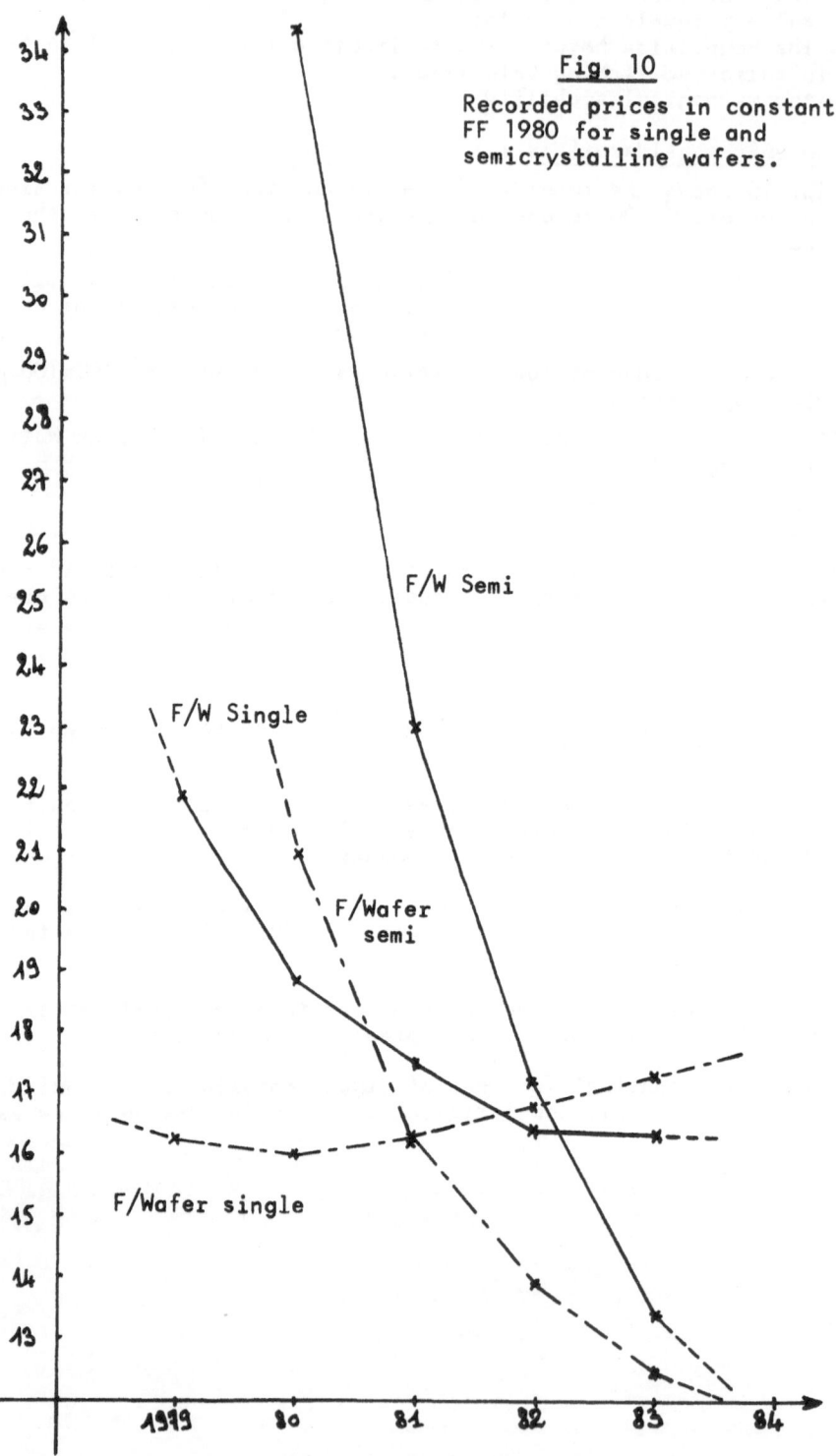

Fig. 10

Recorded prices in constant
FF 1980 for single and
semicrystalline wafers.

OPTIMIZATION OF PROCESSING CONDITIONS OF SOLAR CELLS

VERSUS PHYSICAL PROPERTIES OF RELATIVELY LOW COST SILICON

Authors   : F. DEMONGEOT, J. DONON, H. LAUVRAY, P. AUBRIL

Contract number   : ESC-R-023-F

Duration   : 18 months   1 July 1980 - 31 December 1981

Head of project   : J. DONON (PHOTOWATT) J. LEBAILLY (RTC)

Contractor   : PHOTOWATT International S.A.

Address   : 125, rue du Président Wilson
                     F92302 LEVALLOIS PERRET

Summary

The objectives of this contract concern the research of the optimum
solar cell processing conditions in order to increase the typical
semicrystalline silicon solar cell efficiency. The research programm
was to set up a complete electronic properties analysis, to study the
influence of heat treatments and to fing gettering and grain passivation procedures.

1. Introduction

Lorsque des plaques de matériau semi-cristallin sont utilisées pour la réalisation de cellules photovoltaïques des résultats variables sont obtenus qui dépendent très fortement des structures cristallines des grains, des traitements thermiques effectués et des procédés de fabrication utilisés.

Au cours de cette étude des méthodes de caractéristion sur les matériaux ont été mises au point. Une étude a été réalisée sur des lingots entiers afin de corréler structure et rendement de conversion.

Enfin des études d'évolution des performances soit au cours des traitements thermiques, soit par passivation des joints de grains ont été réalisés.

2. Analyse et caractéristation des matériaux semi-cristallins

Pour caractériser l'influence des procédés de fabrication des cellules avec des matériaux semi-cristallins, les mesures spécifiques effectuées dans cette étude en dehors des mesures classiques (caractéristiques photoélectriques Vco, Isc, FF, η, courants de fuite, et caractéristiques cristallographiques, taille et forme des grains ...) sont les mesures de longueur de diffusion et les mesures effectuées au microscope à balayage en mode EBIC.

2.1 Mesures des longueurs de diffusion
Méthode SVP appliquée au multicristallin

L'éclairement du semi-conducteur induit à sa surface une phototension directement liée à la concentration en surface des porteurs minoritaires en excès (n_o). Si pour différentes longueurs d'ondes on se place à tension de surface constante, la relation entre le flux incident G et L - longueur de diffusion - s' écrit : $G = c'$ $(1 + L)$ où c'est une constante.

La droite de variation de G en fonction de $1/\alpha$ coupera donc l'axe des abscisses en $-L$.

La mesure de la tension de surface est effectuée à l'aide d'une sonde capacitive (Figure 1).

Le matériau polycristallin pose les problèmes suivants pour effectuer correctement la mesure :

. État de surface non homogène à cause de la présence de grains d'orientation différente. Néanmoins pour une même série de mesures, la valeur de L ne sera pas modifiée, seul le niveau constant n_o sélectionné sera obtenu pour une valeur d'illumination plus forte qu'en théorie.

. Modélisation, car le modèle précédent n'est valide que dans la mesure où la surface du silicium est équipotentielle et la densité d'états n_o est la même en surface au voisinage du point illuminé sous la sonde de mesure. Les conditions aux limites de l'équation de base sont modifiées par l'existence de niveaux pièges recombinants aux joints de grains, en particulier la durée de vie τ est reliée à la taille des grains.

Pour effectuer une mesure correcte, il faut se trouver dans une zone où la taille des grains reste constante pour que τ ne soit pas modifiée. Ceci peut être obtenu en réduisant la taille de la sonde à un diamètre de 1 cm, une taille minimale étant imposée pour recueillir un signal détectable. D'autre part, la surface du silicium semi-cristallin ne peut plus être considérée comme équipotentielle. Les pièges aux joints de grains font que la tension mesurée pour un niveau d'illumination donné sera inférieur à la phototension de surface. Un modèle simplifié avec plusieurs grains et pour lequel est pris en compte une chute de tension constante aux joints de grains due à l'occupation des niveaux pièges a été mis au point :

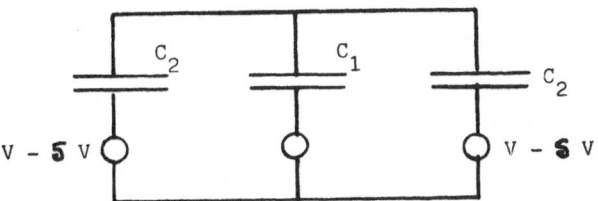

Modèle à 3 zones $C_1 = \dfrac{\varepsilon \, s}{e}$

$C_2 = \dfrac{\varepsilon \, (S - s)}{2 \, e}$

Le schéma équivalent à ce modèle :   avec $C = \dfrac{\varepsilon \, S}{e}$

permet de déduire V' en fonction de V : $V' = V - (1 - \dfrac{s}{S}) \, \delta \, V$ et V' est mesuré
au lieu de V, la valeur de L sera pessimiste.
La diminution de la taille de la sonde permet de faire tendre $1 - \dfrac{s}{S}$ vers
zéro et la mesure vers sa valeur exacte.

Mesure du courant de court-circuit

La longueur de diffusion L peut être déterminée par une technique analogue
au SPV en mesurant le courant de court-circuit pour différentes longueurs
d'ondes. Ceci ne peut se faire que sur des cellules finies. Pour un courant
de court-circuit constant et pour différentes longueurs d'onde, la rela-
tion entre l'éclairement et α s'écrit $[1 - R(\lambda)] \; G = C \, (\, 1/\alpha + L)$. La
droite de variation de l'éclairement en fonction de $1/\alpha$ coupe l'axe des
abscisses en –L. Du point de vue expérimental, le faisceau lumineux de
l'appareil éclairant 1 mm2, aucune correction n'est à effectuer par rapport au monocristallin si le faisceau est centré sur un grain de surface
supérieure à 1 mm2 directement collecté. La mesure sur des grains de
taille inférieure fournira une valeur moyenne de L sur les différents
grains.

Analyse spectrale

Cette méthode est utilisable pour des cellules terminées avec ou sans
anti-reflet.

Le courant de court-circuit dépend d'un certain nombre de paramètres
(profondeur de jonction, vitesse de recombinaison en surface, mobilité
des porteurs, coefficient de diffusion) mais principalement de la longueur de diffusion L. La sensibilité spectrale est le rapport du courant de
court-circuit à l'énergie incidente $\delta = \dfrac{Isc}{E} \, (\lambda)$.

Expériementalement on mesure le courant de court circuit aux bornes
d'une cellule éclairée localement par un monochromateur. L'énergie incidente est connue grâce à la réponse d'une cellule étalon et la comparaison
de la courbe   expérimentale et des abaques permet d'évaluer la longueur
de diffusion. Cette mesure est locale et donc indépendante des joints de
grain si elle est faite au centre du grain ce qui ne pose pas de problème car
le diamètre du faisceau est d'environ 1 mm2.

2.2 Investigation au microscope électronique à Balayage en mode EBIC
(Electron Bean Induced Current)

Cette méthode permet une étude qualitative des grains et de l'activité
des joints sur des cellules terminées en y découpant des échantillons de

de dimension réduite (max 10 x 10 mm2). Deux modes de mesure peuvent être utilisés :

Electrons rétrodiffusés, ceux-ci permettent de visualiser l'image très détaillée de l'échantillon.

Courant induit ; les électrons absorbés par l'échantillon créent des porteurs libres dans un volume très localisé, se propagent par diffusion et atteignent la jonction où ils sont collectés à l'aide des prises de contact de la cellule. Il suffit de balayer ensuite l'échantillon dans une direction pour obtenir la variation du signal et en particulier au niveau des joints de grains.

Cette mesure permet d'interpréter la décroissance du niveau des grains. Deux critères peuvent être retenus :

. Pourcentage de décroissance du signal EBIC dans le joint (niveau zéro sur la métallisation/niveau 100 niveau constant maximum sur le grain).

. Largeur de la zone d'influence du grain.

3. <u>Elaboration de photopiles à partir de matériau semi-cristallin du type Silso (Wacker).</u>

Au court de l'étude de mise au point des méthodes de caractérisation, nous avons constaté une grande dispersion des résultats, un manque de reproductibilité dans un même lot ou entre lots différents, ainsi qu'une variété des structures cristallines rencontrées.

Ceci nous a conduit à étudier l'uniformité d'un lingot.

Les lingots étudiés 100 x 100 x 106 mm3 proviennent de lingots de dimensions 150 x 150 x 106. Ils sont ensuite découpés selon la Figure 2.

Les photographies montrent les différentes structures cristallines observées sur ce type de matériau (Figure 3) ainsi que sur HEM.

Les cellules sont réalisées selon le procédé de fabrication standard suivant :

Décapage acide - rinçage - diffusion $POCl_3$ 850°C 1h30 sérigraphie argent des contacts, dépot de la couche anti-reflet et test.

Les résultats obtenus Figures 4 et 5 ainsi que le tableau I montrent clairement la relation entre structure et propriété photovoltaïques. La qualité électronique du matériau est liée à :

. La position dans le lingot, la longueur de diffusion des porteurs minoritaires croissant de bas en haut.

. La présence de zones contaminées par des impuretés se trouvant aux intersections entre les différentes structures cristallines. En particulier les plaques avec une structure présentant des précipités de SiC n'ont jamais donné de bons résultats.

. Certaines zones comportant d'ailleurs plus ou moins de défauts précédents sont sensibles aux traitement thermiques.

. Il aparait indispensable d'enlever les structures de bord très riches en impuretés sur au moins 2 mm.

. Si ces points ne sont pas observés, il en ressort qu'environ 1/3 du lingot n'est pas utilisable sans modification des conditions de croissance.

Le deuxième lingot Tableau II pour lequel la température de diffusion a été de 800°C au lieu de 850°C donne des résultats nettement supérieurs. De même le matériau HEM (Crystal Systems) donne des différences importantes de rendement en fonction de la structure cristalline.

η = 10,5 % sur matériau monocristallin.

η = 9,5 % sur matériau avec 28 % de zones polycristallines.

3. Influence des traitements thermiques et passivation

Les conditions expérimentales suivantes ont été étudiées :
. Traitement thermique 850°C ou 900°C 1heure en atmosphère neutre.
. "gettering" à l'aide de l'oxyde dopé 800°C à 900°C 1 heure.
Les résultats sont présentés dans le Tableau II suivant :

TABLEAU II		Longueur de diffusion μm		
		Plaque décapée	Traitement Thermique	"gettering"
SILSO	Bord de la Tranche	21 - 37	4 - 14	9 - 32
	Structrure colonnaire	40 - 44		34 51 800°C
HEM	Mono	18 - 42	17 - 51	24 - 51

Pour le Silso la longueur de diffusion est dégradée par un traitement thermique sur tous les types de structures. Le "gettering" restaure partiellement les propriété initiales.
Pour le HEM mono il n'y a pas d'influence du traitement thermique ou du "gettering".
 Des traitement de passivation hydrogène effectués avant ou après diffusion ont été réalisés.
 Une première série d'essais à une température de 600°C n'a pas donné d'amélioration des propriétés.
 Une deuxième série d'essais a été effectuée à plus faible température (420°C) montre que ce traitement est efficace s'il est effectué avant diffusion et avec des durées longues : rendement moyen de 9,8 % supérieur au procédé classique 9,2 %. L'analyse des caractéristiques à l'obscurité Tableau III montre que le courant de recombianaison moyen évolue de façon favorable pour le traitement H_2 avant diffusion. l'amélioration porte sur la diminution des recombinaisons et sur la passivation des joints, ce qui est confirmé par l'analyse en courant induit Tableau IV. De nouveaux essais effectués sur des structures identiques et différentes ne donnent pas toujours les mêmes résultats. Tout se passe comme si les plaques avaient des qualités différentes, des traitements précédents pourront soit améliorer, soit déteriorer le matériau, ceci étant indépendant des structures.

4. Conclusion

L'étude entreprise a permis de mettre au point des méthodes de caractérisations efficaces pour les matériaux semicristallins : mesures de longueur de diffusion sur le matériau ou sur la cellule finie, mesures des réponses spectrales, balayage monochromatique et caractérisation des joints de grains (EBIC).
 Les lingots étudiés ont montré en particulier que des défauts tels structures "grit" étaient inacceptables et que la sensibilité aux traitements thermiques était très importante. Des traitements de passivation hydrogène des joints de grains peuvent dans certains cas améliorer le matériau sur les différents types de structure sans qu'une explication puisse être donnée, puisque dans d'autres cas et sur les mêmes structures aucune amélioration n'est observée.

FIGURE 1

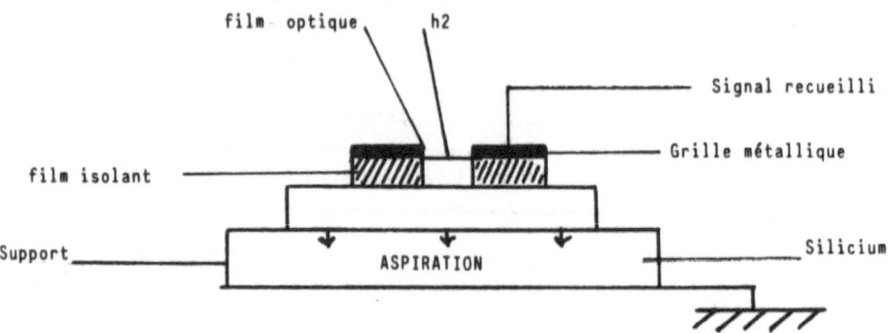

FIGURE 2

DECOUPE DES PLAQUES

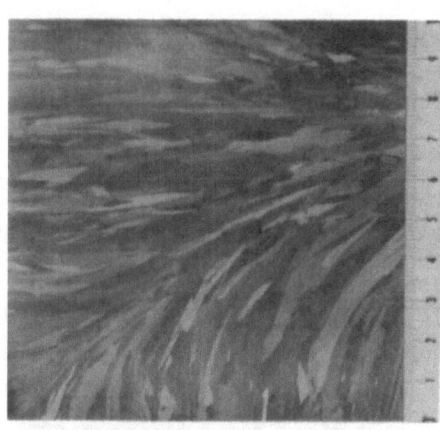

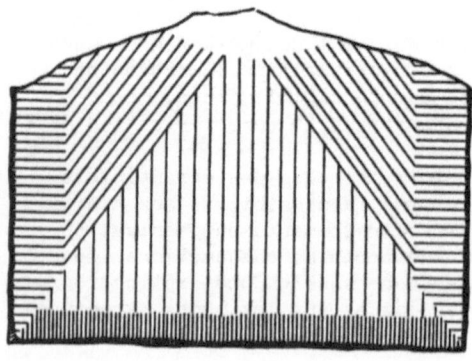

<u>Figure 1</u> Schematic Longitudinal Section of an Ingot

| | | | Columnar STructure
\\\\: Transition Structure
▓▓▓ Edge Structure

|||||||||| Fine Crystalline Structure
——— Transcrystalline Structure

LINGOT N° 1

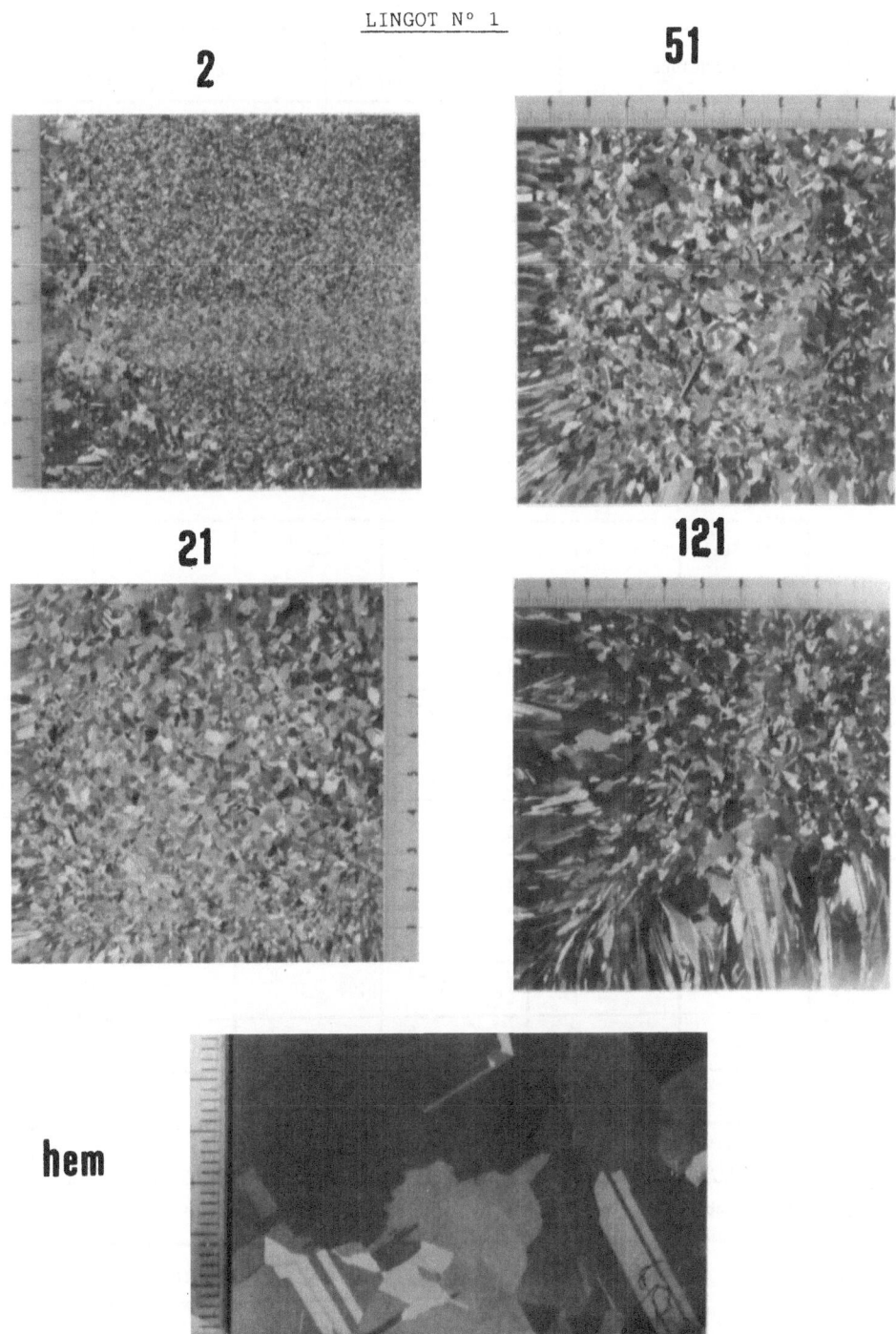

TABLEAU I SOLAR CELL CONVERSION EFFICIENCY

INGOT HOMOGENEITY (SILSO) LINGOT I

N°	STRUCTURE	SPECTRAL RESPONSE MAX(MA/W)	MAX $\overset{\circ}{A}$	L_D (μM)
18	EDGE – COLUMNAR	295	6200	4
38	COLUMNAR TRANSCRYSTALLINE TRANSITION	375	7000	12
65	COLUMNAR – TRANSITION TRANSCRYSTALLINE	480	7500	30
82	"	520	8000	50
104	TRANSITION – COLUMNAR TRANSCRYSTALLINE	415	7500	35
MONOCRYSTALLINE (CZ)		600	8500	100 μM

TABLEAU III

	N° Cell.	Io_2 (A)	$\dfrac{Io_2}{S}$ A/cm2	$\dfrac{n2\,k\,T}{9}$	n_2
Procédé Classique	1/4	10^{-5}	10^{-7}	$4{,}45\ 10^{-2}$	1,73
	1/11	$1{,}38\ 10^{-4}$	$1{,}38\ 10^{-6}$	$5{,}97\ 10^{-2}$	2,32
	1/13	$1{,}43\ 10^{-4}$	$1{,}43\ 10^{-6}$	$5{,}94\ 10^{-2}$	2,31
H_2 avant diffusion	2/3	$9{,}84\ 10^{-6}$	$9{,}84\ 10^{-8}$	$4{,}93\ 10^{-2}$	1,92
	2/5	$13{,}5\ 10^{-6}$	$13{,}5\ 10^{-8}$	$5{,}03\ 10^{-2}$	1,96
	2/15	$13{,}2\ 10^{-5}$	$13{,}2\ 10^{-8}$	$5{,}03\ 10^{-2}$	1,96
H_2 après diffusion	3/8	$4{,}54\ 10^{-5}$	$4{,}54\ 10^{-7}$	$5{,}27\ 10^{-2}$	2,05
	3/4	$1{,}98\ 10^{-4}$	$1{,}98\ 10^{-7}$	$6{,}03\ 10^{-2}$	2,35
	3/11	$3{,}38\ 10^{-4}$	$3{,}38\ 10^{-6}$	$6{,}49\ 10^{-2}$	2,52

k= constante de Boltzmann

$\dfrac{k}{q}$ =constante de Boltzmann

$\dfrac{kT}{q}$ =2.569 10^{-2} à 25°C

TABLEAU IV

N° Cellule et Traitement	Décroissance du signal EBIC au joint (%)		Largeur de la zone d'influence (μm)
Procédé standard	79 %	83 %	60 à 120
Traitement H2 après dif	70 % 86 %	86 % 87 %	70 à 100
Traitement H2 avant dif	90 % 92 % 93 % 55 % joint très actif en recombinaison		60 / 60 / 80 68

FIGURE 4

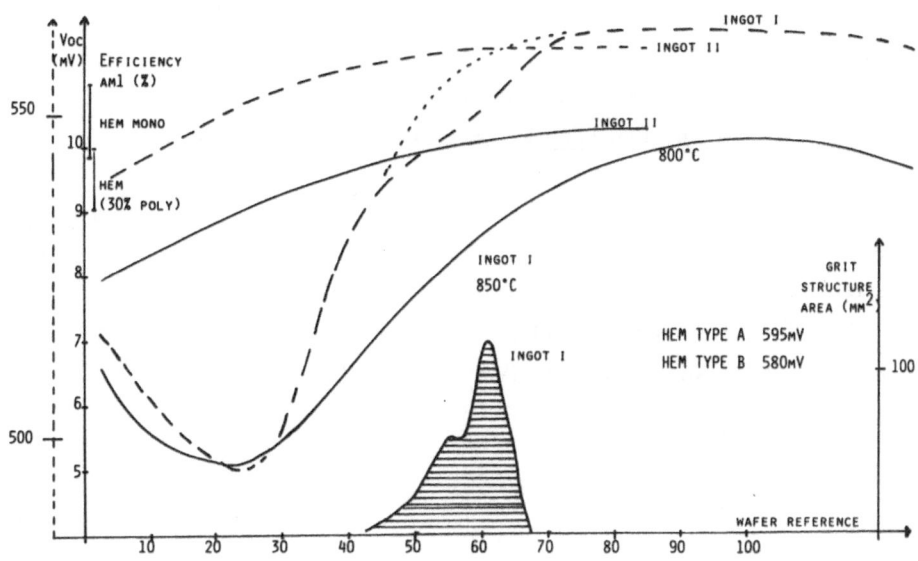

FIGURE 5

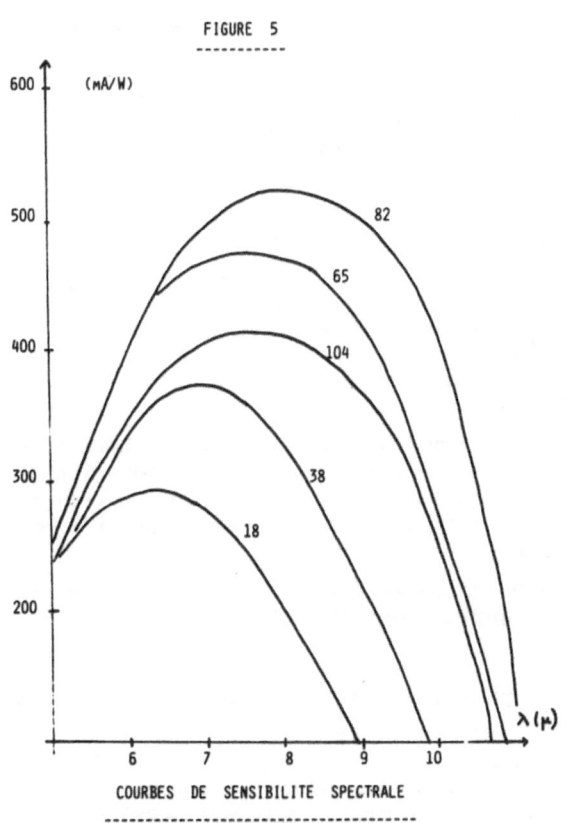

COURBES DE SENSIBILITE SPECTRALE

CONTINUOUS PRODUCTION OF PHOTOVOLTAIC SILICON RIBBONS

BY THE NEW PENDANT DROP GROWTH (PDG) METHOD

Author : J. RICARD

Contract number : ESC R.072.F

Duration : 1.1.81 - 1.1.82

Head of project : M. WINTENBERGER

Contractor : PECHINEY UGINE KUHLMANN

Address : 23, rue Balzac, B.P. 787.08
          75360 PARIS CEDEX 08

Summary -

The PDG apparatus and procedure were significantly improved during this contract. New theoretical considerations enable us to establish the stability of our process as well as the possibility of manufacturing ribbons thinner than the lip thickness, about up to 0,5 mm.

Through the study of materials available for our crucibles, we could work out the ideal crucible, which is made of dense graphite and internaly lined with SiC or pyrocarbon.

We have also markedly improved the crucible powder feeding system involving a vibrating bowl. The rate of output is constant within ± 1.5 % from 1 up to 100 g/hr. The response to a temperature adjustement of the crucible is immediate.

We have manufactured up to 500 mm long and 50 mm wide ribbons at extrusion rate of as much as 90 cm/hr. We could not manage to obtain thinner than 1 mm ribbons without losing width, but now we know the reasons for a such failure new crucibles have machined after the contract for this purpose.

The amount of crystals is convenient in both crystalline and electrical respects. The required annealing of our ribbons prior to submitting them to photovoltaïc test prevented us from recording photovoltaïc outputs exceeding 4.5 % because of impurity from the ribbon surfaces inwards.

Lastly, our actual cost estimates allow us to contemplate a ribbon production at a price lower than 500 FF per square meter for 0.5 thickness.

This contract enabled us to define the subsequent research trends :

- reduction of thickness,
- automation,
- study and control of ribbon cooling,
- reduction of drawing lip external wetting.

1. PRINCIPE DE LA METHODE -

1.1 - Un tube capillaire plonge dans un liquide, se remplit de liquide (fig.1), à condition que le liquide mouille le matériau du tube selon la formule :

$$h = \frac{2 A (1 + K_R)}{R \rho g}$$

avec A = tension superficielle du liquide à la température considérée,
ρ = masse spécifique du liquide,
g = accélération de la pesanteur,
R = rayon intérieur du tube capillaire,
K_R = constante dépendant de la nature du liquide et de la forme de l'extrémité du tube capillaire.

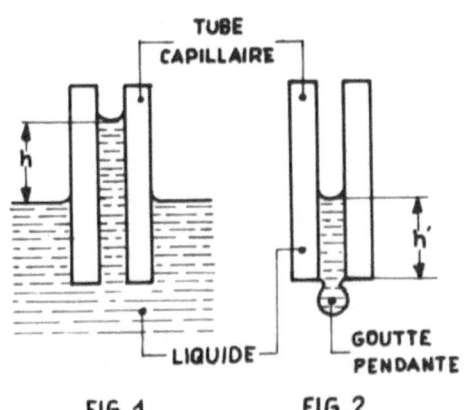

TUBE CAPILLAIRE

h

h'

LIQUIDE

GOUTTE PENDANTE

FIG. 1   FIG. 2

FIG. 1-2 PRINCIPE DE LA GOUTTE PENDANTE

Si l'on retire le tube du liquide, une colonne de liquide de hauteur h' subsiste à l'intérieur du conduit capillaire, à l'extrémité duquel pend alors une goutte (fig.2). h' est de l'ordre de grandeur de h. Cette colonne de liquide et la goutte pendante constituent un système auto-équilibré sans oscillations à l'équilibre.

Si l'on place un germe à l'extrémité inférieure de la goutte (fig.3) et que l'on établit à ce niveau un gradient de température convenable, on peut tirer un cristal vers le bas à condition qu'on alimente, en même temps, le capillaire en liquide. Cette disposition est valable pour les capillaires de section cylindrique (fils, barreaux cylindriques) ou de section rectangulaire (plaque) (1), l'épaisseur des plaques et le diamètre des cylindres dépendant à la fois de la forme de l'extrémité inférieure du capillaire et de son diamètre (ou de sa largeur) interne. En épaisseur, notamment, les plaques peuvent être de 1 à 5 mm environ. La largeur dépend de la largeur du conduit capillaire.

Cette méthode porte le n° 7(000) dans la bibliographie des méthodes préformées établie en 1980 par le Journal of crystal growth (2).

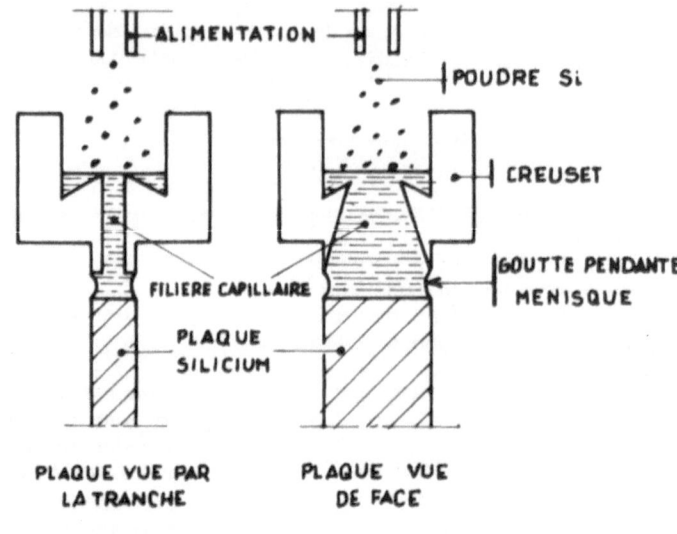

FIG. 3 SCHEMA DU CREUSET

2. ETAT DE LA METHODE PDG AVANT LE CONTRAT -

Nous avons obtenu quelques plaques de 200 x 1,5 x 29 mm avec orientation $\langle 11\overline{2}\rangle$ $(1\overline{1}0)$ donnant une structure colonnaire : le plan (111) est le plan de macle perpendiculaire à la face plane du ruban (3). Les creusets sont en graphite dense.

Nous avions donc à optimiser les diverses parties du procédé de tirage.

3. TRAVAUX EXECUTES PENDANT LE CONTRAT -

3.1 - Enceinte de tirage n° 2 (fig.4) -

Une enceinte de tirage de 200 litres a été utilisée. Elle permet de tirer des rubans jusqu'à 500 mm de long et peut recevoir des creusets pour rubans de 75 mm de large. Elle est équipée d'un groupe vide et d'une alimentation en gaz (argon, hélium, hydrogène).

Un pyromètre optique mesure la température des lèvres du creuset et assure ainsi la régulation de la source de chauffage à induction. L'alimentation commande la vitesse de tirage.

3.2 - Creusets -

C'est le point essentiel de la méthode à la fois pour la forme et le matériau.

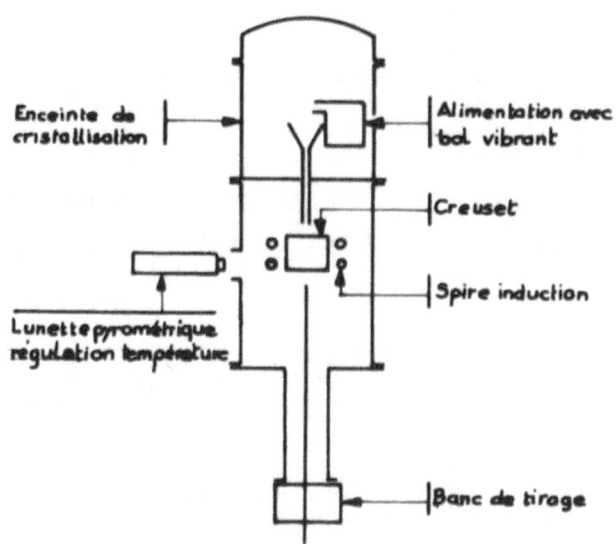

FIG.4 | ENCEINTE DE CRISTALLISATION N°2

3.21 - Matériau -

Nous avons essayé successivement à la fois sous l'angle mouillage et attaque chimique par le silicium :

- graphite chemisé SiO_2 : mouillage défectueux de la partie en SiO_2. abandonné
- graphite chemisé Si_3N_4 : difficulté d'usinage, mouillage difficile
- graphites ATJ, AXF 5 Q (UNION CARBIDE) } conviennent avec
- graphite ELLOR 9, 1116 PT, 1346 PT (CARBONE LORRAINE) } préférence pour le plus dense 1346 PT

- graphite recouvert SiC par CVD : la meilleure solution.

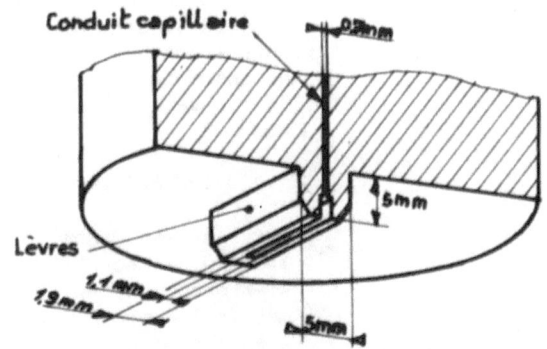

FIG.5 FORME DES LEVRES DU CREUSET N° 20

3.22 - Forme -

Nous avons aminci les lèvres jusqu'à une largeur hors tout de 1,9 mm avec un conduit capillaire de 0,7 mm. Les creusets sont monoblocs (fig. 5).

3.3 - Rubans obtenus -

3.31 - Dimensions -

Grâce à une alimentation en poudre constante ($\pm$ 1,5 %) et à une régulation de température du creuset à $\pm$ 2° au niveau des lèvres, nous avons obtenu des rubans jusqu'à 500 mm de long avec des épaisseurs de 1,5 à 2,0 mm sur 29 mm de large. Ce mode opératoire est au point.

Pour des épaisseurs plus faibles : $\leqslant$ 1 mm, nous n'avons pas réussi à obtenir des rubans aussi longs parce que la largeur du ruban n'était pas conservée pendant le tirage. Après une centaine d'essais de tirage avec différents creusets, nous nous sommes heurtés à plusieurs difficultés :

- la température n'est pas uniforme tout le long des lèvres au bout du capillaire, ce qui provoque le changement de forme du ménisque aussi bien en largeur qu'en épaisseur. Ce phénomène est d'autant plus accentué que le ruban est mince.

- le silicium mouille toute l'extrémité des lèvres, ce qui change la forme du ménisque et épaissit le ruban (fig. 6).

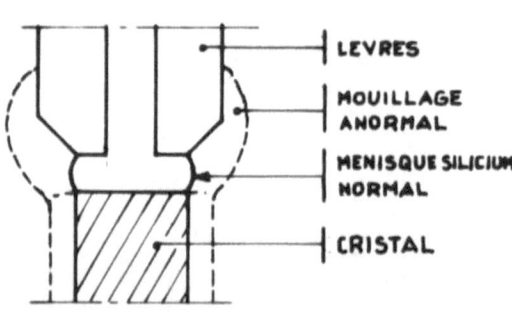

FIG. 6 MOUILLAGE DES LEVRES

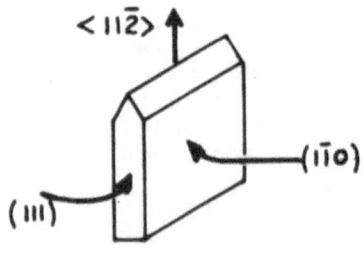

FIG. 7 GERME STRUCTURE COLONNAIRE

- à partir de la vitesse de tirage 1 m/h, la stabilité du processus de croissance est difficile à maintenir.

3.32 - Structure colonnaire -

Des cinq orientations de germe essayées, seule (fig. 7), la structure avec le plan (111) comme plan de côté donne de façon très reproductible une structure colonnaire avec joints de macle et joints de grain. Les autres donnent des grains désordonnés de taille variée, jusqu'à 0,5 cm2.

3.33 - Pureté -

Oxygène - Le dosage de l'oxygène interstitiel s'effectue par spectrométrie infrarouge grâce à une bande d'absorption située à 1105 cm^{-1}. Nous l'appliquons par comparaison avec un silicium Czochralski de même résistivité. L'oxygène est inférieur au seuil de détection de la méthode, soit 6.10^{15} atomes/cm3.

- Carbone - Le carbone substitutionnel est dosé par spectrométrie infrarouge à 5.10^{17} at.cm3 ($3,5\mu$g/g) avec une précision $\pm$ 20 %.

- Le carbone total dosé par combustion est en moyenne de 160μ g/g.
Le taux de carbone dépend du graphite et de la température intérieure du creuset.

- Impuretés métalliques -

Durant la croissance, une certaine purification est produite à cause des impuretés volatiles. Mais, par elle-même, la méthode n'a pas d'effet purificateur. Cependant, grâce à une sonde ionique couplée à un spectrographe de masse, nous avons trouvé que les impuretés sont surtout présentes sous forme d'îlôts à la surface des rubans. Un décapage acide enlève une partie de ces impuretés.

Les principales impuretés sont le fer ($0,4\mu$g/g au lieu de 0,01 admissible), le Mn ($\simeq 10^{-2}$ au lieu de 7.10^{-3}), Ni (1,4 au lieu de 0,2), Ti (1 à 2) et Zr ($\simeq 10^{-1}$).

Ces impuretés proviennent de la poudre de silicium d'alimentation.

3.34 - Caractérisation électrique des rubans -

. Résistivité - Les résistivités mesurées vont de 0,6 à 5,82 Ω/cm et sont fonction de la pureté du graphite utilisé. Les résistivités les plus fortes sont obtenues avec les graphites 1346 PT et 1116 PT du CARBONE LORRAINE.

. Joints de grain - On constate par caractérisation EBIC des joints de grain très actifs en structure désordonnée, plus faible en structure colonnaire. Les impuretés s'y rassemblent par ségrégation.

. Tests photovoltaïques - Ils ont été effectués chez PHOTOWATT et ont donné des rendements de 3,2 à 6,6 %. Les longueurs de diffusion vont de 10 à 25 m.

Ces résultats, insuffisants, sont dus à la présence de certaines impuretés métalliques et à leur diffusion pendant le traitement de recuit.

4. Calculs économiques -

Avec un programme sur calculateur TEXAS TI 59, nous avons calculé le prix de revient du silicium en fonction des divers paramètres. Nous avons pris le prix du silicium à 70 FF/kg.

Nous donnons ci-après la courbe indiquant le prix au m2 en fonction de la vitesse de tirage (fig.8).

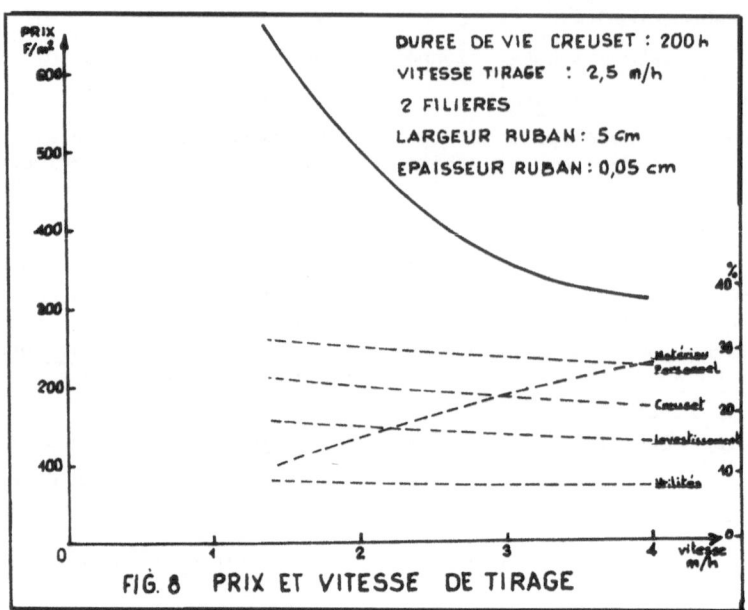

FIG. 8   PRIX ET VITESSE DE TIRAGE

5. TRAVAUX EFFECTUES APRES LE CONTRAT -

. Epaisseur du ruban - Une théorie du ménisque a été établie pour une épaisseur totale des lèvres qui relie l'épaisseur du cristal à la hauteur du conduit capillaire à l'intérieur du creuset et à la hauteur du ménisque (4).

Par exemple, pour une largeur de lèvres de 0,5 mm, on peut atteindre des rubans de 0,4 mm d'épaisseur avec une hauteur de ménisque de 0,25 mm. La théorie a été vérifiée pour des rubans de 1,9 mm d'épaisseur obtenue avec des lèvres de 2,3 mm d'épaisseur. Nous sommes en train de faire usiner des creusets à lèvres de 0,5 à 0,7 mm.

Pour le saphir, nous obtenons aujourd'hui des rubans de 0,8 ± 0,05 mm avec des lèvres de 1 mm.

. Four de recuit - Nous avons modifié les écrans thermiques du creuset pour diminuer l'attaque du carbone par le silicium.

Par ailleurs, nous installons un four de traitement thermique sous le creuset de manière à recuire les rubans en continu en conservant l'atmosphère du four.

. Atmosphère du four - Nous préparons un contrôle d'atmosphère pour jouer sur les concentrations en CO, CO_2 et H_2 en présence d'argon.

. Poudre d'alimentation - Nous avons progressé dans la purification des poudres de silicium pour alimenter la méthode PDG.

6. CONCLUSION -

La méthode PDG a donc prouvé sa faisabilité pour obtenir des rubans de silicium. Les progrès qui lui restent à faire pour l'épaisseur du ruban et la pureté du silicium relèvent d'essais technologiques pour lesquels nous avons des solutions.

Enfin, pour pousser la méthode à son rendement maximal, il faut automatiser complètement le tirage en prenant comme paramètres à tenir constants la hauteur du ménisque et la température des lèvres.

Les prix prévisionnels, calculés selon un modèle industriel, sont favorables pour une production de masse.

BIBLIOGRAPHIE -

(1) J. RICARD - Procédé de fabrication en continu de monocristaux pré-formés - Brevet français 2.321.326 du 8.08.1975

(2) Shaped crystal growth - A selected bibliography - D.O.BERGIN, Journ. of cryst.growth, 50 (1980), p.381-396

(3) J. RICARD - Proceed. Photovolt. Solar Energy - Conférence Luxembourg 1977, p.882

(4) Th. DUFFAR - Cristallisation de rubans de silicium par la méthode de la goutte pendante en vue de l'élaboration de cellules solaires - Thèse Grenoble, 10 novembre 1982.

GROWTH AND SOLAR CELL ASPECTS IN RELATION TO
POLYCRYSTALLINE SILICON RIBBONS GROWN
BY THE RAD PROCESS

Author   : Christian BELOUET

Contract number   : ESC-R-022 F

CEC contribution : 2.7 MF (50 % basis)

                    This work is a part of a 3-year programme also sponso-
                    red by AFME (former COMES) and Société Nationale ELF
                    AOUITAINE (SNEA).

Duration   : 36 months - 1 July 1980/30 June 1983

Head of próject   : J.P. DUMAS, Laboratoires de Marcoussis

Contractor   : Laboratoires de Marcoussis

Address   : Laboratoires de Marcoussis, Research Center of C.G.E.
                    Route de Nozay - F-91460 MARCOUSSIS

SUMMARY

It is reported on the present status of a three-year programme
initiated in June 1980 about polycrystalline silicon ribbons
achieved by the RAD growth process. This presentatioń covers
all the aspects of the study including those which do not expli-
citly pertain to the CEC contract n° ESC-R-022 F.
The results are presented according to the main research areas
which relate to the carbon ribbon, growth and solar cell aspects.
A carbon ribbon structure was chosen in January 82. It is shown
that, although the specifications pertinent to this structure
were individually met, the delayed optimization of this structure slowed down the growth programme. With respect to the
carbon ribbon drawbacks, the solar cell back-contact problem
was obviated by a burn-off step after growth which yields self-
sustained silicon layers.
Continuous growth was demonstrated on the single-ribbon puller
over lengths up to 30 m at a pulling rate of 10 cm.mn^{-1}. The
thickness of each opposite silicon layer thus obtained can be
made larger than the minimal value necessary to prevent the
formation of cracks of thermoelastic origin.
The AM1 conversion efficiencies η of solar cells are in the
9-11% range. Comprehensive studies indicate that η values in
the 11 to 13 % may be envisaged.

Finally, the major goals of this three-year programme should be
reached on schedule by July 1983.

1. INTRODUCTION

It is reported on the status as off october 1982 of a three-year programme running till July 1983 about the RAD polycrystalline silicon ribbon
project.

The main purpose of this programme was i) to define the characteristics
of the carbon ribbon shaper and ii) to demonstrate a number of technical
feasibilities (by July 1983) necessary to establish the economical viability
of the process ; in particular, a number of drawbacks associated with the
use of the carbon shaper had to be overcome.

The original programme was momentarily slown down because of the transfer of the activity from LEP* to LdM**; its content was also modified as
the burn-off of the carbon ribbon, which obviates the solar cell back-
contact problem, was found to be successful.

Following a brief description of the project milestones, the results
obtained in the main research areas, which relate to carbon ribbon (including burn-off), growth and solar cell aspects, are successively presented.

2. PROJECT MILESTONES

The major project milestones as revised in June 1982 in order to take
into account some delays consecutive to the transfer of the activity were
established in view of demonstrating i) deposition lengths of 50 m at
10cm.mn^{-1} using a single-ribbon puller station fitted for ribbons 5 cm in
width and ii) a 10 % AM1 conversion efficiency on A.R. coated cells 50 cm^2
in size. These milestones implied i) the definition and optimization of an
appropriate ribbon structure and ii) the compatibility of the burn-off step
with the achievement of self-sustained sheets at least 50 cm^2 in size. With
respect to the carbon ribbon, the choice of a structure and its optimization
were set for January 1, 1982 and December 31, 1982 respectively.

The feasibility of self-sustained sheets 50 cm^2 in size was set for
June 30, 1982 (this activity was only introduced in September 1981).

3. RESULTS

3.1. Carbon ribbon

The use of a carbon shaper to stabilize the freezing meniscus (Fig.1)
is a unique feature of the RAD growth process which results in reliable
growth conditions, a minimal silicon consumption and a relatively high
throughput rate (1). However, a number of drawbacks associated with this
carbon shaper had to be overcome. They are summarized together with the
appropriate solutions in table I.

The melt contamination by carbon impurities was considerably reduced
using a chlorine purified Papyex*** ribbon base. This treatment resulted in
a reduction of 10^2 to 10^3 in the content of most metallic impurities, and
allowed the deposition of a high quality lamellar pyrocarbon coating****
resistant to molten silicon.

The most significant results which relate to the chlorine purification
step are given in tables 2 and 3. Table 2 shows its efficiency on the
purity grade of the Papyex base (columns 2, 3) and the completed ribbon

----

   * LEP : Laboratoire d'Electronique et de Physique Appliquée.
  ** LdM : Laboratoires de Marcoussis.
 *** Papyex is the tradename of a carbon ribbon manufactured by LE CARBONE
     LORRAINE (LCL).
**** The growth of pyrocarbon was made by LCL ; TEM characterization stu-
     dies were conducted by the Laboratoire Marcel Mathieu (CNRS-ORLEANS).

(columns 4, 5). The consequences in terms of melt contamination appear in
table 3 where melts noted A and B were exposed to the two types of carbon
ribbons in table 2 ; it is seen that with the exceptions of Mo and Na, the
concentration of metallic impurities is reduced by a factor of at least 10^2
for the case of chlorine purified Papyex bases (see other aspects in section
"Growth and Puller Technology").

The thermoelastic stresses developed in the multilayered ribbon, as it
cools down to room temperature in response to the differences of thermal
expansion coefficients of carbon and silicon,may be large enough to generate
cracks in the silicon layers (1). Their magnitude, given the thickness of
the opposite silicon layers, is in direct relation with the values of the
Young's modulus and the thickness of the Papyex ribbon. For the case of
symmetrical growth, of relevance for this work, it was experimentally found
that crack-free silicon layers could be grown down to thicknesses around 50
micrometers on a carbon ribbon made of a Papyex base 200 μm thick having a
pseudo-Young's modulus in the 300-500 MPa range and coated by a thin lamellar pyrocarbon layer. The results of the studies in the above two areas led
to define the list of specifications in table 4, which describe a carbon
ribbon resistant to molten silicon and compatible with the growth of crack-
free silicon layers as thin as 50 micrometers.

The quality of the solar cell back contact was a major concern in the
RAD process as reported in previous works (2). This drawback was definitely
obviated with the burn-off of the carbon ribbon after growth. The burn-off
step, performed at around 1000 °C in oxygen, was found to be compatible
with the achievement of i) large self-sustained sheets and ii) relatively
high efficiency solar cells (see below).

The separation of crack-free silicon layers 100 cm^2 in size (20 x 5 cm)
down to thicknesses as low as 50 micrometers was recently obtained with this
process (the layers were grown at 10 cm.mn^{-1}).

The beneficial consequences of the carbon ribbon burn-off are summarized in table 5. The recovery of the opposite silicon layers results in a
reduction of the effective cost of the carbon ribbon and a two-fold increase
of the throughput rate ; it implies also symmetrical growth conditions with
in turn the advantages shown in table 5. A most important consequence is
that the RAD growth process associated with the burn-off step produces self-
sustained silicon layers with thicknesses eventually below 100 micrometers,
i.e. about equal to the values of the effective minority-carrier diffusion
length L_{eff} (see section "Solar Cells") ; these circumstances should allow
a thin silicon solar cell technology with optimal conversion efficiencies (3).

The geometrical aspects of concern relate to the flatness and straightness of the ribbon (its dimensions being already accurately controlled).
Large deviations from flatness and straightness requirements cause variations of the thickness of each silicon layer to occur in the transversal direction and induce asymmetrical growth conditions. On the basis of the cumula-
ted growth experience, criteria were defined for the deviations to these ribbon characteristics and transmitted to the manufacturer. A careful follow-
up of their occurrence along the fabrication process of the ribbon showed
that they were generated by mechanical imperfections of the kinematics of
the rolling and cutting operations, rather than by the subsequent purification and pyrocarbon deposition steps.

Therefore, the feasibilities of all the specifications pertaining to
the carbon ribbon characteristics were demonstrated individually. However,
most of the ribbons delivered by LCL could not be used for growth experiments since the ribbon structure was definitely chosen on schedule in January 82. The delivery of a reliable product originally planned for December
1982 should take place in the course of the first term of 1983.

3.2. Growth and puller technology

The achievement of a 50 m deposition length (ribbon 5 cm wide) was a major goal of this contrat to be achieved in July 1982. It was also planned to demonstrate the growth of crack-free layers at pulling rates up to 10 cm. mn^{-1}.

The actions concerning the puller technology were expected to result in the concept of a multiple-ribbon puller (preliminary design review by mid-1983) and the demonstration of the feasibility of the laser cutting of the RAD ribbon in simulation experiments (July 1983).

Although growth activities (conducted with the single-ribbon puller in Fig.2) were slowed down by the still hazardous quality of the carbon ribbons, growth goals were closely approached as shown in table 6. The maximum deposition length obtained at a pulling rate of 10 $cm.mn^{-1}$ was 30 m, limited only by excessive deviations from straightness of the carbon ribbon. It appears also in table 6 that the total thickness of the opposite silicon layers could be controlled in the 140 to 170 micrometer range at 10 $cm.mn^{-1}$. These data were obtained in the vicinity of the transition from smooth to dendritic growth (i.e. at small values of the temperature gradient G_L in the liquid at the freezing front)using two types of carbon ribbons having the following thickness and specific weight : 1) 0.30 mm - 0.50 $g.cm^{-3}$ and 2) 0.23 mm - 0.68 $g.cm^{-3}$. These silicon thickness values are significantly larger than predicted on the basis of past experiments (1). This discrepancy is ascribed mainly to a larger contribution of convective heat transfer to the cooling of the ribbon in the present operating conditions. These thickness values were experimentally found to be compatible with the growth of layers free of cracks of thermoelastic origin as predicted if carbon ribbons met the specifications in table 4 (1). Finally, the control of the silicon total thickness in the transversal direction was closely approached by proper adjustments of the heat balance in the silicon pellet injection area, Fig.1.

The result of silicon bulk analyses evidenced a substantial purification effect at the freezing front, the effective partition coefficients K of most metallic impurities being in the 10^{-1} to 10^{-3} range with the exceptions of Cr and Na (table 3, run A). This property of the process is readily ascribed to the growth from a non-confined freezing meniscus, which allows a redistribution of impurities towards the melt.

Besides the continuous improvement of the present puller, the actions relevant of the puller technology were limited thus far to laser cutting experiments using a CO_2 laser. On the basis of the early available results, it may already be concluded that the cutting should be performed on a preheated ribbon. In addition, cracks are generated in large numbers in the silicon layer at the cuts ; their ability to propagate along the ribbon in the subsequent burn-off step has not been studied yet. An increased attention is also going to be devoted to the optimization of the mechanical cutting.

3.3. Solar cells

The main goal in this area is to achieve solar cells 50 cm^2 in size exhibiting AM1 conversion efficiencies $\eta \geqslant 10$ % (A.R. coated, active area).

The solar cells are prepared on self-sustained layers as shown in the flow-chart in table 7. The layers ready from the burn-off step are first submitted to an HF etch to remove the oxide layer at the front face and a chemical lapping to eliminate the thin, discontinuous SiC layer at the back-side. N^+/p homojunctions were made by a classical $POCl_3$ diffusion process at 850 °C (junction depth $\sim$ 0.5 μm) ; a p^+-contact was formed at the back-side by fast annealing at 660 °C of an evaporated aluminium layer and the front contact was a Ti, Pd, Ag evaporated grid. The I(V) curves of these cells were systematically recorded under AM1 illumination (Φ AM1) and for some cases at different η values. Complementary investigations not discussed in this paper involved studies of dark I(V) curves, TEM and SEM observations and resistivity measurements.

The solar cells thus prepared were found to exhibit large trapping effects which impaired their performances. Low-temperature annealings (LTA) proved to be very effective in reducing these deleterious effects. Comprehensive studies aimed at investigating the above phenomena were conducted on complete cells so that the LTA temperature and duration could not exceed 430 °C and one hour respectively in order to obviate front contact degradation problems. The temperature and duration were then set at 430 °C (LTA-430) and 30 minutes respectively.

The overall effect of an LTA-430 is illustrated in Fig.3. The original and broad distribution of conversion efficiency η (noted 1) shifts after an LTA-430 to higher η values, the new Gaussian-like distribution (noted 2) being narrower (η values are given for cells without A.R. coating).

The study of spectral sensitivity curves drawn at different illuminations Φ for solar cells taken before (1) and after (2) an LTA-430 clearly showed that the response of the front "window" and the effective minority-carrier diffusion length L_{eff} were improved in type-2 cells, L_{eff} values up to 100 μm being currently measured at Φ AM1. However, L_{eff} was found to increase with Φ, this effect being stronger for type-1 cells. The increase of L_{eff} with Φ may be interpreted as the result of the saturation of minority-carrier traps (4) ; it implies a superlinear dependence of the photocurrent Jph against Φ :

$$Jph = K. \ [\Phi]^{(1+S)} \tag{1}$$

In relation (1), K is a constant characteristic of each cell and S accounts for the saturation effect, S increasing with the density of active traps. Systematic determinations of S values of type-1 and type-2 cells were obtained from I(V) curves drawn for different Φ values.

It was found that the characteristics of the solar cells (namely, Jph, the open-circuit voltage Voc and the fill-factor FF) were dramatically impaired by a very important trapping effect (S values up to 0.1 were currently observed for type-1 cells). In addition, the activity of traps was systematically and substantially reduced by an LTA-430 of 30 mn only. These findings are examplified in Figs. 4 to 7. Figs. 4, 5 show the respective dependences of Voc and Jph on S for the same ribbon noted "C". The data were indifferently obtained from type-1 and type-2 cells which appear to behave similarly. Least-square linear fittings (broken lines) give an indication on the large sensitivity of Voc and Jph on S ; these figures also show the increments of Voc (ΔVoc) and Jph (ΔJph) following an LTA-430.

Plots of both these increments against the initial values of Voc and Jph measured on type-1 cells are given in figures 6 and 7 for ribbons "C" and "D" (open and full circles respectively). A quasi-linear dependence of ΔVoc on Voc (1) is observed in figure 6 for both ribbons : a least-square linear fitting for ribbon "C" (broken line) gives an extrapolated Voc value of 549 mV for ΔVoc = 0, in perfect agreement with Voc(S = 0) = 552 mV obtained in figure 5 ; the extrapolated Voc value (ΔVoc = 0) is 565 mV for ribbon "D".

In contradistinction to the above results, the dependences of ΔJph on Jph (1) may differ considerably with the ribbons as illustrated in Fig. 7 where a strong and quasi-linear correlation ΔJph(Jph) is obtained for ribbon "D" only. The considerable spread of Jph values noted for ribbon "C" suggests that for some cells, Jph is not primarily limited by trapping effects. The suspected large differences in metallic impurity content along a ribbon, consecutive to a heterogeneous erosion of the pyrocarbon coating, may account for the spread of ΔJph and Jph values (5). Finally, $FF(S)$ and $\Delta FF(FF)$ relations were found to be similar in behaviour to the above ones.

The results stress the important role of traps in controlling both the "bulk" and "surface" electrical properties of the devices as inferred from the $Jph(S)$ and $Voc(S)$ relations respectively.

The reduction of S values with an LTA-430 of 30 mn only and the corresponding increments of Voc, Jph and FF testify to the effectiveness of low-temperature annealings in reducing the trap activity in RAD layers.

The nature of the dominant traps involved in the above phenomena has not yet been determined. They are presumably linked to the presence of a high concentration of both oxygen and carbon in the layers. Detailed investigations of their nature based on I.R. absorption and DLTS measurements have been initiated.

The elimination of the trapping effect would result in fairly high AM1 conversion efficiencies η as indicated by the extrapolated solar cell characteristics at $S = 0$. This is illustrated below for ribbons "C" and "D" (no A.R. coating).

Ribbons	"C"	"D"
Voc (mV)	552	565
Jph (mA/cm^2)	21	23.8
FF	0.73	0.74
η	8.45	9.95

These predictions were experimentally corroborated for a number of A.R. coated solar cells (4 cm^2) which exhibited AM1 conversion efficiencies η between 12 and 13 % and showed weak trapping effects ($S < 0.03$).

This goal could be achieved in practice by a proper adjustment of the diffusion parameters - which definitely influence S values - and the search for optimized LTA treatments in a broad range of temperatures and durations.

The performances of solar cells may also be improved by hydrogen passivation. Recent experiments clearly evidenced that the incorporation of hydrogen in RAD layers by RF plasma treatments at 430 °C (HPL-430) reduced the electrical activity of boundaries and increased L_{eff} in the bulk (6) (Fig. 4 shows an example of a Voc increase consecutive to an HPL-430).

On the basis of these results, the typical performances of the solar cells given in table 8 should be significantly improved.

Finally, the studies on advanced solar cell technologies (ASCT) including "cold junction" formation, printed contact, etc.., will be initiated in January 1983.

4. CONCLUSION

Although the activities of this three-year programme were impaired by the transfer of the research activity and delays in the optimization of the carbon ribbon, the major milestones were met or approached on schedule. The carbon ribbon structure was definitely chosen and it should be optimized early in 1983 ; the drawbacks associated with this ribbon were circumvented. In particular, the burn-off process which obviates the solar cell back contact problem was found to be highly beneficial in all aspects.

Growth results indicate that layers can be currently grown at 10 cm. mn^{-1} and that a strong purification effect takes place at the freezing front.

The solar cells made on the self-sustained layers ready from the burn-off process have AM1 conversion efficiencies in the 9-11 % range. Their performances are primarily limited by trapping effects which should be overcome. Expected performances are in the 11-13 % range.

Finally, the targets of this three-year programme should be met by July 1983.

REFERENCES

(1) Final Report - CEC Contract ESC-R-022 F (01.07.80/31.12.81).

(2) C. BELOUET, C. BELIN, J. SCHNEIDER, J. PAULIN, Proc. IIIrd EC Photovoltaic Solar Energy Conf., CANNES (27-31 Oct. 1980), 558-562.

(3) H.J. HOVEL, Semiconductors and Semimetals, Vol.2, Solar Cells, Academic
Press, p. 100.

(4) E. FABRE, M. MAUTREF, A. MIRCEA, Appl. Phys. Lett. 27, 4 (1975) 239.

(5) J. REVEL, N. DESCHAMPS, J. DEVILLE, C. TEXIER-HERVO, C. BELOUET, Proc.
IVth EC Phot. En. Conf., STRESA (Italy) (10-14 May 1982) - D. REIDEL
Publishing Company, p. 896.

(6) M. AUCOUTURIER, O. RALLON, M. MAUTREF, C. BELOUET, to be published,
Journal de Physique - Proc. Conf. CNRS, "Semiconducteurs Polycristallins", PERPIGNAN (France) -(Sept. 1982).

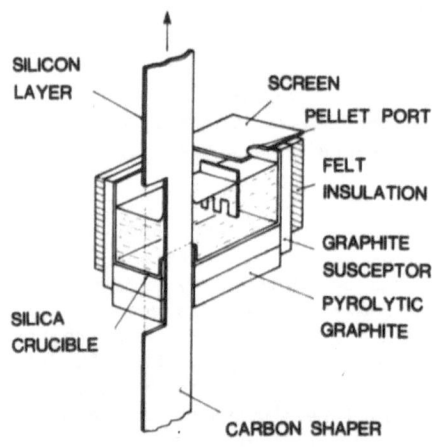

Fig.1 - Scheme of the RAD growth process.

Fig.2 - Puller general view.

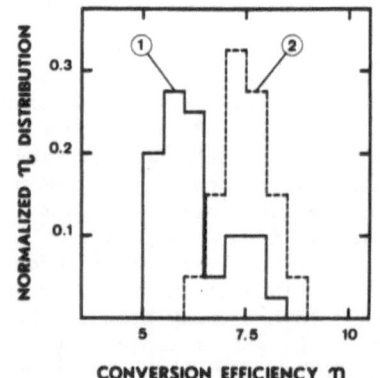

Fig.3 - η distribution before (1) and after (2) a LTA-430.

Table 1

DRAWBACKS	SOLUTIONS
MELT CONTAMINATION	CHLORINE PURIFICATION OF PAPYEX
THERMO-ELASTIC STRESSES	OPTIMISATION OF PAPYEX & FABRICATION STEPS
GEOMETRICAL ASPECT	OPTIMISATION OF PAPYEX ROLLING & CUTTING
SOLAR CELL BACK-CONTACT	CARBON RIBBON BURN-OFF

Table 2 : Neutron activation analysis of the carbon ribbons
(concentrations : $\mu g.g^{-1}$)

CARBON RIBBONS / ELEMENTS	RAW MATERIAL	CHLORINE PURIFIED	CHLORINE PURIFIED PYROCARBON COATED (RUNS B & C)	PYROCARBON COATED (RUN A)
Cl	25	6	7	5
Co	1.5	4×10^{-3}	4×10^{-3}	3×10^{-1}
Cr	40	5×10^{-2}	6×10^{-2}	1
Cu	20	2×10^{-1}	2×10^{-1}	1×10^{-1}
Fe	700	1	1	65
Mn	9.5	5×10^{-3}	1×10^{-2}	1×10^{-2}
Mo	100	2	1.5	35
Na	16	5 to 7	4×10^{-1} to 5	1
Ni	61	$< 5 \times 10^{-1}$	4×10^{-1}	2.5
Ti	10	< 3	< 3	< 5
W	5	1×10^{-1}	1×10^{-1}	2 .
Zn	170	5×10^{-1}	3×10^{-1}	$< 2 \times 10^{-1}$
Zr	2	$< 4 \times 10^{-1}$	$< 5 \times 10^{-1}$	< 1

Table 3 : Neutron activation analysis of the silicon materials
(concentrations : $atom.cm^{-3}$)

SILICON MATERIALS / ELEMENTS	STARTING MATERIALS	RUN A			RUN B	RUN C
		MELT	LAYER	K	MELT	LAYER
As	$< 7.5 \times 10^{12}$	2.0×10^{12}	$< 7.5 \times 10^{11}$	< 0.4	2×10^{12}	1.0×10^{12}
Au	2.0×10^{10}	7.0×10^{12}	7.0×10^{11}	0.1	4×10^{10}	2.0×10^{11}
Co	$< 1.0 \times 10^{13}$	5.0×10^{14}	2.4×10^{13}	0.05	1×10^{13}	1.7×10^{13}
Cr	$< 5.0 \times 10^{13}$	5.0×10^{15}	2.0×10^{15}	0.4	$< 1 \times 10^{13}$	$< 8.0 \times 10^{12}$
Cu	$< 4.0 \times 10^{13}$	3.0×10^{15}	4.0×10^{13}	0.01	$< 4 \times 10^{13}$	6.0×10^{14}
Fe	$< 1.5 \times 10^{15}$	2.8×10^{17}	$< 5.0 \times 10^{15}$	< 0.02	$< 2 \times 10^{15}$	$< 2.0 \times 10^{15}$
Mo	$< 9.0 \times 10^{13}$	2.2×10^{15}	2.6×10^{14}	0.1	1×10^{15}	1.5×10^{13}
Na	-	6.0×10^{14}	1.2×10^{14}	0.2	2×10^{14}	1.0×10^{15}
Ni	$< 1.0 \times 10^{15}$	2.0×10^{16}	$< 5.0 \times 10^{14}$	< 0.3	$< 2 \times 10^{15}$	$< 2.0 \times 10^{14}$
Sb	$< 1.0 \times 10^{12}$	9.0×10^{14}	1.2×10^{12}	0.001	1×10^{13}	3.5×10^{12}
W	$< 5.0 \times 10^{10}$	3.8×10^{14}	$< 5.0 \times 10^{11}$	< 0.001	$< 1 \times 10^{12}$	$< 1.5 \times 10^{12}$
Zn	$< 1.0 \times 10^{14}$	1.0×10^{15}	$< 2.0 \times 10^{14}$	< 0.2	$< 1 \times 10^{14}$	$< 1.0 \times 10^{14}$

Table 4 : Carbon ribbon shaper characteristics

PAPYEX BASE

LENGTH	50 M
WIDTH	5 CM
THICKNESS	200 μM
DENSITY	0.7 G/CM^3
YOUNG'S MODULUS	< 30 daN/MM^2

PYROCARBON COATING

VARIETY	LAMELLAR TYPE
THICKNESS	2 TO 4 μM

Table 5 : Carbon burn-off consequences

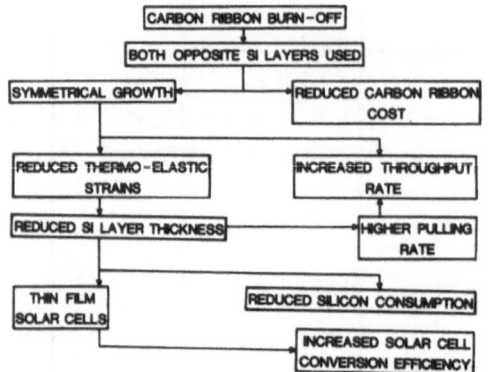

Table 6 : Growth status

SLOT CLEARANCE	0.6 mm
MELT HEIGHT	2 cm
RIBBON WIDTH	5 cm
PULLING LENGTH	30 m
PULLING RATE	10 cm/mn
SILICON THICKNESS (Total)	170 μm

Table 7 : Solar cell fabrication chart

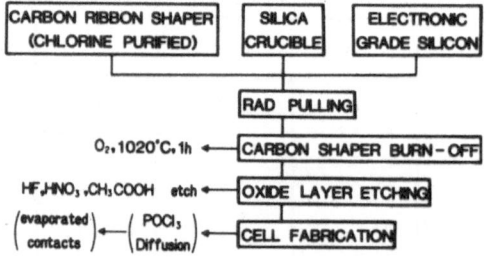

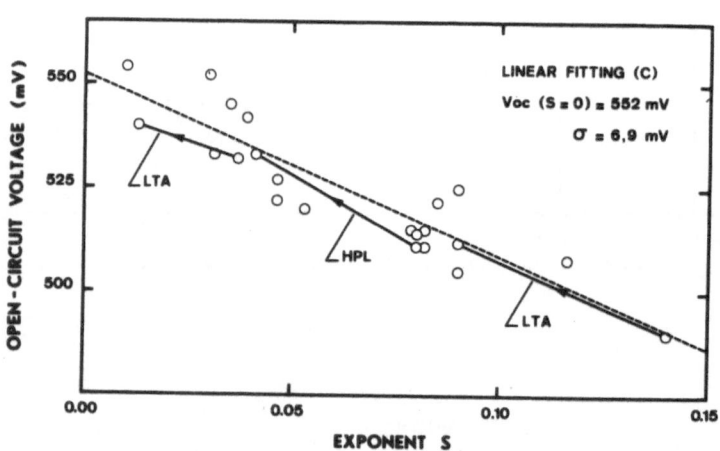

Fig.4 : Open-circuit voltage against S exponent. Voc changes after LTA-430 and HPL-430 treatments are indicated by arrows.

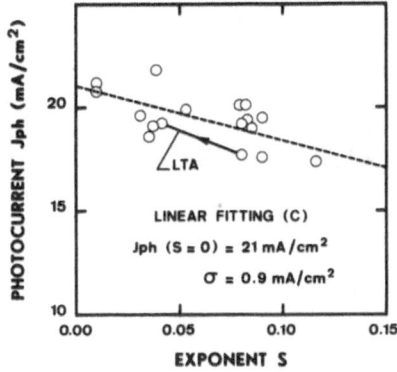

Fig.5 - Photocurrent density against
S exponent. A Jph change after an
LTA-430 treatment is indicated by
an arrow.

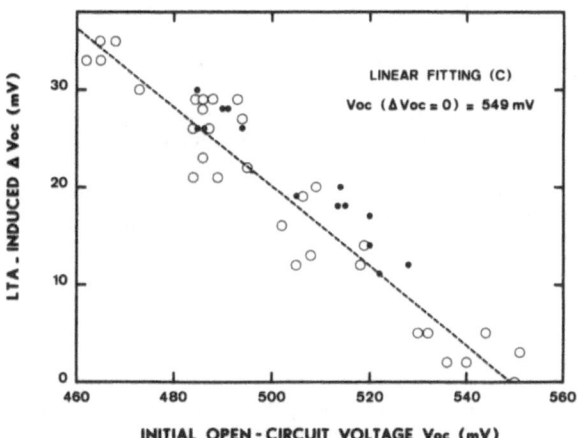

Fig.6 - LTA induced Voc change vs. the initial
Voc value.

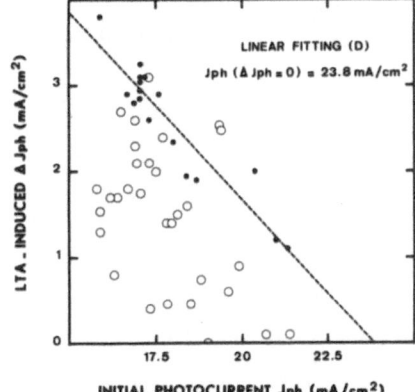

Table 8 : Solar cell typical
performances ; AM1 illumination,
A.R. coating.

SIZE	4 CM²
PHOTOCURRENT DENSITY	24-29 mA/cm²
OPEN-CIRCUIT VOLTAGE	525-555 mV
FILL FACTOR	0.69 - 0.73
CONVERSION EFFICIENCY	9-11%
BEST CELL	12.4%

Fig.7 - LTA induced Jph change vs.
the initial Jph value.

STUDIES RELATING TO NEW ENCAPSULATION MATERIALS

Authors : J. ANGUET, J. DONON*
M.C. MICHEL, J.C. BOBO+

Contract number : ESC-R-076-F

Duration : 18 months   1 January 1981 - 30 June 1982

Head of project : Jérome DONON

Contractors : PHOTOWATT International S.A.*
LABORATOIRES de MARCOUSSIS+

Address : 125, rue du Président Wilson F92303 LEVALLOIS PERRET
Route de Nozay 91460 MARCOUSSIS+

Summary

The studies concern new encapsulation materials for module manufacturing either with cast resin or with laminated resin. A first
category of new material was investigated in the first phase of the
contract, which need only small modifications : polyuréthane resin,
ethylene viny acetate resin. Modules were manufactured with the two
resins and reliability experiments done, showing that the obtained
formulations can be used now for module encapsulation.
During this first phase we also started the investigation of other
new materials like PVC, EPDM, EMA and the first formulations were
tested.

1. Introduction

La présente étude concerne la mise au point de nouveaux matériaux d'encapsulation pour la réalisation de modules photovoltaiques, remplaçant principalement les résines silicones (85-90 F/Kg) et le polyvinyl butyral (55 F/Kg).

Les deux types de modules concernés dans cette étude sont des modules réalisés soit, à l'aide de résine de coulée, soit à l'aide de films laminés.

Deux catégories de matériaux pour ces types d'encapsulation ont été étudiés :

Une première catégorie concerne des matériaux commercialement disponibles (15-20 F/Kg) qui sont pratiquement au point et dont la mise en oeuvre est aisée : résines polyuréthanes pour les résines par coulée et éthylène vinyl acétate pour les films.

Une dernière catégorie concerne des matériaux potentiellement intéressants (inférieur à 10 F/Kg), mais pour lesquels il y a lieu d'améliorer le vieillissement sous UV et sous humidité par incorporation d'additifs appropriés.

Au cours de l'étude et au fur et à mesure de la mise au point des matériaux des modules de grande taille ont été réalisés, permettant ainsi l'étude de la mise en oeuvre des produits utilisés et leur caractérisation du point de vue vieillissement.

Les tableaux I et II donnent pour les différents matériaux étudiés, les ordres de grandeur de prix au kilogramme et les valeurs de transmission totale moyenne entre 400 et 1100 nm.

TABLEAU I

PRIX AU KILOGRAMME DES MATERIAUX PLASTIQUES D'ENCAPSULATION

POUR MODULES PHOTOVOLTAIQUES (Fr/Kg).

SILICONE (RTV)	70 - 80
POLYVINYLBUTYRAL (PVB)	40 - 50
POLYURETHANE	10 - 15
ACRYLIQUE	15 - 25
ETHYLENE VINYL ACETATE (EVA)	10 - 12
CHLORURE DE POLYVINYL (PVC)	5 - 10
ETHYLENE PROPYLENE DIENEMONOMERE (EPDM)	8 - 10
ETHYLENE METHYL ACRYLATE (EMA)	8 - 12
ETHYLENE BUTYLACRYLATE (EBA)	8 - 12

TABLEAU II

TRANSMISSION TOTALE MOYENNE DES DIFFERENTS MATERIAUX D'ENCAPSULATION ETUDIES

TYPE DE MATERIAUX	EPAISSEUR DE L'ECHANTILLON mm	TRANSMISSION MOYENNE ENTRE 400 et 1100 mm %
SILICONE (RTV 615)	0,3	93,8
....	1	93,5
EVA (ELVAX 150)	1	88,4
....	0,25	90,2
EVA (ELVAX 410)	1	88
EVA (Nordel 1320)	1	90
....	0,5	89,5
PU (Qthane - Quinn Co)	0,2	90,7
....	2	88,8
PU (Development Associates)	1	92
PVC (Solvic 372 LC)	1,6	88,4
EMA (2205 Gulf Oil)	1	88,2

2. Les résines mises en oeuvre par coulée

2.1 Résines polyuréthanes aliphatiques

Les résines polyuréthanes aliphatiques appartiennent à la première catégorie de matériau et sont le résultat de l'addition dtunediol et dtun
isocyanate. Trois types de diol ont été étudiés :

a) diol polyether - CH2 - O - CH2 -

b) diol polyesther - CH2 - O - C - CH2 -

$$\begin{array}{cc} R & || \\ | & O \end{array}$$

c) diol branché - HO - C - CH2 -

$$\begin{array}{c} | \\ R \end{array}$$

Le passage de a) à c) améliore lors des essais de vieillissement accé-
léré la stabilité sous UV et la stabilité thermique.

Des polyuréthanes de provenances diverses ont été étudiés en vieillissement (QuinnCo, Bayer, SNPE, Development Associates) et ont donné les
résultats suivants (tableau III).

Tableau III. Vieillissement weather-O-Metre 75°C 80 % RH 1 KW|m2 Xenon Light renforcé UV.

	Direct Transmission		Remarques
	Avant Test	Après Test	
Diol Polyether (Quim)	88 %	81 % 95 h	Fondu après 525 h
Diol Polyether (Quim) entre 2 verres	89 %	88,5 % 95 h	88,5 % après 538 h décollements
Diol Polyesther (Bayer) ...	89 %	90 % 52 h	Fondu après 390 h
Diol Polyesther (Bayer) entre 2 verres	88 %	88 % 180 h	81 % après 500 h
Diol branché (polyether modifié) SNPE	88 %	88 % 700 h	Aucun changement légère présuration
Polyuréthane stabilisé anti UV et anti oxydant (Development Associates) Z2211	92 %	90 % 1400 h	Pas de changement

Les résultats obtenus à l'aide des résines polyuréthanes alophatiques montrent que l'utilisation d'une résine parfaitement stabilisée est possible en tant qu'encapsulant pour module photovoltaïque réalisé par coulée.

2.2 Les chlorures de Polyvinyl PVC

Le PVC brut se présente sous forme de poudre qui peut être mise en solution dans un plastifiant (Plastisol) et ainsi être mis en oeuvre sous forme liquide.

Le PVC étudié est le PVC 372 LC de Solvay stabilisé avec des composés d'étain, de baryum-cadmium, ou calcium-zinc. Le plastifiant utilisé est un dioctylphtalate (DOP) et la composition suivante a été utilisée :

PVC 372 LC 100 parties
Stabilisant 2 parties
Huile de soja epoxydée 2 parties
Plastifiant DOP 120

Le tableau IV suivant donne les résultats de vieillissement obtenus en "weather-o-metre" :

TABLEAU IV	Transmission directe		Remarques
	Avant test	Après test	
PVC 372 LC avec stabilisant Cadmium-Baryum	91 %	81 % 216 h	échantillons déteriorés par l'arrosage
PVC 372 LC stabilisant Calcium-zinc	91 %	75 % 216 h	
PVC 372 LC stabilisant maléate d'étain	74 %	74 % 302 h	485 h 74 % Voile blanc et bulles
PVC 372 LC stabilisant thio étain	86 %	81 % 216 h	551 h 81 % Voile blanc et bulles

L'apparition de bulles constatée est dûe à la migration du plastifiant. Celui-ci a été remplacé par du DIDA qui est moint sujet à la migration que le DOP en raison de sa structure aliphatique. Les échantillons réalisés n'ont pas résisté au test de vieillissement en chaleur sèche à 100°C où après quelques jours il sont devenus bruns puis opaques.

Le mécanisme de vieillissement du PVC est initié par la rupture d'une liaison Cl-C qui entraîne l'élimination d'une molécule HCl. Il y a ensuite formation de liaisons doubles conjuguées par élimination d'autres molécules d'HCl et c'est la formation de ces doubles liaisons conjuguées qui donnent une teinte au matériau.

2.3 Mise en oeuvre de modules de grandes tailles et vieillissement

Des modules de grandes tailles ont été réalisés à l'aide de la résine polyuréthane Quinn La structure était la suivante :
Substrat Aluminium anodisé
Résine
36 cellules Ø 76 mm dans la résine
Pas de face avant verre
La formulation utilisée était la suivante :
525 gr Q621 système aliphatique
136 gr Q626 système polyol
Le mélange une fois réalisé est débullé puis coulé sur les cellules et débullé à nouveau, la polymérisation est effectuée à 60-90°C. Aucun problème n'a été rencontré du point mélange   et dégazage, par contre au cours de la polymérisation, il y a apparition de bulles relativement difficiles à éliminer. Le polyuréthane de Dévelopment Associates permet de résoudre ce problème.

Ces modules ont été mis en essais, mais au bout de six mois et par l'absence de la face avant en verre, la résine est devenue collante. Pour des modules réalisés avec face avant verre ce problème est évité mais des décollements de la résine sont apparus.

Tous ces problèmes se résolvent par l'emploi de la résine Polyuréthane Z2211 de Development Associates dont la tenue   est très nettement supérieure par rapport à la résine QuinnCo.

3. Les résines mises en oeuvre par feuilletage
3.1 Les EVA (ethylène vinyl acetate)
Ces résines appartiennent à la première catégorie de matériau.

Le tableau V donne les propriétés des différents grades d'EVA (copolymer d'éthylène et d'acétate de vinyl). La variation du pourcentage d'acétate de vinyl change les propriétés mécaniques de cette résine : si le pourcentage décroit, la dureté et le module d'élasticité croissent. Avec un même pourcentage d'acétate, lorsque le poids moléculaire augmente, la dureté, le module d'élasticité, la viscosité et point de fusion croissent.

Suivant le taux d'acétate de vinyl, le vieillissement de l'EVA se rapp che soit du vieillissement du polyethylène, soit du vieillissement du polypropylène. Plus le taux d'acétate vinyl est élevé plus on se rapproche du polypropylène. Les mécanisme de dégradation photochimique et thermochimique du polypropylène et du polyéthylène sont bien connus et on est capable de stabiliser chacun d'entre eux. Pour l'EVA l'initiation de la dégradation peut donc se passer soit sur des carbones tertiaires (carbones substitués par des groupements acétates) soit sur des carbones secondaires (carbones non substitués).

Sur un carbone tertiaire le mécanisme de l'attaque est semblable à celui que l'on rencontre dans le cas du polypropylène où l'attaque se fait sur les carbones tertiaires. L'attaque de l'oxygène sur les macro-radicaux tertiaires formés donne des hydroperoxydes tertiaires qui sont photo-instables et qui se décomposent en radicaux alcoxy qui à leur tour donnent des alcools tertiaires.

Sur un carbone secondaire les réactions sont analogues à celles que l'on rencontre dans le polyéthylène. On aboutit à la formation d'alcools secondaires, de cétones, puis d'acides par la réaction de Norrish de type I ou d'insaturations par la réaction de Norrish de type II. Ces réactions sont des réactions photo-chimiques sur des composés carbonylés aboutissant à des ruptures de liaison C-C.

La protection des EVA contre de tels processus de dégradation se fait en piégeant à l'aide de molécules spécifiques, les produits intermédiaires formés au cours du processus :

    absorption des UV - anti UV
    réducteurs - anti-oxydants
    capteurs de radicaux - stabiliants
Différentes formulations ont été testées :

	FORMULATION CLASSIQUE A	FORMULATION AMELIOREE B
Péroxyde de réticulation	1,5 pcr	1,5 pcr
UV absorbeur	0,25 "	0,9 "
Stabilisant	0,1 "	0,4 "
Anti-oxydant	0,2 "	0,2 "

Les résultat du vieillissement sur les deux formulations montrent que le système de protection est nettement amélioré par l'augmentation des anti UV et des stabilisants. En particulier l'augmentation du pourcentage d'anti UV permet d'avoir une densité optique supérieure à 2 entre 300 et 350 nm. La tenue est supérieure à 2000 h en "wheather-o-mètre" sans dégradation apparente ou mesurée. De la même façon le passage d'un ELVAX 150 à un ELVAX 420 permet une amélioration très limitée de la tenue.

Les propriétés d'adhérence des EVA sur le verre ou d'autres matériaux font qu'il est nécessaire d'utiliser des primaires d'accrochage ou systèmes équivalents. Deux solutions peuvent être utilisées : primaire déposé

TABLEAU V

PROPERTIES OF ELVAX (DUPONT)

	Vinyl acétate (%)	Molecular weight	Melt index (g/10 min.)	Hardness shore A	Elasticity modules (MPa)	Melting point (°C)
ELVAX 40 P	40.5		57	40	2.1	104
ELVAX 150	33		43	65	4.8	110
ELVAX 210	28		402	62	4.1	82
ELVAX 220	28		150	69	5.9	88
ELVAX 230	28		110			
ELVAX 240	28		43	73	7.6	110
ELVAX 250	28		25	75	9.0	127
ELVAX 260	28		6	80	11	154
ELVAX 265	28	increase	3	83	14	171
ELVAX 310	25		402	70	7.6	88
ELVAX 350	25		19	80	12	132
ELVAX 360	25	increase	2.0	85	18	188
ELVAX 410	18		502	80	14	88
ELVAX 420	18		150	84	19	99
ELVAX 450	18		8.0			
ELVAX 460	18		2.5	90	24	199
ELVAX 470	18	increase	0.7			
ELVAX 550	15		8	92		
ELVAX 560	15		2.5			
ELVAX 565	15	increase	1.5			
ELVAX 660	12		8	93		
ELVAX 660	12		2.5			
ELVAX 670	12	increase	0.3			
ELVAX 750	9.0		7	95		
ELVAX 760	9.3		2			
ELVAX 770	9.5	increase	0.8			
ELVAX 3120	7.5					

directement sur les matériaux, primaire incorporé dans la résine.
Les résultats sont présentés dans le tableau suivant.

TEST	PEEL STRENGTH
1. Formulation A coated with primer	2,8 Kg/cm
2. Formulation B coated with primer	Rupture du plastique pour 4 échantillons sur 6
3. Formulation A with 1 g. of primer incorporated........................	Rupture du plastique pour 5 échantillons sur 5
4. Formulation B with 1 g. of primer incorporated	Rupture du plastique pour 6 échantillons sur 6
5. (1) and (2) after 630 h 95°C 100 % RH	0 Kg/cm
6. (3) and (4) after 630 h 95°C 100 % RH	2,94 Kg/cm

Ces résultats montrent que la solution avec primaire incorporé est la
seule devant être retenue. Il est à noter que l'incorporation du primaire
dans la résine ne modifie pas les vieillissements des formulations
utilisées.
    Les propriétés rhéologiques des EVA sont très importantes pour l'ex-
trusion des films et pour les cycles d'encapsulation.
    Pour l'extrusion une qualité d'EVA à faible "melt index" est la
mieux adaptée, mais pour l'encapsulation une valeur plus élevée est favorable pour réduire les temps de cycles.
    Ceci montre l'importance de l'optimisation de la qualité d'EVA à
utiliser des deux points de vue : statibilité et utilisation. Les résines
suivantes ont été extrudées avec les températures optimum suivantes
d'extrusion.
    ELVAX 150   80°C
    ELVAX 420   75°C
L'ELVAX 420 plus stable que le 150 est plus difficile à extruder à cause
de son "melt index". Pour le feuilletage des modules l'ELVAX 420 devenant
très liquide nécessite des modifications importantes du cycle de feuilletage.

3.2 Les EMA (éthylènes methyl acrylate)
    L'EMA copolymère de l'éthylène a une bonne résistance au vieillisse-
ment et présente l'avantage par rapport à l'EVA de s'extruder plus faci-
lement en films d'épaisseur bien contrôlable. De plus sa stabilité thermique est supérieure. Après 1400 h de "wheatheromètre" aucune dégradation
n'a été constaté. Des essais plus poussés sont en cours avec d'autres
grades d'EMA.

3.3 Les EPDM (éthylènes propylène diène monomère)
    Ceux-ci sont surtout connus pour leur très bonne tenue en température.
Les premiers travaux ont porté sur le NORDEL 1320 qui est un copolymène
à 53 % d'éthylène, 3,4 % d'hexodiène - 4 et 43,7 % de propylène. Les
premiers problèmes rencontrés ont concerné l'agent de réticulation qu'il
a fallu optimiser. A partir de là des formulations ont été réalisées et
sont en cours de vieillissement.

3.4 Mise en oeuvre et vieillissement de modules de grandes tailles

Deux types de modules ont été réalisés : module avec cellules ayant des contacts étamés et module avec cellules ayant des contacts sérigraphiés.

Les tableaux suivants VII et VIII donnent les évolutions du facteur de forme dans les différents tests accélérés.

Tableau VII

TEMPS	VAPEUR D'EAU BOUILLANTE				150°C				100°C			
	$\Delta I_{M\%}$	$\Delta FF_\%$	$\Delta I_{M\%}$	$\Delta FF_\%$	$\Delta I_{M\%}$	$\Delta FF\%$	$\Delta I_{M\%}$	$\Delta FF\%$	$\Delta I_{M\%}$	$\Delta FF\%$	$\Delta I_{M\%}$	$\Delta FF\%$
336 h	0	0	0,7	0,7	29,7	5,91	4,1	1,1	0	2,3	0	0
750 h	0	0	6,5	7,8	36,2	5,6	3,5	2	0	2,9	0	0
1000 h	0	0	7	7,2	45,6	9,1	4,9	2,9	0	2,3	0	0
2200 h	1,4	3,1	4,3	6,9	cellules cassées		27,4	20,9	0	3,5	0	0
3000 h	7,8	8,1	5,1	6,1	"		47,8	38,2	0,7	4	0	0
4000 h	14,1	11,8	10,8	7,2	"		Arrêt		2,1	3,2	0	0
	Formulation classique avec ELVAX 150		Formulation améliorée avec ELVAX 420		Classique		Améliorée		Classique		Améliorée	

Contacts étamés. Formulations après primaire déposée

TEMPS	VAPEUR D'EAU BOUILLANTE 85° 85%								150°C				100°C			
	II		VI		VII		V		III		IV		I		VIII	
	ΔI_M	ΔFF	ΔI_M	ΔFF	ΔI_M	ΔFF	ΔI_M	ΔFF	ΔI_M	ΔFF	ΔI_M	ΔFF	ΔI_M	ΔFF	ΔI_M	ΔFF
168 h	0	0	0	0	0	0,4			0	0	0	0				
336 h	0	0	0	0	0	0,4			0	0	0	0				
500 h	1,6	1,2	0	0,4	1,5	0,7	0,5	0	0	0	0	0,3				
700 h	8	0	0	0	0,5	0,9			0	0	2,6	1,7				
1000 h	9	0	0	0,4	2,6	2,2			0	0	2,6	1,7	0	0	0	0
1500 h	10	2,6	0	0	2,1	0,9	1	0	1,1	0	7,4	3,3				
2000 h			4,2	2,1	4,7	1,8			9,3	4,8	13,9	8				
3000 h	18	6,9	6,3	2,1	10,6	8			21,5	14,7	31,5	21	5	4	3	4

Contacts sérigraphiés. Formulation améliorée avec primaire incorporée

Ces résultats montrent le très bon comportement des EVA. Sur le plan de l'aspect il y a un avantage très net pour la formulation améliorée. De même la formulation avec primaire incorporé est meilleure car aucune délamination n'apparait, ce qui n'est pas le cas avec le primaire déposé au bout d'un certain temps.

4. Conclusion

Au cours de cette première phase de l'étude, deux matériaux se sont plus particulièrement dégagés pour une utilisation en tant qu'encapsulant pour cellules photovoltaïques.

. Résine polyuréthane aliphatique de Development Associates pour des encapsulations par coulée.

. Résine EVA pour encapsulation feuilletée. Cette dernière résine avec une formulation améliorée et un primaire incorporé est tout à fait intéressante, en particulier avec un taux d'acétate de vinyl de 33 % où les problèmes de réticulation sont résolus.

Il reste pour l'utiliser de façon industrielle à optimiser les conditions d'extrusion sous forme de film d'épaisseur contrôlée et en de grande quantité.

D'autres matériaux potentiellement intéressants nécessitent encore des améliorations : il s'agit principalement des PVC, EMA et EPDM pour lesquelles les premières formulations élaborées montrent un comportement qui est déjà assez satisfaisant sauf pour le PVC, mais sur lequel il y a lieu de poursuivre pour obtenir une stabilité, des conditions de mise en oeuvre et des adhérences équivalentes aux résines polyuréthanes ou EVA.

ENCAPSULATION OF PHOTOVOLTAIC SOLAR CELL MODULES

Authors   : Bernd Melchior

Contract number   : ESC R - 026 - D (B)

Duration   : 12 Months   1 July 1980 - 30. June 1982

Head of project   : Bernd Melchior

Contractor   : JMC JMCHEMIE Kunststoff GMBH

Address   : JMCHEMIE Kunststoff GMBH
                     Adolf Flöring Str. 22

                     D 5632 Wermelskirchen

Summary

The comercial and technical encapsulation of silicon solar cells in an acrylate system via extruded acrylate gel layers having different qualities and the subsequent polymerisation pressing has been stopped.

There were fundamential difficulties during the production of elastic pearl polymerisate, which was necessary for the embedding gel layer via the extrusion processing.
A special development of acrylic pearls with a soft core and hard shell are producable but in a very expensive way, so it is no longer economical for silicon soler cell encapsulation.

Other technical difficulties arised from the extrusion-process with gel acrylates. The normal extrusion machinery can not press out a geled acrylate layer. For these main reasons JMC changed the encapsulation technologie.

1. Acrylate polymer concrete technologie as a casting technologie which can be converted to molded plates, roof bricks or roof tiles dyed in any desired colour.

2. Acrylate mineral concrete. Very cheap mineral concrete plates can be completely perfused with acrylates up to core via the JMC soaking prozess.

3. Acrylate mineral concrete with coefficient of expansion alpha = $2,33 \times 10^{-6}$. Formed by special ceramik and glass powder with a mineral binder and then acrylated by JMC soaking prozess.

4. Combination polymer concrete plates with enpressed solar cell moduls.

Introduction

The development of an encapsulation system for photo-voltaic solar cells on the basis of acrylates is to be considered as having failed inasmuch as the encapsulation system consists completely of a hard upper layer, an elastic embedding layer an a hard lower layer of sandwich design.

The specific and good qualities of acrylates, such as best transparency with a good resistance to ultraviolet rays, best resistance to weather conditions at a low weight, easy machining and good electrical insulation, easy processing by casting or injection moulding respectively with an elegant shaping possibility do not supersede the negativ basis qualities of an encapsulation system purely built up of acrylates.

Acrylates have by nature a very high coefficient of expansion which is in the region of 70×10^{-6}.
A silicon solar cell, however, has a linear coefficient of expansion of $2,33 \times 10^{-6}$, which is nearly 30 times smaller. Strains due to changes in temperature will result in very strong mechanical forces acting on the cells and on the contact system respectively which will lead to a breakage of the cells and of the contact paths respectiveley. All attempts to reduce the coefficient of expansion by a chemically and mechanically successful incorporation of glass fibres did not result in a drastic improvement. The difference of tension between the cells and the hard acrylate layers had to be compensated via three to four mm thick elastic intermediate acrylate layers. This is uneconomical.

The low hardness of the acrylate surface leads to dulling by scratches and consequently to a strong decrase of performace of the moduls. A cleaning from finest dust, which is frequently necessary, especially in sunny but dusty and windy regions, such as for instance on the Canarian Islands in Fuerteventura would soon deaden the surface.

A scratch-resistant coating is also very problematic. The siloxane layers would very soon lead to crackled fissures under strains due to changes in temperature subsequently to flakings under the influence of humidity. Also the hard acrylate interlace system which was polymerized into the surface during the same curing process by JMC did not show an appreciable improvement of scratch resistance during long-time tests.

Another negative quality is the electrostatic attraction of dust to be noted with acrylates.
Also a treatment with antistatic agents will bring only a temporary improvement of short duration, and moreover such a treatment is rather complicated.
Each wiping of dust, spraying with antistatic agents and the subsequent rubbing-in of these agents into the acrylate surface will result in a gradual dulling of the acrylate surface. Moreover this additional maintenance and care ist absoluteley unsuitable for a large area photo-voltaic system.

The low inherent strength of acrylates will lead to skeleton-strengthened modular system. An additional torsion-resistant bordering with aluminium or stainless steel profiles respectively ist absolutely necessary especially for large moduls. This, however, is expensive with regard to material and effort.

Moreover acrylates are sensitive to selvents and may dull very soon due to inexpert cleaning with cleansers containing solvents. Interlaced acrylate material which is resistant to solvents can normally not be worked into a complete sandwich acrylate system.

A further negative point ist the price of acrylates compared with silicate glass systems.
Today the extruded acrylate plate or the profile extruded acrylate plate costs between DM 8,-- and DM 10,--, the better cast acrylate plate between DM 12,-- and DM 15,-- per kg. This means a price of DM 18,80 to DM 23,60 for 2 mm thick covering layer.

These prices are increased by another 50 % when glass fibres are incorporated. Thus the price for a glass fibre reinforced acrylate cover plate is in the region of DM 25,-- to DM 30,--. Contrary to these figures today's price for a 2 mm thick float glass amounts only to DM 8,-- to DM 10,-- per square meter. A pure acrylate system is only suited for the encapsulation of individual cells in the range of toy manufacture, for a responsible manufacture of moduls for the use in a large-area system a pure acrylate system has to be rejected.

During the development work under the research contract 449-78 the elastic acrylate embedding material has prooved to be successful, it is higly elastic, resistant to weather conditions and to ultraviolt rays, diffusion-proof, and its price of DM 6,-- to DM 8,-- per kg is much lower than for instance the price of silicone caoutchouc which cost between DM 20,-- and DM 25,-- per kg.

Moreover this elastic acrylate material can also be produced as fusion-adhesive foil, and thus offers itself for the compound glass technique for the encapsulation of solar cells.

The commercial encapsulation of silicon solar cells in an acrylate system via extruded acrylate gel layers having different qualities and the subsequent polymerization pressing in a fixes operation cycle has been stopped for the reasons explained before. EG Nr. ESC-R-026-D (B)

Moreover there were fundamental difficulties during the production of elastic pearl polymerizate which was necessary for the embbeding gel layer via the extrusion processing.
A special development of acrylate pearls with a soft core and a hard shell did result on principle in a progress for the extrusion gel layers processing, however a commercial produktion of these special pearls is very costly and for reasons of expenses it is no longer economical for a silicon solar cell encapsulation.

Also for this reason a change of approach to the encapsulation
of solar cells became necessary.

Now JMC develops and researches the following encapsulation
systems:

New encapsulation technologies by JMC

1. Acrylate polymer concrete technology

Filling materials as for instance sand, chalk etc. are very
cheap.

A mixing ratio of 70 - 75 parts by weight of filling materials
and 30 - 25 parts by weight of acrylate binding agents will
produce an acrylate concrete mass which can be cast and
pumped, and which can be converted to moulded plates, roof
bricks or roof tiles in a kind of injection moulding process.

Such a moulded profiled base plate is non-porous and
therefore diffusion-proof. It can be dyed in any desired color,
for instance in the colours of roof bricks or roof tiles black
to red/yellow/brown, and thus forms a phantastic building
element for the production of low-priced photo-voltaic modules
of all kinds.

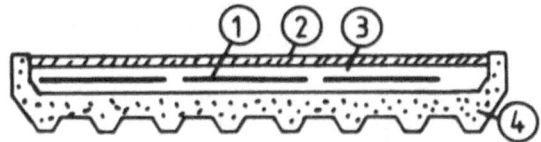

1.1.

Figure 1

1. Solar cells
2. Silicate glass pane
3. Elastic tension-compensating acrylate layer
4. Acrylate concrete body

2. Acrylated mineral concrete

Steam-hardened, high-strength mineral concrete plates are
normally not suited as base plate for a solar cell module
because of their high water vapour diffusion permeability.
On the other hand, however, then can be commercially produced
in a simple and low-priced way.
Such mineral concrete plates can be completely perfused with
acrylate up to the core via the JMC soaking process, and thus
they will get a higher stability and they become chemically
resistant to weather conditions and water vapour diffucion-
proof.

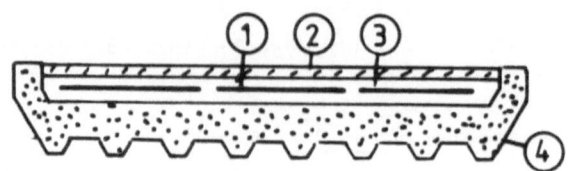

2.1.

Figure 2

1. Solar cells
2. Silicate glass pane
3. Elastic tension-compensating acrylate layer
4. Acrylated mineral concrete body

3. Acrylated mineral concrete with coefficient of expansion
Alpha = $2,33 \times 10^{-6}$

A casting compound which had the same coefficient of expansion
as the silicon solar cells would be ideal.
Unfortunately this is not possible in view of the very low
coefficients of expansion of silicon.
However, in the ceramic-and glass industry there is a material
which has the coefficient of expansion 0 - 0,3.
This material occurs in larger quantities as swarf in some
companies.

To use a polymer material as binding agent in this case will
not even result in the desired low coefficient of expansion
of the mixture when the addition of binding agent is limited
to the necessary 10 - 20%. All plastic binding agents have a
coefficient of expansion of $60 - 70 \times 10^{-6}$.

Only by using mineral binding agents with a coefficient of
expansion near $10 - 12 \times 10^{-6}$ and via the added quantity
of this mineral binding agent to the core material having
a coefficient of expansion 0, the desired coefficient of
expansion of $2,33 \times 10^{-6}$ can be achieved.

Since such a mineral plate is not water vapour diffusion-proof,
it is necessary to subject also this material subsequently
to the JMC soaking acrylate process.
The acrylate medium which fills up the pore capillary system
does not participate in the total coefficient of expansion of
the mineral concrete plate, but it has only a sealing function.

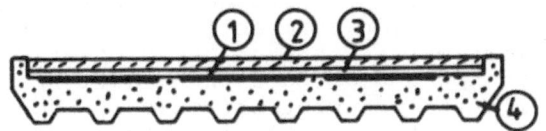

3.1.

Figure 3

1. Solar cells
2. Silicate cover plate
3. Elastic acrylate layer
4. Acrylated mineral concrete with a coefficient of expansion 2,33 x 10 $^{-6}$

4. Acrylated mineral concrete with a coefficient of expansion 2,33 x 10 $^{-6}$ with a highly transparent cover foil which heals scratches.
Meanwhile highly tough transparent foils are available which have the quality to yield to an impression lead, and to return to the original position after this impression load no longer exists.
These foils are already successfully used for crash helmets worn by motor-cylists.
Such a system could replace the silicate glass which is susceptible to shocks.

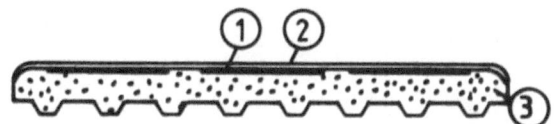

4.1.

Figure 4

1. Solar cells
2. Protective foil healing scratches
3. Acrylated mineral concrete with a coefficient of expansion 2,33

5. Mineral glass with low - priced protective foil as encapsulating system.

A nonferrous, highly transparent tempered textured silicate glass has proved to be the optimum solution for a cover element.

Such a silicate glass pane is extremely stable and firm, it has a hard, scratch-resistant surface, an in spite of this its price per square meter is lower than that of an acrylate glass plate. In order to save weight and costs, JMC now applies the method to replace the normally used basic glass pane by a protective foil.

However, this method involves difficulties because nearly all plastic foils which come into consideration for this application, are still diffusion permeable. In order to avoid any diffusion of noxious matter through the protective foil to the solar cell, JMC is now embedding a thin gastight glass foil or a thin Al-foil unter the protective foil. Such a glass or Al-foil is highly resistant to tearing and is easy to incorporate.

Waterdamp - diffusion-coefficient

Typ	Wd	(g $/$ m $\cdot$ day $\cdot$ bar)
Tedlar	0,01	
Teflon	0,0004	
EVA	0,04	
PC	0,02	
PMMA	0,01 - 0,04	
PE	0,0004	
Tefzel	0,01	
PUC	0,0004	
Polyester	0,01	

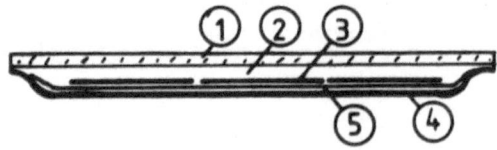

5.1.

Figure 5

1. Solar cells
2. Silicate glass pane
3. Elastic acrylate layer
4. Glass foil
5. Protective foil

At the moment JMC favours this system and is building up a suitable encapsulating automation.

<u>6.</u> The JMC hybrid solar system represents the practical application of the last encapsulating system with polymer-concrete.

The solar cell module is either pressed into the polymer concrete body 1 via a surrounding sealing band, or it is cemented into the roof brick with an elastic fusion-adhesive foil.

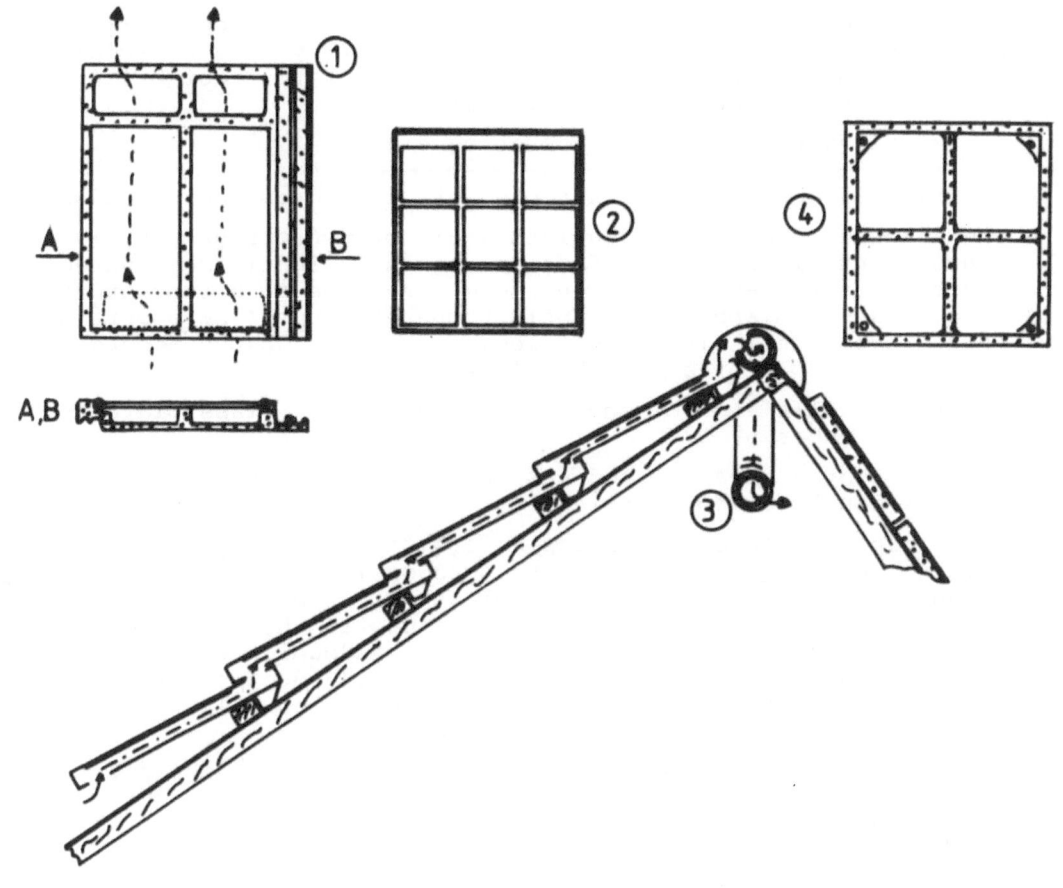

As it can be seen from section A/B, the acrylate concrete body with the solar module will then form a hollow under the cells.

Since the solar cells will get hot during exposure to sunlight, the air behind the cells will heat up and due to the stack effect it will ascend, and thus it will always attract cooler air from below, so that the solar cells are cooled. By this special arrangement of the cooling channels a complete hybrid roof system can be built up, in which the solar cells will supply the electic current and in which the ascending heated air can be additionally thermically used via a heat pump.

In order to avoid that the solar module is fixed one-sided on the polymer roof brick body, JMC has additionally developed a polymer concrete frame which can be nailed to some kind of a foundation and which can be arranged side by sider and one upon the other in any desired number of pieces.

Regarding the roof bricks it has still to be remarked that the now used square mono-crystalline solar cells are provided with a dull black antireflex coating and that the pattern is also surprinted in the same dull black shade.

Thus I have a building element which incorporates optically and architectonially into an existing roof system, especially because I have shaped the outer geometry of the roof bricks like that of a standard mineral roof brick.

7. JMC all-glass solar roof covering element

This all-gass roof brick represents a lower-priced version.

The silicate glass pane consists again of nonferrous, highly transparent glass. The surface is again sligntly corrugated, textured, the glass was given the outer shape of a roof brick in a press-form and subsequently tempered, Thus I have obtained an extremely resistive glass element which is shaped according to the principle of a roof brick.

The solar cells are now encapsulated under the glass plate as described under 5.

A large - area encapsulated suspension element serves simultaneously as contact lead.

Such solar module elements copied from the overlapping roof bricks can of course also be produced in larger or longer dimensions respectively. Such elements will then allow to cover very quickly enormous surfaces. Due to this shaping of the solar modules I have simultaneously a double function of a roof element. However, since the ventilating duct is dropped with this element, the temperature at the solar cells may rise much higher than with the composition system No. 5.

Of course, an all-glass roof system is essentially lower priced than the combination with a polymer concrete roof brick body.

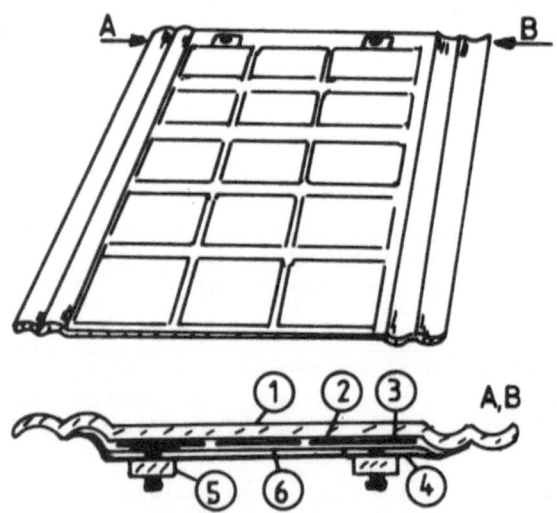

7.1.
Figure 7

1. Silicate glass pane
2. Solar cells
3. Elastic acrylate layer
4. Protective foil
5. Contact- and suspension elements
6. Diffusion glass foil

Because of the necessary discontinuance of the complete-
acrylate encapsulation technique and because of giving up
the development of this technology to a mass production
system according to the research contracts EG 23 and EG 24,
a realistic change of approach for new encasulation
technologies of solar cells became necessary.
EG Nr. 449 - 78 - 1 ESD
EG Nr. ESC-R-026-D (B)

I hope, that the proposed encapsulation techniques will bring
about new impulses for a lower-priced and better production
of solar cell modules with the variants for roof system
and stellage system.

R & D WORK ON THE ENCAPSULATION OF SOLAR CELLS WITH IMPROVED POTTING
AND COVER MATERIALS

Author: E. GRUBER

Contract number: ESC - R - 027 - D

Duration: 12 months, 1st April 1981 - 1st April 1982

Head of project: Dr. E. Gruber, RESART-IHM AG

Contractor: RESART-IHM AG

Address: Gassnerallee 40
                         D-6500 Mainz

Subcontractor: BATTELLE-Institut e.V.
                         Am Römerhof 35
                         D-6000 Frankfurt/Main

Summary:

Encapsulation materials for a full-plastic-panel consisting of a scratch
resistant coated PMMA-cover, an acrylic pottant, a PMMA-substrate and
an acrylic sealant were optimized. The main problems of inhibition of
the redox-initiation-system, of adhesion between pottant and cover
plates, and of water take up by the pottant could be solved. The optical
properties of panels were improved by scratch resistant coating and by
applying a diffusely reflecting substrate. Only 60° C could be reached
as a longterm service temperatur, tests about performance under practical
weathering conditions are made.

1.0 Introduction

Due to its physical properties (see tab. 1). Poly(methylmethacrylate)
(PMMA) is probably the best choice among plastic materials for making
photovoltaic panels. Even when compared to inorganic glass as a cover
material, PMMA shows a number of advantages, e.g. lower density,higher
impact strength and better transparency. Nevertheless up to now plastics
in general have not been used successfully for encapsulation of solar
cells.

Our work aimed at solving the most important specific problems, which
are the causes for failing of a PMMA-system. These problems are:

- The surface of a PMMA-panel is sensitive to scratching and soiling.

- Due to differences in thermal expansion delamination between cover
  sheet and pottant, or between pottant and solar cell takes place.
  The situation is worst, when Si-cells are embedded directly into
  PMMA without using a separate pottant.

- Acrylics are not completely tight against permeation of water and
  atmospheric gases, which may lead to corrosion.

For trials and checks we used simple, handmade sandwich-panels, which
are shown in fig. 1.

Polymer class		Poly(methylmethacrylate)s	Poly(carbonate)s	Celluloseesters	Poly(amide)s	Epoxi Resins	Poly(styrene)s	
Property of a typical polymer	Unit	PMMA	PC	CAB	PA	EP	PS	inorgan. glass
Density	kg/dm^3	1,18	1,20	1,19	1,10	1,20	1,05	2,25
Mod. of elasticity	N/mm^2	3300	2400	1500	2000	3200	3500	$7 \cdot 10^4$
Impact strength	kJ/m^2	20	n.m.	45	n.m.	15	20	1
Abrasive hardness	Mohs	3	2	1	3	2	3	7
Softening point (VICAT)	$^\circ C$	125	150	80	200	160	88	520
Heat capacity(20°C)	kJ/kgK	1,5	1,2	1,5	1,7	0,8	1,3	0,7
Lin. expansionscoef.	K^{-1}	$8 \cdot 10^{-5}$	$6 \cdot 10^{-5}$	$1,2 \cdot 10^{-4}$	$1,1 \cdot 10^{-4}$	$7 \cdot 10^{-5}$	$8 \cdot 10^{-5}$	$6 \cdot 10^{-9}$
Heatconductivity (20°C)	kJ/mhK	0,6	0,7	0,9	1,0	0,8	0,5	5
Index of refraction n_{D20}		1,49	1,59	1,48	1,53	1,58	1,58	1,48
Light transmission (3mm, 460 nm)	%	92	80	82	78	68	90	80
Permeability to water vapor	g/m^2d	0,8	0,7	0,9	0,5	0,5	0,5	0,1
Waterabsorption (DIN)	mg	45	10	70	300	90	3	
Weatherability	qual.	good	bad	bad	medi.	medi.	bat	excellent
Price	qual.	medi.	high	high	high	medi.	low	low

Tab. 1 Physical properties of various transparent plastic materials pared to inorganic glass

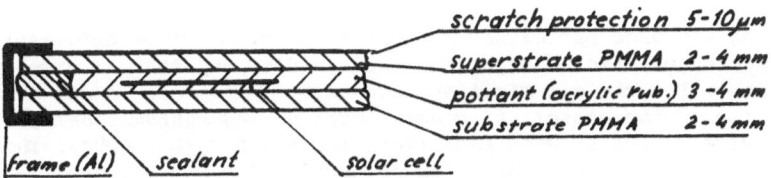

Fig. 1 Sandwich structureof the test panels

2.0 Developing an acrylic potting system

For embedding two procedures are possible: laminating or casting (potting).
As laminating needs more elaborate machinery and is not easily applicable
to shaped panels (which may be of interest for special applications) we
decided on using the casting process for developing the casting materials.
By slight modifications however, these materials may be adjusted to the
laminating process.

For casting a solution of polymers in monomers is used, which polymerizes
and cures in the casting chamber. The basis component is acrylic rubber,
formed by various acrylates and methacrylates, which are permanently
crosslinked by diacrylates. A certain amount of special comonomers is
added to guarantee for adhesion onto the surfaces of substrate, superstrate, silicon cells, and wiring.

It took us several months to find an appropriate composition. The best
elastic properties are furnished by poly (butylacrylate) (PBA), which
has the lowest glass transition temperature ($T_G = -56^oC$). PBA shows very
good adhesion to PMMA, medium adhesion to Si and poor adhesion to antireflection coatings of solar cells. PBA monomer could not be used as
the main component of pottant, as the monomeric butylacrylate attacks
PMMA. It can be used in form of a prepolymer, but there is only a small
window of allowed degree of polymerisation, due to severe polymer-monomer
incompatibility. Higher acrylates do not attack PMMA anymore, but this
polymer has a higher T_G-value and poorer adhesion to PMMA. Polar comonomers improve adhesion to Si but increase T_G considerably.

The tricky problem of optimization of casting syrups, using T_G and inter-
facial tensions towards PMMA and cell-materials as aim function and co-
monomer composition and degree of polymerisation spectrum as variables,
could not be solved theoretically. So it was done empirically by a series
of trials. We finally arrived at a recipe furnishing optimum adhesion to
all components and good low-temperature elasticity.

Another problem consisted in the redox-initiating system, which works via
a complimentary electron-proton-transfer. Under normal conditions the
reducing component is not able to reduce protons to hydrogen due to
a phase transition barrier. This barrier is lowered considerably by copper and its alloys, especially when it is contacted with another metal,
forming a local galvanic. So normally at the wires and the welding dots,
H_2 is evolved in the form of bubbles. This problem had to be solved also
empirically by changing the proton donator (complexed acid) and selecting
the wiring and welding material.

Crosslinking was also optimized in order to obtain maximum deformability
(low shear modules, high elongation) and avoiding any thermal creep at
the same time. For the optimization concentration, spacer length and
functionality of the crosslinking agent were varied systematically. The
specimens were tested between PMMA (shearloads 3,5 kg/m^2, 60^oC). The
optimum was found at a concentration of appr. 0,2 % and a spacer length
of 2 to 3 -CH_2O- units.

In the course of investigations it became evident that the hydrophylic
properties of the pottant are still more important than those of the
cover material. After reformulation, the tendency to take up water by
the cured pottant was reduced considerably (30 %). This was done by
using less polar monomers. Further reduction was not possible because
more hydrophobic materials did not adhere so well to PMMA.

3.0 Developing an acrylic sealant

A thermoplastic, which is elastic for a short time, with good adhesion
to PMMA was developed . It was made on the basis of uncured poly (butylacrylate). It is produced in a melt-polymerization process and extruded to a cylindrical cable. In the panel the sealant forms an inter-
penetrating network with the pottant, producing a stable junction. It
is completely transparent.

4.0 Work on the panel

The pilot panels were assembled as shown in fig. 2, the cells were supported each by one fixed and two movable points in order to allow for
differences in thermal expansion. The casting syrup was poured in latter-
ally and was cured under an IR-reflector-heater. This procedure guaran-
tees that polymerization starts first at the surface of the heat ab-
sorbing solar cells, leading to good adherence and little shrinking
stresses. Photocurrent: The photocurrent and efficiency where increased
by embedding by 10 to 30 %, depending on the original level of efficiency. This increase is based on total reflection of light, which acts
according to a light trap, and the action of a diffusely reflecting
back cover.

Effect of coating

The polysiloxane top coating (RESARIX SF) augments the amount of incident
light appr. by 1 % only but it protects the surface of the panel effec-
tively against scratching, which is mainly caused by cleaning the panel.
Fig. 3 shows, that transparency of PMMA-pane as measured by a photo-
meter decreased considerably when exposed to blowing sand due to defects
in the surface. But these damages have not a severe effect on the performance of a panel because the solar cells also make use of diffused
light (fig. 4). The main practical effect of coating is to diminish
soiling and to ease washing. A half years test showed that uncoated
panels loose more in efficiency than coated ones and washing cannot
fully compensate for the loss (fig. 5).

Influence of water

An elaborate study was made to evaluate penetration of water into coated
and uncoated PMMA-panes as well as into the embedding material. It was
found that water penetration in PMMA via intact surfaces is very slow
and follows a self hindering mechanismus type II.

The concentration profiles of water after storing thick PMMA in water
are shown in fig. 6 (calculated from measurements of panes of different
thickness). Fluctuations of external moisture are thus damped effective-
ly by the cover material. The moisture content inside the panel at the
surface of solar cells is determined nearly exclusively by the long term
level of relative humidity in the surrounding atmosphere. Water take up
of the less hydrophylic pottant is only about one third. It has to be

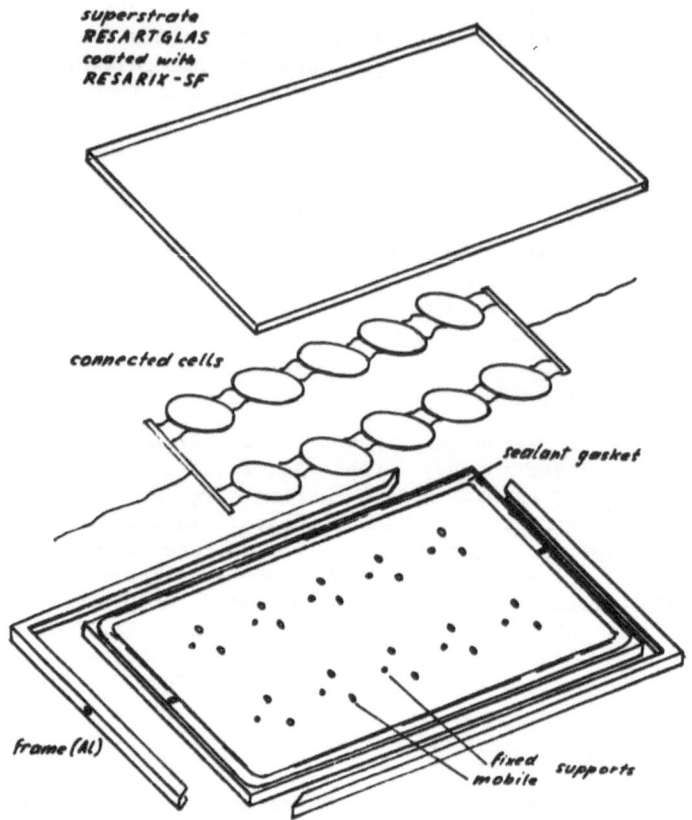

superstrate
RESARTGLAS
coated with
RESARIX-SF

connected cells

sealant gasket

frame (AL)

fixed supports
mobile

substrate RESARTGLAS white   Fig. 2   Assembly of test panels

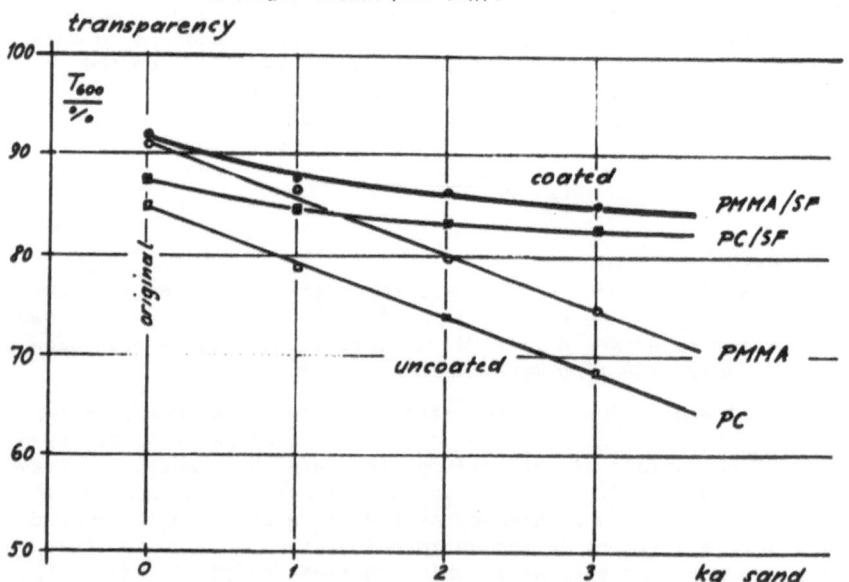

Fig. 3   Decrease of transparency of PMMA caused by blowing sand

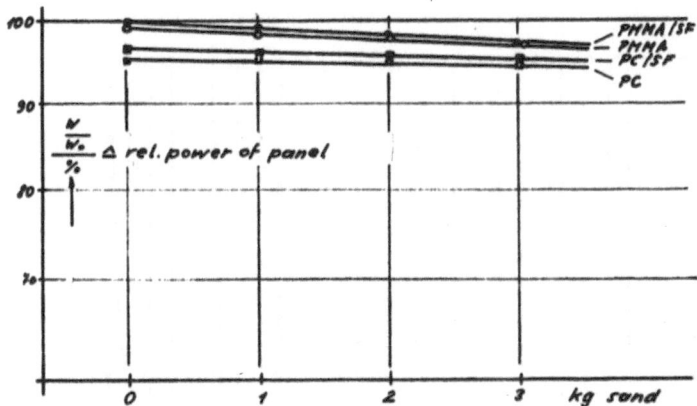

Fig. 4   Influence of blowing sand on the performance of panels

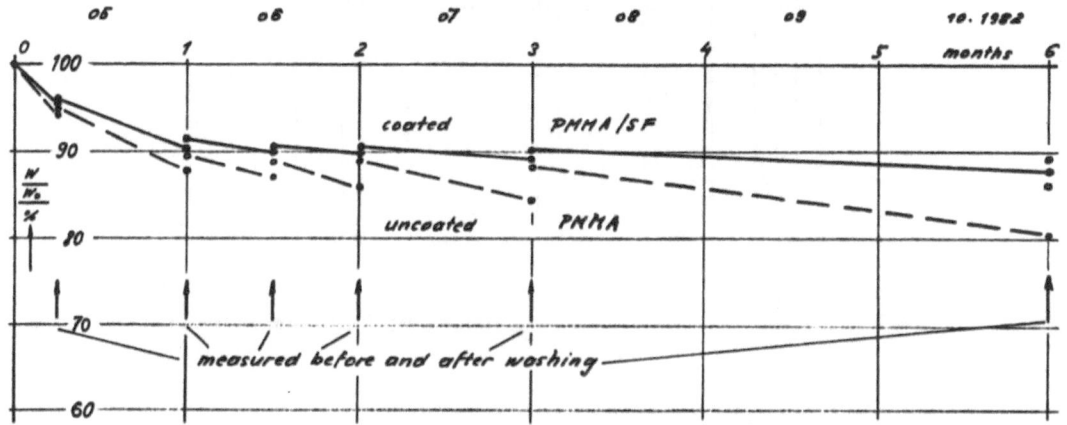

Fig. 5   Influence of soiling and washing on the performance of panels

tested whether their residual moisture may cause any damage of the cells.
Tests were made, storing panels at 100 % rel. humidity: within three
months, there was no decrease in photovoltaic efficiency. Under these conditions only the sealant was swelling, so the panels had to be framed.

Influence of temperature

Test panels were exposed to temperatures in the range of $-40^{\circ}C$ to $90^{\circ}C$,
either at constant temperature or in cold-warm cycles (total cycle 24 hrs).
Low temperatures did not cause any difficulties. On the other hand, high
temperatures are generally very critical. At temperatures about $80^{\circ}C$
normally bubbles were formed beginning at the surface of the cells and
delamination took place. Though we tried a great number of modifications,
we could only reach service temperatures of $60^{\circ}C$ for long term and
$80^{\circ}C$ for short term. At normal exposure, $60^{\circ}C$ should not be exceeded,
which was shown by heating the panel by a radiating heater (appr. 2 suns.)
The hottest point inside the panel (surface of the cell) did only reach
$50^{\circ}C$). So it can be expected, that thermostability of the panels will be
sufficient for practical purposes (fig. 7).

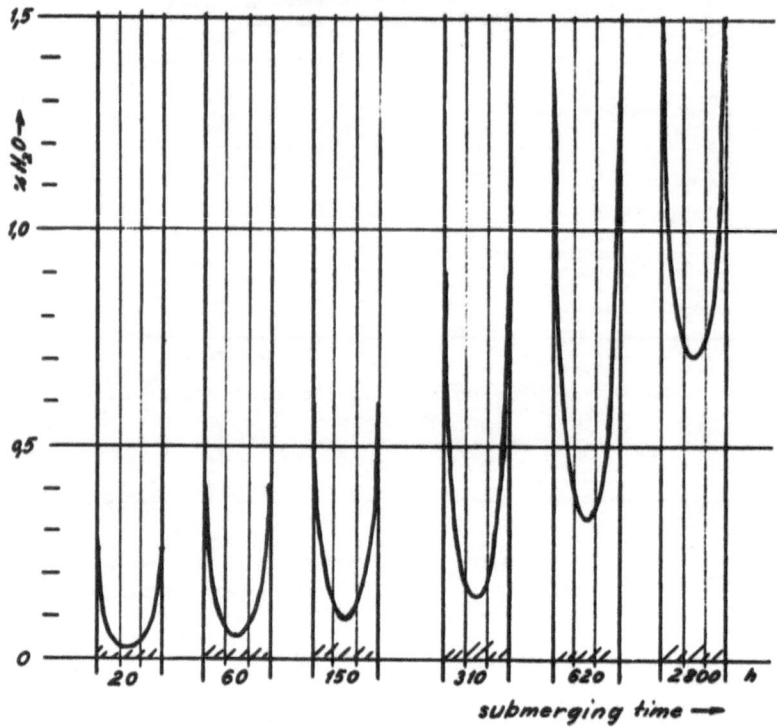

Fig. 6   Profiles of water concentration in PMMA submersed in water

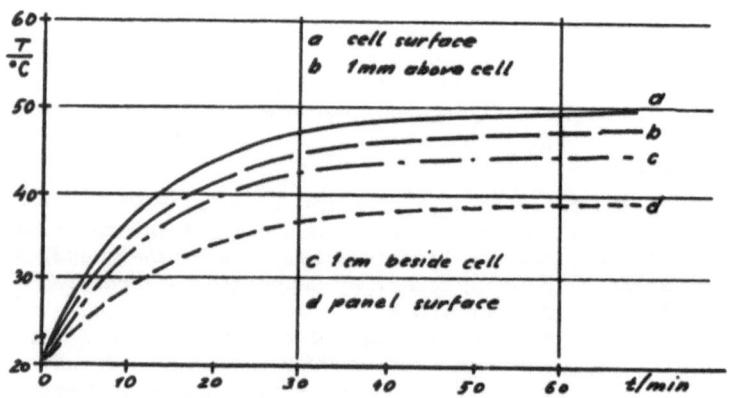

Fig. 7   Increase of temperature inside the panel when irradiated by
           appr. 2 suns

ALTERNATIVE CELLS

a) α - Si

Photogeneration in amorphous silicon solar cells

Atomic transport in and stability of amorphous silicon solar cells

Generateur photovoltaïque au silicium amorphe hydrogène

Development of sputtered thin-film a-Si solar cells

Preparation, study and characterization of hydrogenated amorphous silicon for photovoltaic cells

b) CdS - Cu_2S

Studies to improve the effifiency of Cu_2S-CdS spray solar cells

Electrochemical preparation of cuprous sulfide

Electrophoretically deposited thin films for low cost solar cells

c) CdSe

Development of a CdSe thin film solar cell

PHOTOGENERATION IN AMORPHOUS SILICON SOLAR CELLS

Authors   : W.E. SPEAR and F. CARASCO

Contract number   : ESC/003/UK/D

Duration   : 36 months, 1 July 1980 - 30 June 1983

Total Budget   : £93,280   )
                     DM538,550) CEC contribution 50%

Heads of Project   : Professor W.E. Spear, FRS ,   Dr. S. Kalbitzer

Contractors   : University of Dundee   Max-Planck Institut
                                                  für Kernphysik

Address   : Carnegie Laboratory of Physics, 6900 HEIDELBERG 1,
                     The University,   Postfach 10 39 80,
                     DUNDEE DD1 4HN, Scotland, U.K.   W. Germany.

Summary

The work of the Xerox group, published in 1981, suggested that the
efficiency of photogeneration in a-Si was fundamentally limited to
about 0.5 by geminate recombination. If correct, these results would
throw considerable doubt on the applicability of a-Si as an efficient
photovoltaic material. It appeared therefore of considerable importance to carry out a detailed study of photogeneration in a-Si photovoltaic junctions in order to assess the validity of the Xerox work.
The results are reported in the present paper which deals with photoexcitation both into the extended states and the tail states of the a-
material. It is shown from direct measurements of the generation
efficiency η and from the analysis in terms of the Onsager theory that
for undoped glow discharge Si $\eta \gtrsim 0.95$ at photon energies $\varepsilon_{ph} \gtrsim 1.6eV$.
It is therefore concluded that, contrary to the Xerox results geminate
recombination does not impose any fundamental limitation in photovoltaic applications. At $\varepsilon_{ph} \gtrsim 1.5eV$, transitions into electron tail
states begin to predominate. η decreases rapidly and shows a field
dependence in agreement with the Onsager theory. Geminate recombination therefore becomes important in this spectral range. The temperature dependence of η now shows a rapid rise in activation energy,
suggesting that photogeneration involves a thermal activation step
into the extended states. These results are discussed on the basis of
a simple model.

1. Introduction

The efficiency of photogeneration in a material is of critical importance for its application in a photovoltaic device. In 1981 the Xerox
group (1) reported results on the quantum efficiency of photogeneration η
which suggested that in amorphous silicon (a-Si) prepared by the glow discharge process η was limited to values between 0.44 and 0.55. It was
estimated (2) that on this basis the collection efficiency of an a-Si
Schottky barrier solar cell should at most be 0.65 in the green part of the
spectrum. The results of the Xerox workers implied that the low η in a-Si
imposes a fundamental limitation on the applicability of this material in
photovoltaics. These conclusions were challenged by a number of groups
working in the field (3,4,5) and in view of the basic importance of this
parameter to the a-Si solar cell development we decided to carry out a
detailed investigation of photogeneration and efficiency in undoped a-Si.

The work described in this paper was carried out on a series of reverse
biased a-Si p-i-n junctions, which form ideal structures for photogeneration
measurements. The experiments have been extended to photon energies ε_{ph}
above and below the optical gap of a-Si (1.65eV) and led to important
conclusions about photogeneration and the contribution of geminate recombination in these two spectral regions.

2. Theoretical Basis

The extensive work on photogeneration and recombination during the last
decade has shown that the process of geminate recombination can play an
important role in both inorganic and organic semiconductors. The first
detailed approach to this mechanism was given by Onsager (6) and the theory
has subsequently been used with considerable success to interpret photogeneration measurements on both crystalline and amorphous semiconductors
(e.g. 7).

Fig.1a illustrates the model describing the excitation of an electron-
hole pair with $\varepsilon_{ph} > \varepsilon_{opt}$, predominantly involving transitions between
extended states below ε_v and above ε_c. In undoped glow discharge Si

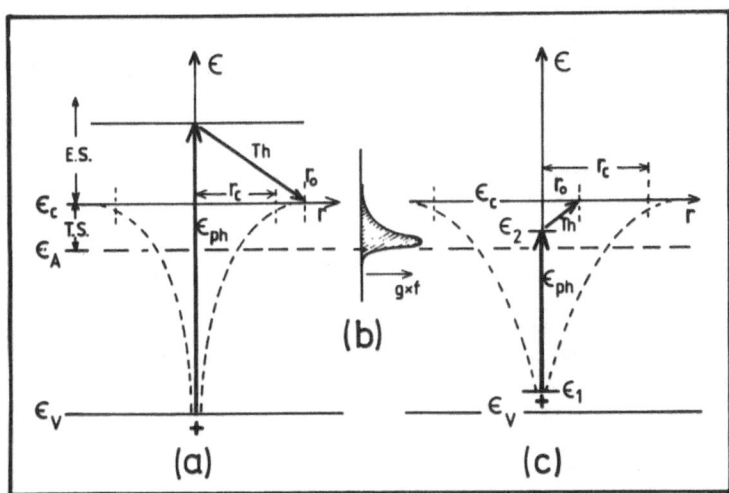

Fig.1   Models for photoexcitation of an excess electron (a) into
the extended electron states (E.S.), and (c) into the localised
tail states (T.S.). r_c: Coulomb radius, r_0: thermalisation
distance. (b) illustrates the thermalised electron distribution.

investigated in this paper the observed photocurrent above 250K is almost entirely carried by the generated excess electrons, so that in the following we shall mainly consider effects above and below ε_c. As indicated, the transition takes place in the Coulomb field of the hole and the electrostatic interaction of the charges is expressed in terms of the Coulomb radius r_c. For a-Si, $r_c \simeq 50\text{Å}$ at room temperature. The generated electron and hole will rapidly thermalise through phonon interactions and also change their relative separation by diffusion. We shall assume that the thermalising electrons populate the lowest extended states, just above ε_c, at an average separation r_0 from the hole.

This is the starting point for the Onsager theory which gives an expression for the probability $\Omega(E,T,r_0)$ that the thermalised electron at a temperature T and in the presence of an applied field E should escape geminate recombination in the Coulomb well. The overall quantum efficiency (or yield) η, defined as the number of electron-hole pairs per absorbed photon which are separated and can take part in the subsequent conduction process, is then given by

$$\eta(E,T,r_0) = \eta_0\, \Omega(E,T,r_0) \tag{1}$$

η_0 denotes the efficiency of the initial generation process leading to the transition in fig.1a.

Instead of a single electron-hole pair it is more relevant to consider the totality of generated excess electrons. Their energy distribution at $r > r_0$ is sketched in fig.1b (8), showing that in the a-semiconductor a large fraction of the electrons will thermalise into localised tail states. During transport the latter will remain in quasi-thermal equilibrium with the extended states (e.g. 9) and, provided $r_0 \gtrsim 4r_c$, most of the distribution will escape geminate recombination, irrespective of whether the carriers are in tail or extended states.

As ε_{ph} is reduced below about 1.5eV, transitions into the electron tail states become very likely. In fig.1c it is assumed that the transition takes place from an energy ε_1 just above ε_v, to ε_2 in the electron tail states. There are several important differences between the excitation into states above and below ε_c. First, the diffusive separation during thermalisation will be much smaller for an electron in the tail states, as the diffusion constant is only about 1/50 of that above ε_c. Secondly, the thermalised distribution of fig.1b is now produced at $r_0 \gtrsim r_c$, leading to a rapid decrease in η through geminate recombination.

If, in a simple model calculation, one assumes that only electrons in extended states have a chance of escaping geminate recombination it can be shown that

$$\eta \propto \left[g(\varepsilon_c)/g(\varepsilon_2)\right]\ \exp\left[-(\varepsilon_c-\varepsilon_2)/kT\right]\Omega(E,T,r_0) \tag{2}$$

for excitations into an electron tail state at ε_2. $g(\varepsilon_c)$ and $g(\varepsilon_2)$ denote the corresponding density of states. We shall return to this model in section 5.

3. The Experimental Approach
3.1 Specimen Preparation and Properties

P-i-n junctions, similar to those used in a-Si photovoltaic cells, were deposited on to stainless steel substrates by the glow discharge technique. Fig.2a shows a typical potential profile through a junction. The undoped (or 'intrinsic') i-layer, about 0.5μm thick, is sandwiched between thin ($\simeq 100\text{Å}$), highly doped n^+ and p^+ regions. The light enters through a transparent indium-tin oxide (ITO) top electrode, evaporated on to the n^+

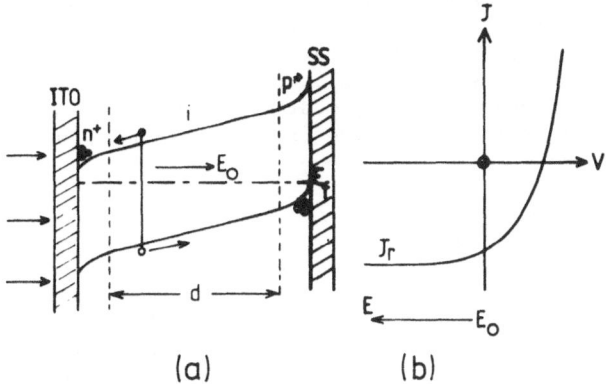

Fig.2   (a) potential profile through a p-i-n junction.   SS:
stainless steel substrate, ITO: indium-tin oxide top electrode,
E_0: built-in field.   (b) typical J-V characteristic under
illumination.

layer.   An important feature of the junction is the 'built-in' field E_0.
Drift mobility or transient pulse height measurements (10) were used to
determine its magnitude, which ranged between 15kV cm^{-1} and 20kV cm^{-1} for
the specimens used.   The internal field can be increased by applying a negative potential V to the stainless steel substrate and the   resultant
reverse field is approximately

$$E \simeq E_0 + \frac{|V|}{d} \tag{3}$$

The specimen is mounted in a temperature controlled vacuum chamber and
N_{ph}, the number of photons s^{-1} falling on to the area of the electrode, is
determined to within an accuracy of $\pm$ 2% by a specially calibrated crystalline Si photovoltaic cell.   In most experiments phase-sensitive detection is
used.   The load resistor, connected form the ITO electrode to ground, is
much smaller than the junction resistance and therefore does not limit the
observed current.

3.2   Determination of η
The advantage of a reverse-biased junction in quantum efficiency
measurements lies in the fact that one deals with a primary photocurrent.
It is evident from the potential profile in fig.2a that the injection of
excess carriers from the electrodes is prevented by the reverse field E.
In the present experiments, the reverse current J_r has been investigated as
a function of E at a given photon flux N_{ph} and ε_{ph}.   Fig.2b shows a typical
J_r-V characteristic.   If all the electron-hole pairs escaping geminate
recombination reach the electrodes, then for the primary photocurrent,

$$J_r(E) = e\,\eta(E)\,N_{ph}\,(1-R) \tag{4}$$

N_{ph} is corrected for reflection at the ITO surface and it is assumed that
all photons are absorbed in the specimen.   Eqn.(4) requires the complete
collection of separated electron-hole pairs.   This condition will be approached closely at fields at which the drift length, μτE, of the carriers
exceeds the specimen thickness by a factor of 3 to 4.   τ describes the volume lifetime with respect to trapping and/or recombination.   With

$\mu\tau \gtrsim 10^{-8} \text{cm}^2 \text{V}^{-1}$ for both electrons and holes (10,11) and thickness d ≈ 0.5μm we would expect eqn.(4) to be applicable for $E \gtrsim E_O$.

Two methods, based on eqn.(4), have been used to obtain η from the measurements. In the first η is evaluated directly by applying the appropriate correction factors. However, most of the results were obtained by fitting the experimental $J_r(E)$ to the field dependence predicted by the Onsager theory.

4. Results

4.1 Direct determination of η

In these experiments J_r was measured at $E \simeq E_O$ in the range of photon energies between 1.9eV and 3.0eV, and fig.3 shows some typical results. The directly measured J_r/eN_{ph} curve (denoted by (a)), leads to a value of 0.81 at $\varepsilon_{ph} = 2.3\text{eV}$, which is increased to 0.87 after correcting for surface reflection (curve (b)). The drop below $\varepsilon_{ph} \simeq 2.3\text{eV}$ arises from increasing loss of photons through transmission. So far we have not yet considered the absorption in the 100Å thick n+- layer of the junction (fig.2a). It is well established (12) that because of the high phosphorus doping level, this region represents a 'dead layer' of thickness δ as far as photogeneration is concerned. In order to obtain a realistic estimate for η of the intrinsic (undoped) part of the junction, we shall correct for this effect by writing $N_{ph} \exp(-\alpha\delta)$ for the photon flux incident in the i-region. α was taken from published absorption data and with δ = 100Å η values between 0.95 and 1 ((c) and (d) in fig.3) were obtained in the highly absorbed range of ε_{ph} from 2.4eV to 3.0eV. Although this correction procedure depends critically on the choice of δ, it was found that rapid variations of η in the above spectral range (which are unlikely) could be avoided only with δ values between 96Å (points(c)) and 110Å (points(d)), which lie close to the δ deduced from the deposition rate. We therefore conclude that a quantum efficiency between 0.95 and 1 is a characteristic feature of undoped glow discharge Si. In this case it follows that the efficiency for the production of thermalised pairs, η_O, (eqn.1), must also lie close to unity.

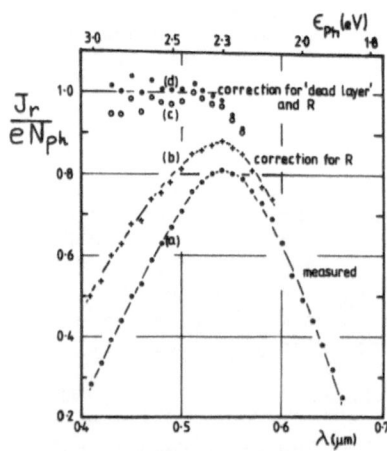

Fig.3 Plot of J_r/eN_{ph} against photon energy. Curve (a) measured points, (b) corrected for reflection at ITO, (c) corrected for 'dead layer' of thickness δ = 96Å, (d) for δ = 110Å.

4.2 The Field Dependence of the Reverse Current, η from Onsager Theory

Fig.4 shows a typical set of J_r vs E curves for fields from 2.5kV cm^{-1} to 150kV cm^{-1}. The results refer to the stated photon energies lying above and below the optical gap. To facilitate the comparison of the field dependence, J_r is plotted in relative units and the curves have been

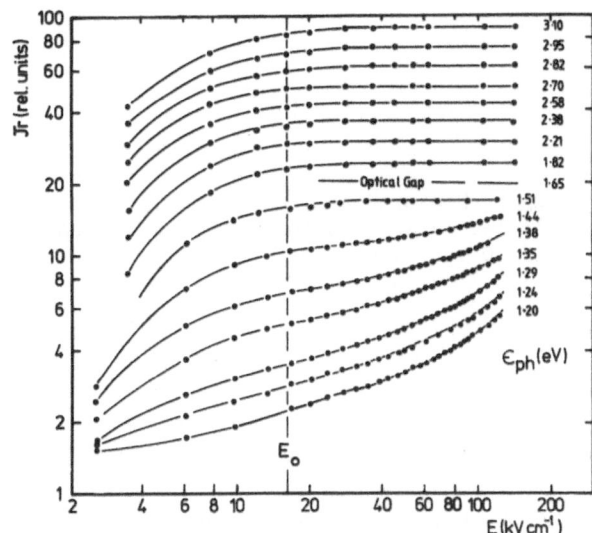

Fig.4 Plot of the reverse current J_r against the internal electron field E at the stated photon energies. To facilitate comparison the curves have been shifted and J_r is in relative units.

suitably shifted.   The results can be divided into two field regions.   At E < E_O one observes the volume life-time limitation in the extraction of photogenerated carriers from the i-region, discussed in section 3.2.

For E > E_O we suggest that the primary photocurrent J_r becomes a true measure of $\eta(E)$.   The important result is that for photon energies above 1.5eV, J_r shows complete saturation with field up to the highest E that could be used in the junctions.   Comparing these results with η vs E curves calculated for a-Si from the Onsager theory (eqn.1) it is evident that $r_o \gtrsim 400Å$ and $\eta \gtrsim 0.95$.

The character of the field dependence changes for photon energies below 1.5eV.   Analysis of the $J_r(E)$ curves shows good agreement with the Onsager theory.   For example at ε_{ph} = 1.38eV the fit leads to $r_o = (45 \pm 5)Å$ and to $\eta(E_O) \simeq 0.37$.

5.   Discussion
The first, and perhaps most important, conclusion that can be drawn from the results in the last section is that for photoexcitation into the extended states of glow discharge Si the quantum efficiency lies close to unity.   This is supported by both the direct determination of η (E_o) des-cribed in sect.4.1 and by the results of fig.4 showing a complete absence of any field dependence in $\eta(E)$ for E > E_o.   It is therefore most unlikely that geminate recombination can impose any limitation on the photogenerat-ion in this material for photon energies above 1.6eV, which is the range of interest in the photovoltaic applications of a-Si.

The results of the Xerox group (1) which led to η values between 0.44 and 0.55 and r_o between 45Å and 80Å, are difficult to understand in the light of the above conclusion.   The reason for the disagreement could,at least partly,lie in the electronic quality of the material investigated, as the properties of glow discharge Si depend critically on the plasma con-ditions during deposition.

Finally, it is of interest to test the model proposed in section 2 for photoexcitation into the electron tail states.   Eqn.(2) suggests that η should possess a thermal activation energy $\varepsilon_c-\varepsilon_2$ for $\varepsilon_{ph} \gtrsim 1.5eV$.   The temperature dependence of J_r has been investigated at E $\sim$ 100kV cm^{-1} between

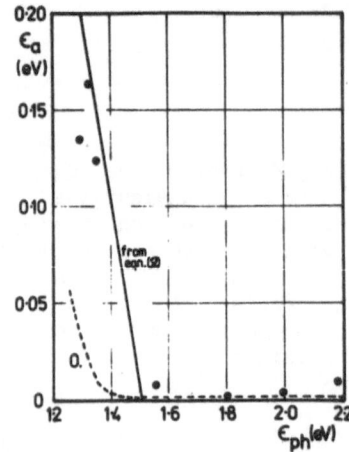

Fig.5 Measured activation energy ε_a of J_r plotted against photon energy. Broken line O.: calculated from Onsager theory, solid line calculated from eqn.(2).

ε_{ph} = 2.2eV and 1.3eV. The observed activation energy ε_a is plotted against ε_{ph} in fig.5. For transitions into extended states (ε_{ph} > 1.5eV), $\varepsilon_a \gtrsim$ 10meV, in agreement with the Onsager theory (see broken line O. in fig.5). At ε_{ph} < 1.5eV a remarkable increase in ε_a is observed, which cannot be explained by the Onsager theory, but fits well to the solid line calculated from eqn.(2). It shows that the predicted thermal activation energy plays a decisive part in the photogeneration process. Further support for the model comes from a comparison of the directly measured η curve with that calculated from eqn.(2).

Acknowledgments: The authors would like to thank Dr. R. Gibson and Mr. S. Kinmond for preparation of the junctions and Dr. P.G. LeComber for helpful discussions.

References:

1.   Mort, J., Troup, A., Morgan, M., Grammatica, S., Knights, J. and Lujan, R. 1981, Appl. Phys. Letters 38, 277.

2.   Chen, I. and Mort, J. 1980, Appl. Phys. Letters 37, 952.

3.   Madan, A., Czubatyj, W., Adler, D. and Silver, M. 1980, Phil.Mag. B42 257.

4.   Okamoto, H., Yamaguchi, T. and Hamakawa, Y. 1980, J. Phys. Soc. Japan 49, Suppl.A. 1213.

5.   Spear, W.E. Third Photovoltaic Solar Energy Conference Cannes 1980, (ed. W. Palz, Reidel Publishing Co. Dordrecht), p. 302.

6.   Onsager, L. 1938, Phy. Rev. 54, 554.

7.   Pai, D.M. and Enck, R.C. 1975, Phys. Rev. B11, 5163.

8.   Spear, W.E. and LeComber, P.G. 1972, J. Non-Cryst.Solids 8-10, 727.

9.   LeComber, P.G. and Spear, W.E. 1970, Phys.Rev.Letters 25, 509.

10. Spear, W.E., et al. 1981, J.de Physique 42, C4-1143.

11. Recent Dundee measurements, also R. Crandall, 1981, RCA REview 42, 441 and R. Street (Xerox) private comm.

12. Carlson, W.E. 1980, Solar Energy Mat.3, 503.

ATOMIC TRANSPORT IN AND STABILITY OF AMORPHOUS

SILICON SOLAR CELLS

Authors   : S. KALBITZER and M. REINELT

Contract number   : ESC/003/UK/D

Duration   : 36 months, 1 July 1980 - 30 June 1983

Total Budget   :£ 93,280   )
                     DM538,550) CEC contribution 50%

Heads of project   : Professor W.E. Spear, FRS   Dr. S. Kalbitzer

Contractors   : University of Dundee   Max-Planck-Institut
                                                   für Kernphysik

Address   : Carnegie Laboratory of   6900 HEIDELBERG 1,
                     Physics,   Postfach 10 39 80
                     The University,   W. Germany
                     DUNDEE DD1 4HN, Scotland,
                     U.K.

Summary

By using high resolution nuclear spectroscopy we have measured distribution profiles of a variety of impurities in amorphous silicon. Heavy
elements, i.e. those with atomic numbers higher than that of silicon,
have been studied by using the Rutherford backscattering technique,
whereas hydrogen has been detected by a nuclear resonance reaction employing a beam of 15N ions. Temperature induced changes of the initial
profiles, as obtained after film deposition or by an additional ion implantation step, have been evaluated in terms of diffusion coefficients.
The general finding is that all impurities studied so far diffuse much
more slowly in a-Si than in c-Si, or are practically immobile in the
temperature regime below the recrystallization temperature of a-Si.
Nevertheless, there are a number of fast diffusors, with hydrogen among
them, which can be critical as regards redistribution processes during
fabrication and operation of thin film solar cells based on amorphous
silicon.

1. Introduction

Solar cells based on glow discharge amorphous silicon (gd-a-Si) are diodes consisting of different thin film regions. Fig. 1 shows a typical structure of this kind. In view of the rather thin layers the question arises whether critical redistribution processes, by chemical diffusion and/or ion drift in the built-in electrical field, may take place during certain fabrication steps at elevated temperatures and also during the actual operation as a solar cell under ambient conditions. Clearly, these unwanted changes have to be restricted to a fraction of 100 Å. This will be the case, if during a fabrication time of about one hour the diffusion constant is kept below $D \sim 1E-18$ cm^2/s and during an operation time of ten years below $D \sim 1E-23$ cm^2/s. Also, as may be seen from Fig. 2, the corresponding drift lengths of charged impurities will be negligible at the existing maximum field strength of $E \sim 1E5$ V/cm.

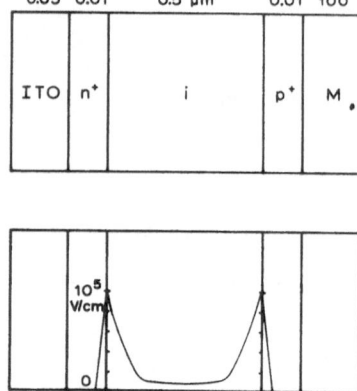

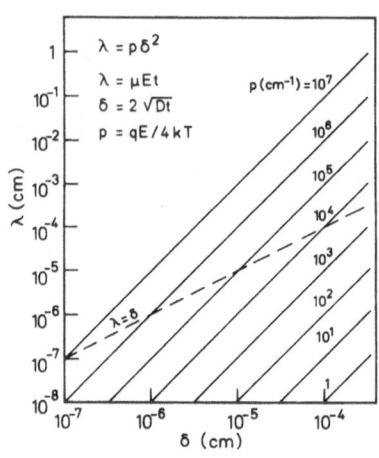

Fig. 1: a) Schematic diagram of a typical solar cell deposited onto a metallic substrate and coated with a conductive and antireflecting layer of indium-tin-oxide. b) Electrical field distribution in this cell (7).

Fig. 2: Relation between drift length λ and diffusion length δ for various drift strength parameters p. The dashed line separates the two regimes of predominant diffusion and drift, respectively.

2. Experimental

The amorphous samples have been prepared either by bombarding c-Si with heavy ions, such as Si or Ne, or, as in the case of H effusion measurements, glow discharge samples have been provided by Prof. Spear, Dundee, and by Dr. Müller, MBB Ottobrunn. The impurities of interest have been introduced by a second implantation process (1).

The nuclear techniques, as described in detail elsewhere, are capable of resolving depth differences of the order of 10 Å, which is adequate for the investigation of the problem as outlined above (2). In terms of diffusion coefficients, values of $D \sim 1E-20$ cm^2/s can be measured.

Figs. 3 and 4 give examples for the Rutherford backscattering and the 15N reaction technique, respectively.

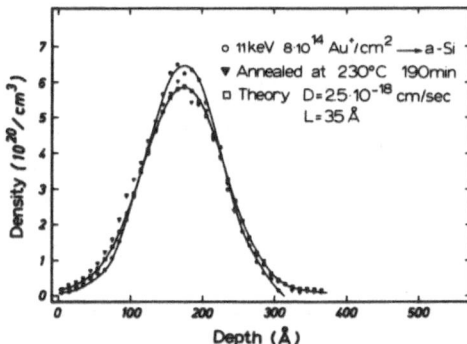

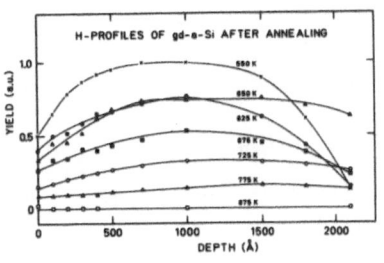

Fig. 3: Gold profiles in a-Si, as implanted and annealed. The diffusion length as evaluated from the broadening of the distribution amounts to 35 Å (4).

Fig. 4: H concentration profiles for annealed gd-a-Si: H samples from which diffusion coefficients at the respective temperatures have been derived. Sample thicknesses about 2000 Å.

3. Results

In the following we shall first report results obtained for the diffusion of H, which is the key impurity for solar grade a-Si. Then we shall turn to other impurities which are needed as dopants, or which are potential poisons for solar cells.

3.1 Hydrogen diffusion

As is definitely known by now, H is incorporated into glow discharge material with typical concentrations of 1 to 10 at% depending on deposition temperature and plasma conditions (3). Also, the addition of gas phase dopants to the plasma discharge will modify these figures to some extent (3).

Photoconductivity measurements have clearly established that H losses due to effusion during post-deposition heating treatments give rise to a severe deterioration of the opto-electronic quality of the material, as soon as the deposition temperature of typically 280°C is exceeded markedly. (5). Thus, it is of considerable interest to know the distribution profiles of H as exactly as possible.

Systematic measurements on a large number of gd-a-Si:H samples have revealed that a subsurface layer of up to several 100 Å in depth may be depleted from H by a considerable fraction (3). This fact may readily explain several observed phenomena, such as excessive front dead layers of solar cells, doping deficits in films and anisotropic conduction in them.
Fig. 5 demonstrates the surface depletion phenomenon for a number of glow discharge samples.

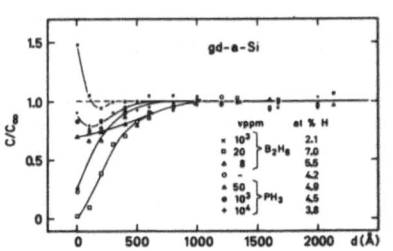

Fig. 5: Normalized hydrogen concentration vs. depth with the doping ratio as parameter. Note that in some cases deviations from the bulk value (c_∞) extend from the surface to depths of about 500 Å. Deposition temperature 280°C.

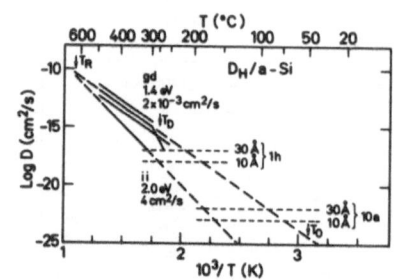

Fig. 6: Schematic overview for H diffusion in glow discharge (upper curve) and ion implanted (lower curve) amorphous silicon. The range of measurements is given by the full drawn lines. Note that at about 300°C a low temperature release process may be operative. The horizontal dashed lines denote critical diffusion parameters.

The surface droop of H in gd-a-Si:H, prepared at a deposition temperature of 280°C, becomes understandable, when we consider the H diffusion data as displayed in Fig. 6 and Table 1.

Table 1: Hydrogen diffusion parameters for amorphous silicon

Author	Material	Q(eV)	$D_o(cm^2/s)$
CM (78)[10]	gd	1.5	1E-2
ZGSBFD (81)[11]	gd	1.5	5E-3
RK (81)[2]	gd	1.1	6E-3
RKM (82)[12]	gd	1.4	1E-3
RKM (82)[12]	ii	2.0	4E-0

First we note that all diffusion data on gd material are based on effusion type experiments, where either the released or retained gas has been measured, be it by conventional mass spectrometry or by nuclear reaction analysis and secondary ion mass spectrometry, respectively.

Secondly, due to the small temperature interval accessible to experiment, the quoted activation energies cannot be very precise and are estimated to be uncertain by at least 10%. The concomitant errors in the prefactors are about a factor of 10.

The experimental data on gd material can be presented by an average activation energy <Q> ∿ 1.4 eV and prefactor <D_o> ∿ 1E-3. As is indicated by the upper and lower parallel lines in Fig. 6, the data of all three groups fall into this bracket of a factor of about 3 for temperatures T≳300°C. At lower temperatures, we believe that a change in the release mechanism takes place, which is characterized by a steeper temperature dependence. The data by the RCA group do not really contradict this interpretation, since they could also be fitted by two straight lines instead of a single one. This matter is under further investigation with the aim of clarifying the nature of this second mechanism.

Whereas it cannot be ruled out that some structural differences exist in the gd material prepared by different laboratories, by which the "minor" differences in the diffusion data may be caused, the situation is grossly different for ii material.

The experimental technique of implanting H into the a-Si matrix allows for measuring diffusion coefficients in a quasi-infinite body. This means that boundary reactions, such as bimolecular recombination of H and the following desorption of H_2, cannot disturb the kinetics of diffusion-controlled redistribution of H. This may explain the observation of a single activation energy over the measured temperature interval. The values of the characteristic parameters Q and D_0 considerably differ from the gd values indicating major structural differences between the two kinds of material.

We now turn back to the explanation of the H surface droop in gd-a-Si. In making a rough estimate, we find the diffusion length to be about 100 Å for a deposition time of about one hour and a diffusion coefficient of 1E-16 cm^2/s. This is of the observed order of magnitude. We further note that a temperature difference by only 20°C with respect to this standard deposition temperature of 280°C would strongly change the magnitude of this effect, provided the steep temperature dependence of D on T is true. Thus, small changes in deposition temperature - not unlikely to occur for runs under "constant conditions" and for large area samples as a lateral gradient - are by no means uncritical as regards the fabrication of optimum performance solar cells on the basis of this material.

At ambient temperatures of 50°C, not unrealistic for cell operation conditions, the average diffusion parameters of Fig. 3 would lead to D(50°C) $\sim$ 3E-25 cm^2/s. This would be sufficient as to guarantee a life time of 10a for a solar cell. In contrast to the situation at deposition conditions, the existence of a slower effusion mechanism appears to be irrelevant here.

Since a solar cell is actually operated under illumination, the present results do not allow to make predictions for light induced effects, as known in literature (13).

3.2 Interstitial diffusion

In c-Si a variety of fast diffusors occupy interstitial sites, e.g. alkali atoms and some noble metal atoms. Alkali impurities have been found to be efficient donors in a-Si:H also (14). They are easily and reproducibly introduced into the material by ion implantation, and the process can, in principle, be tailored such as to form ultra-thin windows of less than 100 Å thickness. Here also the question is whether the stability of the junction is sufficient with respect to atomic redistribution processes.

Among the group Ib noble metals Cu is an extremely fast diffusor in c-Si and therefore, as an interstitial impurity with deep energy levels, also a potential poison for a-Si:H. Thus, a study of its diffusion appears worthwhile.

The results, to be published in full detail elsewhere, may be generalized as follows. The diffusion in a-Si as compared to c-Si is dehanced by orders of magnitude. Initially freely diffusing, the interstitials undergo trapping and diffusion slows down considerably. It is likely that the electrical activity of interstitial dopants changes in the trapped state. The trapping process, characterizable by an apparent diffusion coefficient changing with time, is demonstrated in Fig. 7 for the system K/a-Si. (2). The proposed structural model is shown in Fig. 8 (2), where average size and spacing of voids, interspersed into the amorphous bulk, are given.

While the heavy alkali impurities K, Rb and Cs may safely be used for solar device technology, the diffusion of Cu in ii-a-Si at temperatures of 300°C would be fast enough as to penetrate into depths of the order of 1 μm in relatively short times, since D $\sim$ 1E-13 cm^2/s. In gd-a-Si:H, however, the diffusivity of Cu appears to be strongly retarded for reasons not yet known.

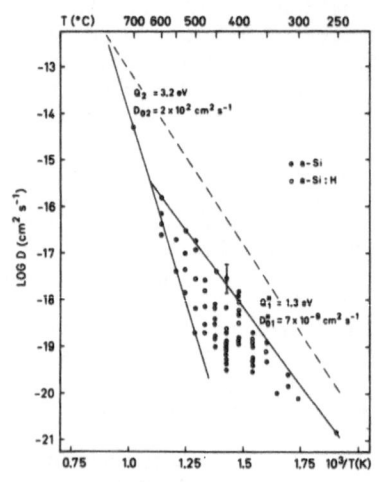

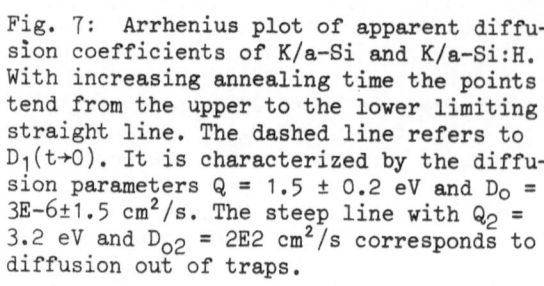

Fig. 7: Arrhenius plot of apparent diffusion coefficients of K/a-Si and K/a-Si:H. With increasing annealing time the points tend from the upper to the lower limiting straight line. The dashed line refers to $D_1(t \to 0)$. It is characterized by the diffusion parameters $Q = 1.5 \pm 0.2$ eV and $D_0 = 3E-6\pm1.5$ cm^2/s. The steep line with $Q_2 = 3.2$ eV and $D_{02} = 2E2$ cm^2/s corresponds to diffusion out of traps.

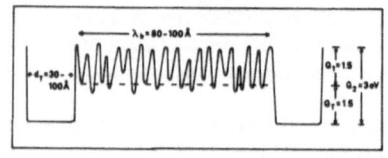

Fig. 8: Model scheme for the diffusion in an amorphous two-phase solid consisting of a disordered bulk with fluctuating atomic potentials into which voids are embedded. After diffusing a certain length in the bulk the particles become trapped at the voids.

3.3 Substitutional diffusion

Tetrahedrally bonded impurities of group IIIa and Va, e.g. B and P, act as acceptors and donors in both c-Si and a-Si:H. There were reported SIMS measurements by the RCA group (8) on the diffusion of these substitutional impurities among others in a-Si:H. The results were such that within deposition time B and P would have diffused through the whole pin cell structure.

By comparing our diffusion results on Sb at 400°C with those data, i.e. 2E-19 vs. 2E-15 cm^2/s, we have concluded, by using scaling arguments derived from c-Si data on diffusion, that D(B) and D(P) should also be orders of magnitude smaller in a-Si than the corresponding RCA data. In addition, we have found that In and Tl are practically immobile over the entire accesible temperature range in a-Si. Recently, $D(B) \sim D(P) < 3E-17$ at 400 and 450°C, respectively, was reported (9), which is uncritical for the above preparation conditions.

4. Conclusions

a)   a-Si solar cells with long term stability against atomic redistribution processes can be made by using suitable group III, V and I dopants.

b)   Similar requirements can be satisfied for hot processing steps over typical fabrication times of 1h duration.

c)   H effusion is critical during deposition at temperatures of 300°C; a reduction to 250°C should help to largely suppress the surface droop phenomenon.

d)   Regular diffusion of H at an ambient temperature of about 50°C appears to be unproblematic over a period of 10a.

e)   Drift effects are negligible under the present conditions of an electrical field strength of 1E5 V/cm and a tolerated maximum diffusion length of about 30 Å.

References

(1) S. Kalbitzer, M. Reinelt and W. Stolz, Proc. IV. E.C. Photovoltaic So-
     lar Energy Conference, Stresa, W.H. Bloss and G. Grassi, eds., Reidel
     Publishing Company (1982) p. 1059
(2) M. Reinelt and S. Kalbitzer, J. Physique, Suppl. 10, 42 (1981) C4-843
(3) F.J. Demond, G. Müller, H. Damjantschitsch, H. Mannsperger, S. Kalbit-
     zer, P.G. LeComber, W.E. Spear, J. Physique, Suppl. 10, 42 (1981)C4-779
(4) W. Stolz, Diplomarbeit Heidelberg 1982
(5) D.I. Jones, R.A. Gibson, P.G. LeComber and W.E. Spear, Solar Energy Ma-
     terials 2 (1979) 93
(6) H. Fritzsche, Thin Solid Films 90, (1982) 119
(7) H. Ohamoto, T. Yamaguchi, S. Nonomura and Y. Hamakawa, J. Physique,
     Suppl. 10, 42, (1981) C4-507
(8) D.E. Carlson, C.R. Wronski, J.I. Pankove, P.J. Zanzucchi and D.L. Staeb-
     ler, RCA Review 38, (1977) 211
(9) D.E. Carlson, J. Vac. Sci. Technol. 20 (1982) 290
(10) D.E. Carlson and C.W. Magee, Appl. Phys. Lett. 33(1), (1978) 81
(11) K. Zellama, P. Germain, S. Squelard, B. Bourdon, J. Fonteuille, R. Da-
     nielon, Phys. Rev. B23 (1981) 6648
(12) M. Reinelt, S. Kalbitzer and H. Mannsperger, to be published
(13) D.L. Staebler and C.R. Wronski, Appl. Phys. Lett. 31 (1977) 292
(14) W.E. Spear, P.G. LeComber, S. Kalbitzer and G. Müller, Phil. Mag. 39
     (1979) 159

GENERATEUR PHOTOVOLTAIQUE AU SILICIUM AMORPHE HYDROGENE

Auteurs   : L. VIEUX-ROCHAZ, R. CUCHET, A. CHENEVAS-PAULE

Numéro du contrat : ESC.R.028 F (S)

Durée   : 18 mois (1er juillet 1980-31 décembre 1981)

Tête de projet   : Dr D. RANDET

Contractant   : Centre d'Etudes Nucléaires de Grenoble
                      LETI/Laboratoire des Composants Electroniques

Adresse   : CEA/CEN.G - LETI/CE
                      85X - 38041 GRENOBLE CEDEX - FRANCE

Summary

The purpose of this work is to realize by R.F. sputtering two stacked photovoltaic structures in view of increasing the conversion efficiency of hydrogenated amorphous silicon solar cells. The dual Schottky structure must enhance the short-circuit current J_{cc} by an electrical parallel connection of two diodes, while the twin (dual) p-i-n structure must double the open circuit voltage V_{oc} by in-series connecting two junctions. We have first elaborated elementary Schottky and p-i-n diodes which have been electrically investigated. P-i-n diodes offer better V_{oc} but weaker J_{cc} than Schottky diodes. Good tunnel junctions have been realized and have allowed to double the value of V_{oc} in stacked p-i-n structures. On the other hand, the attempt to make dual Schottky structures was not successful, resulting in low V_{oc} values without any increase of J_{cc}. These results prove that an effective enhancement of the efficiency will be possible only with multispectral stacked junctions.

1. Introduction

Dès la parution de l'article de Carlson et Wronski (1), de nombreuses équipes (U.S.A., Japon, G.B., France) ont commencé à étudier le silicium amorphe hydrogéné (a-Si:H) en vue de son application à la conversion photo-voltaïque de l'énergie solaire en grande surface. Notre équipe, pionnière dans le domaine, a choisi dès le départ pour élaborer ce matériau, la pulvérisation cathodique réactive. Ceci essentiellement parce que ce procédé présente un rendement matériau déposé/matériau de cible (silicium monocristallin en l'occurrence) voisin de 1 contre 0,2 environ pour les procédés qui mettent en oeuvre la décomposition d'un gaz (SiH4 par exemple). Cette méthode tout à fait adaptée à la réalisation de panneaux de grande surface (style baie vitrée) doit aussi permettre ultérieurement de doper à partir des cathodes et donc d'éviter la manipulation de gaz toxiques tels que PH3, ASH3, B2H6, pour réaliser des photodiodes de type Schottky ou p-i-n. Mettant en oeuvre cette technique, notre groupe a couvert dans ce sujet un domaine assez vaste, concernant tant la physique du matériau que l'étude et la réalisation des dispositifs.

Il ressort de la littérature (2, par exemple) que la longueur de collection maximale dans ce matériau ne peut excéder 3.000 Å (pour un gap optique de 1,9 eV) et décroît exponentiellement avec la teneur en hydrogène. Il paraît donc inutile d'élaborer des diodes dont l'épaisseur dépasse 3.000 Å. Par ailleurs, le spectre du coefficient d'absorption de ce matériau est tel que cette épaisseur est insuffisante pour absorber l'énergie solaire dès 1,5 eV. Une solution pour améliorer le rendement de conversion des photopiles à base de a-Si:H semble être de multiplier ces zones de collection et de les placer optiquement en série, en faisant éventuellement varier le gap optique du matériau pour augmenter encore l'efficacité d'absorption (structure multispectrale).

Cette approche constitue précisément l'objectif de cette étude, en excluant dans un premier temps l'aspect multispectral. Notons que cette approche a été suggérée simultanément par plusieurs laboratoires (Hamakawa, Université d'Osaka (2), Hanack (3) RCA), notre propre équipe ayant déposé un brevet sur les structures type Schottky empilées, électriquement en parallèle dès 1978 (brevet n° 78/22826).

Il convient de remarquer que l'épaisseur des dispositifs élémentaires est plus faible que la taille des dispositifs de la VLSI ! Cela laisse augurer des difficultés technologiques liées à cette solution.

Notons que lorsque ces dispositifs sont électriquement en série, la clef du problème est de réaliser une diode tunnel (p$^+$n$^+$ ou cermet style Pt/SiO$_2$ (3)) de moins de 200 Å d'épaisseur (afin de ne pas induire une absorption parasite) pour éviter un redressement parasite entre diodes.

2. Réalisation des dispositifs
2.1. Appareillage et conditions de dépôt

Nous possédons trois bâtis de pulvérisation cathodique "diode radio-fréquence". Les premières diodes p-i-n ont été réalisées dans deux bâtis différents, le dopage se faisant dans l'un, le matériau intrinsèque dans l'autre. Pour éviter la pollution due à la remise à l'air entre les dépôts, nous avons fait l'acquisition d'un bâti multicathode. Deux cathodes de silicium monocristallin ainsi que deux cathodes de platine et de chrome permettent de déposer dans un même cycle de vide les différentes couches d'une structure. Le dopage se fait grâce à l'adjonction de phosphine ou de diborane dans le flux d'argon-hydrogène. Les couches d'oxyde d'indium transparent et conducteur sont déposées par pulvérisation cathodique continue magnétron. Un spectromètre de masse raccordé à l'enceinte permet d'analyser les gaz résiduels avant et au cours du dépôt.

La température de dépôt est généralement 250°C. La pression partielle d'hydrogène est de 1,4 Pa, celle de l'argon de 0,6 Pa. La vitesse de dépôt est 3 A/s. Les gaz dopants sont dilués à 3 % dans de l'argon et le rapport pression partielle/pression totale dans l'enceinte varie de 10^{-5} à 10^{-2}.

Par ailleurs, grâce à un appareillage type LBIC (Light Beam Induced Current) constitué d'une binoculaire, de diaphragmes, d'une lampe à mercure et d'une fibre optique, nous avons pu effectuer une topographie très fine du courant de court-circuit des cellules qui nous a permis de déceler des défauts aux points singuliers (bords, poussières, trous...) et de mesurer avec précision la surface active des diodes (section du faisceau 20 µm x 20 µm).

2.2. Dispositifs réalisés
2.2.1. Diodes élémentaires

Plusieurs configurations ont été réalisées et testées pour les diodes Schottky ou p-i-n.

Structure normale	Schottky	Pt /	a-Si:H (i) /	n^+ /	Cr /	substrat verre
	$\leadsto$ hν $>$	8 nm	700 nm	20 nm	300 nm	
	p-i-n	p /	a-Si:H (i) /	n^+ /	Cr /	substrat verre
		30 nm				

structure inverse	Schottky	Al /	n^+ /	a-Si:H (i) /	Pt /	In₂O₃ /	substrat verre hν
		300 nm	20 nm	700 nm	8 nm	300 nm	$\leftharpoondown$
	p-i-n	Al /	n^+ /	a-Si:H (i) /	p /	In₂O₃ /	substrat verre
					30 nm		

La structure inverse présente le double avantage de pouvoir être éclairée par le contact transparent et d'avoir une autoprotection des couches actives du Pt et du p par le substrat.

2.2.2. Jonctions tunnel p^+n^+

La réalisation des structures doubles p-i-n/p-i-n (diodes en série) requiert la présence d'une jonction tunnel p^+n^+ reliant les deux diodes, pour éviter l'apparition d'une tension parasite et la chute du photocourant due à de la recombinaison dans ces couches. Sachant que dans a-Si:H, la présence d'états de queue de bande ne permet pas de faire de véritables diodes tunnel de type Esaki (5) dans lesquelles les deux types de porteurs sont dégénérés, nous avons néanmoins obtenu des diodes dont le comportement est similaire, précisément grâce à la position du niveau de Fermi situé dans les états conducteurs de queue de bande.

2.2.3. Structures empilées

Nous savons que la densité d'états au milieu du gap mesurée pour a-Si:H est très faible (6)(7)(8) et donne lieu à des régions de déplétion étendues (de l'ordre de 500 nm pour 10^{16} cm⁻³). En fait, la zone efficace de collecte est plus courte, ce qui équivaut à associer aux porteurs une certaine longueur collecte L_C (pour notre matériau, L_C = 300 nm) (9). Ceci signifie que seules les quelques centaines de nanomètres adjacents à la jonction sont utiles bien qu'insuffisants pour collecter toute la partie énergétique du spectre solaire. Pour améliorer le rendement des diodes, nous avons choisi de multiplier les zones de collecte en empilant plusieurs jonctions dont les épaisseurs devront être adaptées selon le montage électrique choisi et permettront d'augmenter le courant de court-circuit ou la tension de circuit ouvert. Pour obtenir une meilleure efficacité de conversion, les cellules associées devraient être colorées, offrant des gaps différents et compatibles avec le spectre solaire. Dans ce travail, nous avons seulement démontré la faisabilité de multijonctions de même gap.

- structure Schottky double (montage parallèle - figure 1)

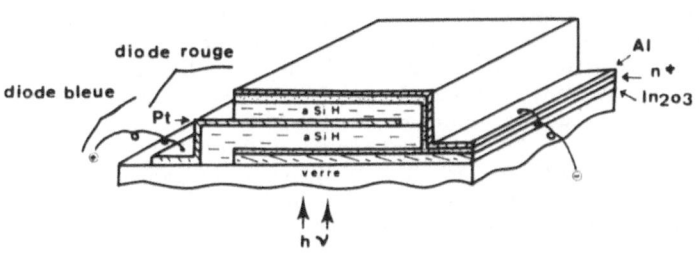

Figure 1 : Double structure Schottky

Le montage parallèle doit conserver la tension de circuit ouvert Voc d'une diode élémentaire et augmenter le photocourant grâce aux deux zones de collecte situées de part et d'autre du platine médian. Voc est optimal quand les valeurs correspondantes des deux diodes sont identiques. Cette condition ne semble pas, a priori, trop critique à réaliser dans la mesure où la valeur de Voc ne dépend que de façon logarithmique du flux absorbé. La difficulté de cette structure est de "sortir" le contact Pt médian.

- structure p-i-n double (montage série - figure 2)

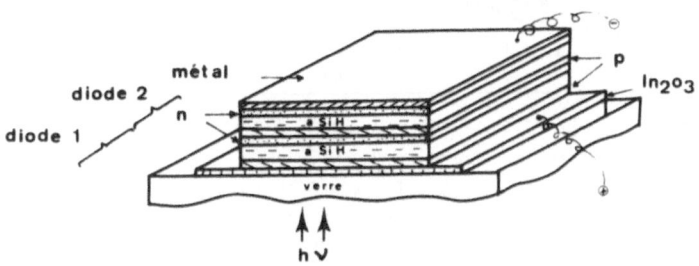

Figure 2 : Double structure p-i-n

Grâce au montage série, la tension de circuit ouvert devrait être multipliée par 2, mais le photocourant ne peut être, au mieux, que légèrement supérieur à la moitié du photocourant d'une diode simple de même épaisseur. Comme le photocourant est égal au plus faible des deux diodes, l'optimisation consiste à égaliser les deux photocourants et donc à respecter la règle de l'égalité des flux lumineux absorbés, sachant que le photocourant dépend linéairement du flux absorbé. Le flux élémentaire de photons absorbés à une distance x dans une tranche dx de matériau et pour un rayonnement de longueur d'onde λ est :

$$d\phi_{abs}(\lambda) = \phi_{inc}(\lambda)\alpha(\lambda)\exp(-\alpha(\lambda)x)dx$$

avec $\alpha(\lambda)$ = coefficient d'absorption du matériau.

Après intégration sur l'épaisseur et pour l'ensemble du spectre solaire, l'égalisation des flux donne les valeurs x_1 et x_2 des épaisseurs représentées dans la figure 3.

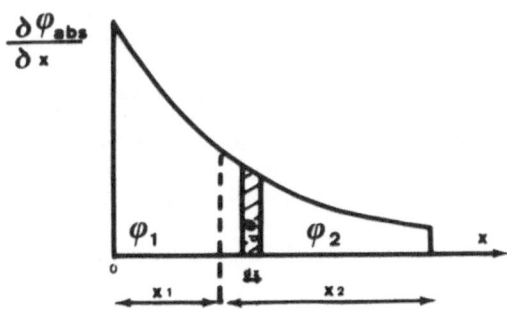

Figure 3 : Schéma de la densité de flux absorbé suivant l'épaisseur

3. Expériences et résultats
3.1. Caractérisation électrique des diodes élémentaires
3.1.1. Caractérisation I(v) dans le noir

Pour cette étude, nous avons étendu aux diodes a-Si:H la théorie des diodes cristallines ; aussi devrons-nous quelquefois considérer avec prudence les résultats obtenus.

Sous polarisation V, le courant noir des diodes Schottky suit la loi de l'émission thermo-ionique et celui des diodes p-i-n vérifie l'équation de diffusion de Shokley qui se traduit par la même expression $I = I_o (\exp \frac{qV}{nkT} - 1)$ avec I_o = courant de saturation, q = charge électronique, k = constante de Boltzmann, T = température en Kelvin, n = facteur d'idéalité (égal à 1 si la diode est parfaite). Les valeurs élevées de n (comprises entre 1,2 et 1,5 pour Schottky et entre 1,3 et 2 pour p-i-n) témoignent d'un régime mixte mettant en jeu la génération-recombinaison (due aux états dans le gap). A haute température, l'émission thermo-ionique pour les Schottky et la diffusion pour les p-i-n dominent. Par des mesures en température, nous déduisons la hauteur de barrière Φ_B des diodes qui est de l'ordre de 1 eV pour les Schottky et de 0,9 eV pour les p-i-n.

3.1.2. Caractérisation sous éclairement

L'évaluation du rendement des diodes a été effectuée au soleil sous des conditions valables d'illumination (> 70 mW/cm^2). La puissance lumineuse incidente a été mesurée avec un solarimètre KIPP et ZONEN type CM 5.

TABLEAU I

Structure	N° diode	Surface (mm^2)	Epaisseur (μm)	Voc (V)	Jcc (A/cm^2)	FF	n %
Schottky	M 530 C2	1,3	0,75	0,66	$7,6.10^{-3}$	0,44	2,2
	M 534 E3	4,0	0,75	0,62	$6,2.10^{-3}$	0,42	1,6
	M 541 B4 (inverse)	3,2	0,75	0,46	$4,6.10^{-3}$	0,37	0,78
p-i-n	E 301 E3	4,0	0,5	0,73	$1,87.10^{-3}$	0,49	0,7
	M 532 B4	2,5	0,9	0,86	$7,8.10^{-4}$	0,40	0,27
	M 601 D2 (inverse)	4,0	0,8	0,76	$3,75.10^{-3}$	0,54	1,5

De la lecture du tableau I il ressort que les diodes Schottky ont une meilleure densité de courant Jcc que les p-i-n, mais un plus faible Voc. Le courant recueilli dans la p-i-n inverse ainsi que le facteur de remplissage FF sont nettement améliorés grâce à une meilleure transmission du rayonnement au travers de la couche In$_2$O$_3$. Ceci n'est pas vrai pour la Schottky inverse dont le photocourant et la tension ont chuté. Nous avons en effet remarqué que la barrière Pt/a-Si:H était toujours moins bonne dans les structures inverses, entraînant un Voc et un champ interne plus faibles (donc une mauvaise collecte du photocourant).

Notons qu'à la différence des autres diodes, la p-i-n inverse a été faite dans un même cycle de vide, ce qui a certainement contribué à l'amélioration observée. Cependant après plusieurs dépôts alternés de matériaux dopés et de matériau intrinsèque, nous avons constaté une nette dégradation des performances des dispositifs que nous avons attribué à une pollution par les impuretés dopantes résiduelles.

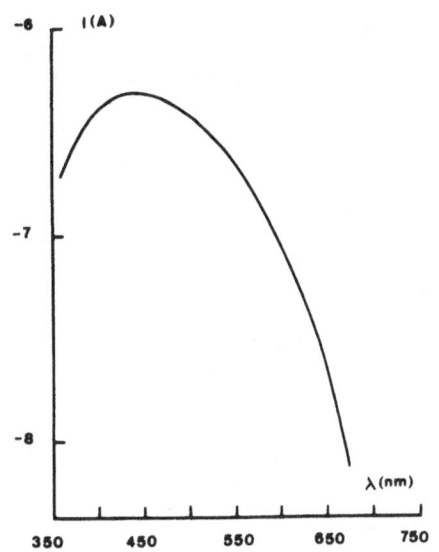

Figure 4 : Réponse spectrale d'une diode p-i-n

La figure 4 montre l'allure de la "réponse spectrale type" des diodes que nous fabriquons, p-i-n ou Schottky. Elle résulte de la combinaison du coefficient d'absorption du matériau et de la longueur de collecte L_C des porteurs.

Nous constatons que le courant recueilli pour $\lambda > 500$ nm décroît très vite, reflétant à la fois la valeur élevée du seuil d'absorption et la faible valeur de L_C. Ceci explique la faible densité de courant des diodes qui ne recueillent que le courant dû aux courtes longueurs d'onde.

3.2. Caractérisation des jonctions tunnel p^+n^+

Faisant varier le dopage des couches n^+ et p^+, nous avons mesuré la conductivité apparente de ces jonctions et comparé celle-ci aux conductivités séparées des couches n^+ et p^+ établies en fonction du dopage. Nous en avons déduit que c'est le matériau p^+ qui pilote le fonctionnement de la jonction, quel que soit le dopage n. Une bonne p^+n^+ doit d'abord posséder une caractéristique I(v) symétrique par rapport à l'origine, ce que nous avons constaté seulement sur des diodes dont la couche p était fortement dopée. D'autre part, Marfaing (10) a montré que les jonctions tunnel p^+n^+ pouvaient être considérées comme telles si elles satisfaisaient certains critères dont notamment une dépendance linéaire du rapport des dérivées d(LnI)/d(LnV) vis à vis de V et également un comportement linéaire de LnI en fonction de T à très faible polarisation. Nous avons vérifié que certaines de nos diodes satisfaisaient ces critères.

3.3. Résultats sur les structures empilées
3.3.1. Double structure p-i-n

Tableau II

n° diode	épaisseur i_1(nm)	épaisseur p^+n^+(nm)	dopage $p(p^+n^+)$ (P_{B2H6}/Ptotale)	épaisseur i_2(nm)	Voc(V)	Jcc(A/cm^2)	FF	n %
M 647	150	20	$1,6\ 10^{-3}$	300	1,45	$2,3\ 10^{-3}$	0,44	1,46
M 653	150	20	$7,5\ 10^{-3}$	250	1,59	$1,65\ 10^{-3}$	0,41	1,07
M 657	120	20	$1,6\ 10^{-3}$	300	1,7	$1,67\ 10^{-3}$	0,40	1,13
M 658	120	20	$7,5\ 10^{-3}$	300	1,2	$1,95\ 10^{-3}$	0,40	0,94
M 675	55	30	$1,6\ 10^{-3}$	300	1,3	$9,2\ 10^{-4}$	0,35	0,42
M 685	80	30	$7,5\ 10^{-3}$	300	1,6	$1,4\ 10^{-3}$	0,36	0,81

Le tableau II montre que, dans l'ensemble, une tension double de Voc se retrouve aux bornes de la structure. Les caractéristiques I(V) sous AM1 des p-i-n/p-i-n ont pour la plupart l'allure de celle montrée sur la figure 5. Pour expliquer cette forme, nous devons analyser le fonctionnement de la structure.

Quand on applique une tension directe à une p-i-n/p-i-n, cette tension se retrouve toute entière aux bornes de la diode 2 (voir fig. 2) tant qu'elle n'a pas atteint la valeur "coude" de cette diode. Sous l'effet de cette tension, la zone de collecte de cette même diode diminue, entraînant une décroissance quasi-linéaire de son propre photocourant et un mauvais FF. Si ce photocourant se révèle être le plus faible des deux diodes, alors ce sera celui de la structure et la forme de I(V) aura l'allure donnée par cette diode (fig. 5). Par contre, si le photocourant de la structure se trouve être celui de la diode 1, comme celle-ci ne subit pas tout de suite une chute de tension, sa zone de collecte va rester constante, donc également le courant, et le FF sera bon (Fig. 6). Ce fonctionnement se vérifie également quand on éclaire la structure du côté opposé à la fenêtre d'entrée, car on se trouve alors dans le cas où c'est la diode 1 qui a le plus faible courant ; FF passe alors de 0,36 à plus de 0,50.

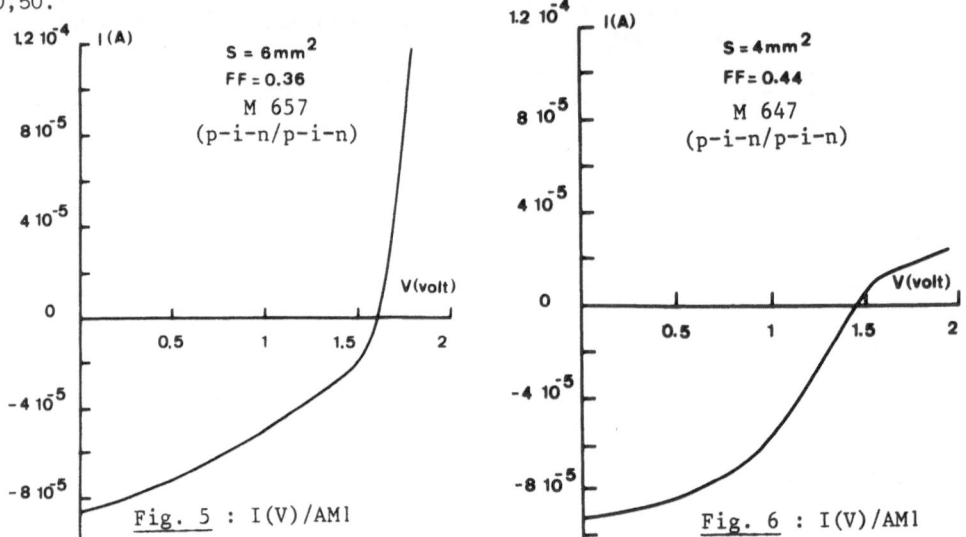

Fig. 5 : I(V)/AM1   Fig. 6 : I(V)/AM1

Dans le tableau II, nous remarquons que si les couche p^+n^+ sont trop épaisses, cela diminue encore le courant ainsi que FF parce que le rayonnement est en partie absorbé dans cette double couche (M 675 et M 685). De plus, les résultats sur les jonctions tunnel p^+n^+ ont montré que celles-ci étaient d'autant meilleures qu'elles étaient plus dopées. Si ceci est favorable au passage du photocourant collecté, en revanche cela nuit à la bonne transmission optique de ces couches à cause du surplus de défauts introduit par le dopage. Il faudra donc trouver des compromis pour l'épaisseur et le dopage des jonctions p^+n^+. Sur la figure 6, on remarque une saturation du courant direct due à une trop forte résistance série. L'emploi d'une couche n^+ plus épaisse a définitivement éliminé cette anomalie. Les valeurs de Jcc montrent que le photocourant atteint pratiquement la moitié de celui des meilleures p-i-n (M 601 D2). Les rendements obtenus sont du même ordre de grandeur.

Nous pensons donc qu'avec une très bonne optimisation des épaisseurs nous serons en mesure d'augmenter le rendement. Néanmoins, il ressort de cette étude qu'il est essentiel d'utiliser des jonction colorées pour lesquelles notre matériau pourrait convenablement servir de matériau de fenêtre d'entrée.

3.3.2. Double structure Schottky

Malgré plusieurs essais, nous n'avons pu obtenir qu'une seule structure apte à fonctionner. Comme nous l'avions prévu, la réalisation de la barrière de Pt dans le plan médian s'est avérée délicate. La structure obtenue, dont les épaisseurs de matériau intrinsèque sont voisines de 200 nm et 500 nm, possède un courant de court-circuit inférieur au mA/cm^2 et un Voc de 0,3 V. Ces valeurs sont inférieures à celles trouvées pour la diode 1 ou "diode bleue" (Jcc = 1,1 mA/cm^2 et Voc = 0,42 V) que nous avons pu étudier séparément grâce à un substrat qui a été mis lors du dépôt puis retiré après le dépôt de Pt. Cette différence de résultats prouve que non seulement la diode 2 (ou diode rouge) était mauvaise, mais qu'en plus la diode bleue s'est dégradée en fin de fabrication de la structure.

4. Conclusion

Cette étude a permis d'atteindre en plusieurs étapes le but que nous nous étions fixé, à savoir de montrer qu'il était technologiquement possible de réaliser par pulvérisation cathodique des multijonctions qui permettent d'améliorer le rendement photovoltaïque des diodes.

Dans une première étape nous avons montré que différents types de diodes élémentaires pouvaient être réalisés sans problème technologique particulier, notamment que des diodes aussi minces que 1500 Å étaient capables de fonctionner sans subir de court-circuit. La comparaison des résultats photovoltaïques sur diodes Schottky et p-i-n conduit à dire que d'ores déjà les p-i-n sont plus intéressantes que les diodes Schottky. L'étape intermédiaire a vu la réalisation de jonctions tunnel, passage obligé des multijonctions p-i-n. Des expériences caractéristiques ont attesté de leur véritable comportement tunnel bien qu'elles ne soient peut-être pas optimisées au pdv optique, notamment à cause de l'absorption trop importante du matériau p$^+$. La phase finale a servi de synthèse aux précédentes par la réalisation de structures empilées, les unes (Schottky) dans un montage électrique en parallèle, les autres (p-i-n) en série. S'il est un peu prématuré d'affirmer que la structure Schottky double n'est pas à retenir, parce qu'elle n'a pas donné satisfaction dans notre réalisation, en revanche on peut dès à présent affirmer que la structure p-i-n/p-i-n répond à nos espérances technologiques. Les p-i-n sur In$_2$O$_3$ confirment que le matériau est bien adapté pour la partie bleue des cellules multispectrales.

REFERENCES

1) D.E Carlson and C.R Wronski, Appl. Phys. Lett. 28 (1976) 671
2) Y. Hamakawa, H. Okamoto, Y. Nitta, J. Non Cryst. Solids, 30 (1980) 749
3) J.J Hanak, J. Non Cryst. Solids, 30 (1980) 775
4) W.E. Spear, P.G Le Comber, Solid State Comm., 17 (1975) 1193
5) L. Esaki, Phys. Rev. 109 (1953) 603
6) A. Madan, P.G. Le Comber and W.E. Spear, J. Non Cryst. Solids, 20 (1976) 239
7) P. Viktorovitch, D. Jousse, A. Chenevas-Paule and L. Vieux-Rochaz, Rev. Phys. Appliquée, 14 (1979) 201
8) J. Beichler, W. Fuhs, H. Mell and H.M. Welsch, 8 th International Conf. on Amorphous and Liquid Semicond., Cambridge (USA) Aug. 27-31, 1979
9) P. Bouchut, thèse (Université de Grenoble 1981)
10) W.U. Zhong Yan, Y. Marfaing and J. Dixmier, 9th International Conf. on Amorphous and Liquid Semicond., Grenoble (FR) July 2-8, 1981

DEVELOPMENT OF SPUTTERED THIN-FILM a-Si SOLAR CELLS

Authors   :   J. ALLISON,   D.P. TURNER

Contract Number   :   ESC-R-040-UK

Duration   :   36 months from July 1st 1980 to June 30th 1983

Head of Project   :   Professor F.A. Benson

Contractor   :   Dr. J. Allison

Research Assistant :   Mr. D.P. Turner

Address   :   The University of Sheffield,
                        Department of Electronic &
                            Electrical Engineering,
                        Mappin Street,
                        SHEFFIELD. S1 3JD
                        U.K.

Summary

    The technique of magnetron sputtering has been thoroughly researched
and a-Si:H displaying similar characteristics to the best glow-discharge
material has been produced.   The field effect density of states has been
reduced below $10^{17} cm^{-3} eV^{-1}$ and in these films the defect luminescence band
at 0.9eV is completely undetected.   The normalised photoconductivity
$(n\mu\tau)$ is $4.10^{-8} v^{-1} m^2$, an order of magnitude higher than that for films
bias-sputtered from a conventional target.

    Initial trials have shown that magnetron sputtered material may be
doped n or p type by the inclusion of phosphine or diborane in the
sputtering gases.   Activation energies of 0.37eV and 0.5eV respectively
have already been achieved and more efficient doping is expected from the
optimisation of sputtering conditions.

1.   Introduction

Amorphous silicon has attracted intense interest as a material for thin-film solar cells.   Several different deposition techniques are being developed, such as chemical vapour deposition (1), evaporation (2), ion beam deposition (3) and sputtering (4).   The most established technique is by decomposition of silane in a glow discharge (5).   The high quality material this produced provides a yardstick for other methods.

The technique of sputtering a-Si has potential advantages including large area and low cost production, but so far sputtered films, although comparable, have not matched their glow discharge (g.d.) counterparts in solar applications.   Typical solar cell efficiencies for p-i-n junction cells incorporating glow discharge a-Si:H are $\approx$ 8% (6).

Our research programme is in three sections.   The first aim is to improve the quality of intrinsic sputtered films to have similar properties to their glow discharge counterparts.   Then we require efficient n and p type doping by the incorporation of phosphorous and boron.   Finally we will combine these layers in a p-i-n structure to form a cell.   At present we are midway through the second phase.

2.   Intrinsic Materials

In order to improve the intrinsic a-Si:H material a magnetron target has been used instead of a conventional one without a magnetic field.   The main feature of the magnetron is a reduction in target voltage, assuming the same power level, thus lowering substrate bombardment by secondary electrons and high energy neutrals.   The introduction of electronic flow control and gas purification has also greatly improved reproducibility.

Magnetron material has been produced at four deposition temperatures (170, 200, 230 and 260°C) over a range of hydrogen partial pressures, P_H, 0-1 m.Torr, at a fixed argon partial pressure, P_{Ar}, of 5 m.Torr.   At the substrate temperature, T_S = 170°C two additional series of films have been prepared at P_{Ar} 2.5 and 10 m.Torr over a range of P_H 0-1 Torr.   The conditions for optimum intrinsic materials were found to be P_H 0.6 m.Torr, P_{Ar} 5 m.Torr, T_S 230°C for a target power of 100W, voltage 400V and electrode spacing of 9 cm.

Many film properties have been monitored during the optimisation procedure.   Photoconductivity has been found to be very sensitive to preparation details and varies in a way representative of film quality. Figures 1 and 2 show the variations in photoconductivity for the series of depositions varying T_S and P_{Ar}.   The peak photoconductivity is found at the stated optimum conditions.   We do not observe the two orders of magnitude increase reported by Anderson et al (7) as P_{Ar} is raised from 5 to 10 m.Torr for P_H 0.6 m.Torr.   The luminescence efficiency is also highest at P_{Ar} 5 m. Torr.

Some properties of an optimised intrinsic film are listed in table 1. The low values of the density of states < $10^{17} cm^{-3} eV^{-1}$ from field effect and < $10^{16} cm^{-3} eV^{-1}$ from C-V measurements also indicate the quality of the films.

A direct comparison has been made of the photoluminescence, hydrogen content and hydrogen bonding (by infra-red spectroscopy) in sputtered samples and g.d. films prepared by the Dundee group.   In the best material of both types the photoluminescence emission band at 0.9eV, attributed to an intrinsic defect, is absent.   The photoluminescence efficiency of the main band at 1.3eV approached unity in both cases and the hydrogen content and percentage of single bonded hydrogen become comparable.

Field effect transistors have been fabricated on sputtered films having an on-off drain-current ratio greater than 10^4 with turn on

voltages of 6 volts (8).   These and other device characteristics are
comparable with those prepared on g.d. silicon (9) underlining the improvement in our intrinsic material.

3.1   Doping Studies

We have now completed our preliminary trials for both n and p type
doping.   The optimum conditions for intrinsic material were taken as a
starting point.   Phosphine and Diborane 1% mixes in argon were added to
the sputtering gases at flow rates 0-10 S.C.C.M.   This is equivalent to a
maximum dopant partial pressure of 1×10^{-5} m.Torr and P_H was maintained
at 0.5 m. Torr.

3.2   n-Type Doping

The arrhenius plots of phosphine doped samples are shown in figure 3.
All the samples were found to be singly activated across the temperature
range 200-400K with activation energies ranging from 0.85 to 0.44eV, and
room temperature conductivities 9.9×10^{-9} to $2.4 \times 10^{-4} S.m^{-1}$.

Films sputtered from a conventional target have also been doped, using
a phosphine partial pressure 2.5×10^{-6} Torr.   An activation energy of
0.4eV with room temperature conductivity $1.2 \times 10^{-4} S.m^{-1}$ was obtained.
Gas phase doping of conventionally sputtered silicon was first reported by
Paul et al (10) with sputtering conditions : P_{Ar} 5 m.Torr, T_S $250^{\circ}C$ and P_H
in the range 0.5-0.7 m.Torr.   An activation energy of 0.42eV with
$4 \times 10^{-4} S.m^{-1}$ room temperature conductivity was reported.   Subsequently
(11) the activation energy was reduced to 0.25eV for the same dopant level
by a small reduction in P_H to 0.4 m.Torr.   This value is lower than our
original result, but it is anticipated that by further varying the
sputtering conditions more efficient doping may be possible.

From i-r spectroscopy we find that the introduction of phosphine at a
1 S.C.C.M. flow rate increases the hydrogen content, C_H, from 3.3% to 4.3%
suggesting that a reduction in P_H may lead to higher doping.   The deposition rate also increases with the introduction of phosphine, from 7.1 to
7.6 nm. min^{-1}.

The phosphorous content, C_p, of these films has been measured by
energy dispersive X-ray spectroscopy and figure 4 shows the remarkable
linearity between C_p and the phosphine flow rate. C_p ranges from 0-12%,
compared with 1% and 40% reported by Anderson (11), for films prepared at
1×10^{-5} and 2×10^{-4} Torr phosphine.   A C_p of 1% has also been reported
by R.C.A. (12) for optimum p-i-n devices and by Matsumura (13) for g.d.
layers.

The high phosphine content in the sputtered films is constant with an
observed fall-off in luminescence intensity in the doped films (14).

3.3   p-Type Doping

The arrhenius plots of diborane doped samples are presented in figure
5; two intrinsic samples "i" and "o" are also included.   Sample "i" was
prepared between the phosphine and diborane runs and has similar properties to an intrinsic sample; no phosphorous contamination was detected.
The sample "o" was produced after the diborane doped runs and the high
activation energy and low photoconductivity it exhibits are typical for
several runs following indicating probable diborane cross-contamination.
We intend to reduce the possibility of contamination in the future by
introducing argon and hydrogen through a separate line which will never be
exposed to toxic gases.

As will be seen from figure 5 none of the diborane doped samples are
singly activated.   A good fit was found by separating the conductivity

into two parallel components

$$\sigma = \sigma_o \exp\{-\frac{E_a}{kT}\} + \sigma_h \exp\{-\frac{W}{kT}\} \qquad (1)$$

The activation energies E_a and W for the two components are shown in figure 6 for diborane flow rates O-5 S.C.C.M. The boron content of these films has not yet been determined. The pre-exponential factor σ_o ranges from 1.9×10^5 to 1.5×10^3 S.m^{-1} and is associated with extended state transport. Thus the Fermi level has been shifted by a $\sim$ ·2eV to O.5eV above the valance band.

Infra-red measurements show that the hydrogen content of p-doped films drops to below 1% for a-Si sputtered with 5 S.C.C.M. diborane, compared to 3.3% in our optimum material. This is also reflected in the change in optical gap from 1.8 to 1.3eV. In material with such a low hydrogen concentration a high density of states is expected at midgap. Green-baum (5) has demonstrated from NMR measurements that boron in a-Si:H is largely incorporated in a three-fold configuration which Freeman (16) suggests will give rise to a dual acceptor state deep in the gap. Both of these factors may be contributing to the hopping conduction found in these samples.

The properties of our most heavily doped films are presented in table 3. The photoconductivity falls off with increasing doping and the peak wavelength shifts to lower energies as the energy gap is reduced as seen from figure 7. Also shown in figure 7 are the two intrinsic films ("i" "o") clearly displaying the contaminated film with low photoconductivity.

As with the n-type doping we expect to improve the doping efficiency by changing the deposition conditions, especially the hydrogen pressure. Moustakas (17) has reported an activation energy of O.3eV and room temperature conductivity of 10^{-2}S.m^{-1} by conventional sputtering.

3.4 Bias Doping

The reduction in the band gap with diborane doping is undesirable for solar cell applications, as the absorption in the p region is increased without generating photocurrent. Various methods have been used to increase the band gap without reducing the σ_{RT}, by high power g.d. deposition (18) or by introduction of Methane to the plasma to produce Si:C:H (6).

We have previously demonstrated that the application of a suitable substrate bias can reduce the density of states and boost the photoconduct-ivity of conventionally sputtered material by an order of magnitude (4) ; the ratio of singly bonded to double bonded hydrogen is also enhanced. We have applied this technique to boron doped films and found a partial recovery of the hydrogen content to $\approx$ 2%, using a substrate potential -25 volts and a flow rate of 5.O S.C.C.M. of diborane; the bandgap is then 1.64eV. Figure 8 shows the shift in E_a and W with various bias voltages and it will be seen that for -25V bias the hopping energy is a maximum involving a lower density of states; the activation energy shifts to higher values as the energy gap widens but is still $\sim$ O.17eV below the midgap position. The $n \mu \tau$ product also recovers to $\sim 1 \times 10^{-12}$m^{2}V^{-1} for the 5OV bias. Table 3 summarises the properties of the 25 volts 5.O S.C.C.M. boron sample.

4. Conclusion

We have shown that the magnetron sputtering technique is capable of producing intrinsic material with a low density of states ($< 10^{16}$ cm^{-3}eV^{-1}

using C-V measurements) and properties comparable with its g.d. counterpart.

Our initial doping trials indicate that the material can be doped both n- and p- type by inclusion of phosphorous or diborane. More efficient doping will be sought by varying preparation conditions, particularly hydrogen pressure, which is affected by the dopant gasses. The technique of Magnetron-Bias sputtering has been shown to have a pronounced effect on material properties consistent with previous findings and may prove important when the p-doped films have been optimised.

Measures are being taken to evaluate and eliminate cross-contamination from dopants.

We are now in a position to produce good intrinsic a-Si:H and p-n junction devices; the fabrication and evaluation of solar cells using such films, particularly p-i-n structures, which will commence shortly.

REFERENCES

1. M. Tanguchi, Y. Osaka and M. Hirose : J. Elect. Mat. 8 (5),689 (1979)

2. D. Kaplan, N. Sol, G. Velases and P.A. Thomas : Appl. Phys. Lett. 33, (5), 440 (1978)

3. G.P. Ceasar, K. Okumura and S.F. Grimshaw : Proc. 9th Int. Conf. Amorphous and liquid semiconductors, 627 (1981)

4. D.P. Turner, I.P. Thomas, J. Allison, M.J. Thompson, A.J. Rhodes, I.G. Austin and T.M. Searle : Proc. Tetrahedrally bonded Amorphous Semiconductors (Carefree, Arizona), A.I.P. Conference 73. 47 (1981)

5. W.E. Spear : Advances in Physics 26 (6), 811 (1977)

6. Y. Tawada, M. Kondo, H. Okamoto, V. Hamakawa : Solar Energy Materials, 6, 299 (1982)

7. D.A. Anderson, G. Noddel, M.A. Paesler and W. Paul : J. Vac. Sci. Technol. 16 (3), 906 (1979)

8. M.C. Abdulrida, J. Allison : Submitted for publication

9. M.J. Powell, B.C. Easton and O.F. Hill : Appl.Phys.Lett.38, 794(1981)

10. W.J. Paul, A.J. Lewis, G.A.N. Connell and T.D. Moustakas : Solid State Communications, 20, 969 (1976)

11. D.A. Anderson and W. Paul : Phil. Mag. B45 (1), 1, (1982)

12. D.E. Carlson, R.W. Smith, C.W. Magee, P.J. Zanucchi : Phil. Mag. B45, (1), (51), (1982)

13. A. Matsuda, S. Yamasaki, K. Nakagawa, K. Tanaka, S. Iizima, M. Matsumura and H. Yamamoto : Jap. J. Appl. Phys. 19. L305,(1980)

14. R.A. Street : Advances in Physics, 30, (5), 593 (1981)

15. S.G. Greenbaum, W.E. Carlos and P.C. Taylor : Solid State Communications 43 (9), 663 (1982)

16. E.C. Freeman and W. Paul : Phys. Rev. B20 (2), 716 (1979)

17. T.D. Moustakas : J. Elec. Mat. 8 (3), 391 (1979)

18. A. Matsuda, M. Matsumura, S. Yamakasi, H. Yamamoto, T. Imura, H. Okushi, S. Iizima and K. Tanaka : Jap. J. Appl. Phys. 20 (3) L186 (1981)

TABLE 1 : Properties of optimised i-layer.

Photoconductivity $n\mu\tau$ $(V^{-1}.m^2)$	3.96	E-8
Energy Gap (eV)	1.8	
Room temp. Conductivity σ_{RT} $(S.m^{-1})$	9.9	E-9
Activation Energy E_a (eV)	0.85	
Conductivity Prefactor σ_0 $(S.m^{-1})$	3.5	E 6
Hydrogen Content C_H (%)	3.3	

TABLE 2 : Properties of an n-type film prepared with
           10 S.C.C.M. Phosphine.

Energy Gap (eV)	1.74	
Room temp. Conductivity σ_{RT} $(S.m^{-1})$	2.5	E-4
Activation Energy (eV)	0.44	
Conductivity Prefactor σ_0 $(S.m^{-1})$	9.4	E 3
Hydrogen Content C_H (%)	3.92	
Phosphorous Content C_p (%)	11.6	

TABLE 3 : Comparison of p-doped films prepared with diborane
            flow 5.0 S.C.C.M. showing the effect of substrate bias.

Substrate bias. (Volts)	Zero		25	
Photoconductivity $n\mu\tau$ $(V^{-1}m^2)$	Zero		5.5	E-13
Energy Gap (eV)	1.32		1.64	
Room temp. Conductivity σ_{RT} $(S.m^{-1})$	7.5	E-6	7.5	E-6
Activation Energy E_a (eV)	0.50		0.65	
Conductivity Prefactor σ_0 $(S.m^{-1})$	1.5	E 3	2.8	E 5
Hopping Energy W (eV)	1.4		0.24	
Conducting Prefactor σ_H $(S.m^{-1})$	8.2	E-4	6.1	E-2
Hydrogen Content C_H (%)	0.9		2.05	

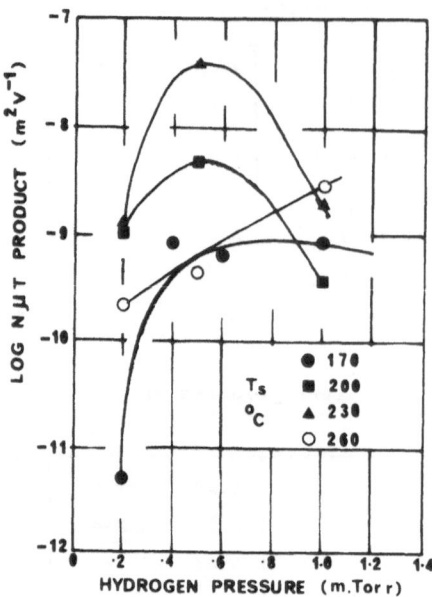

Fig.1 : Normalised photoconductivity of
films prepared over a range of
deposition temperatures with
fixed argon pressure 5m.Torr.

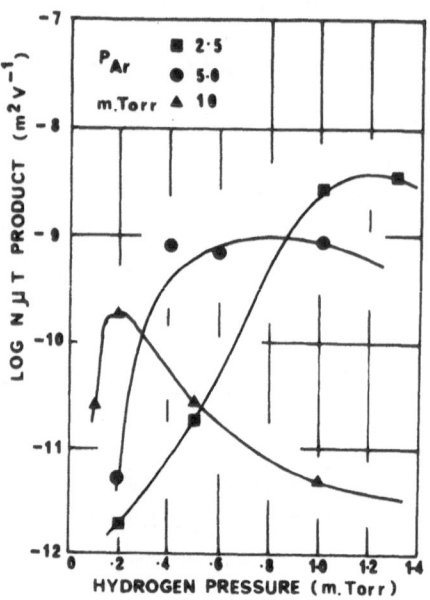

Fig.2 : Normalised photoconductivity of films
prepared over a range of
argon pressures at a
fixed deposition temp-
erature 170°C.

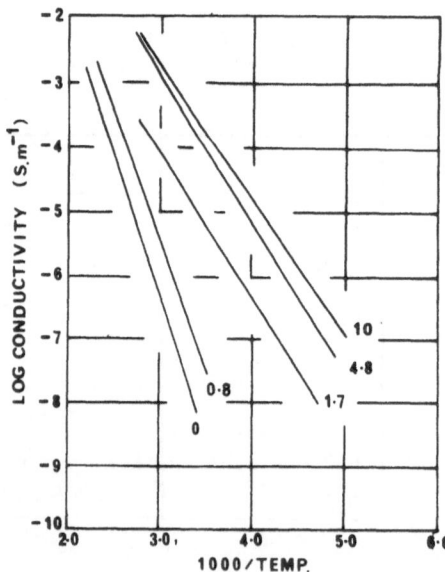

Fig.3 : Temperature dependence of
conductivity for n-doped films
prepared with phosphine flows
0-10 S.C.C.M.

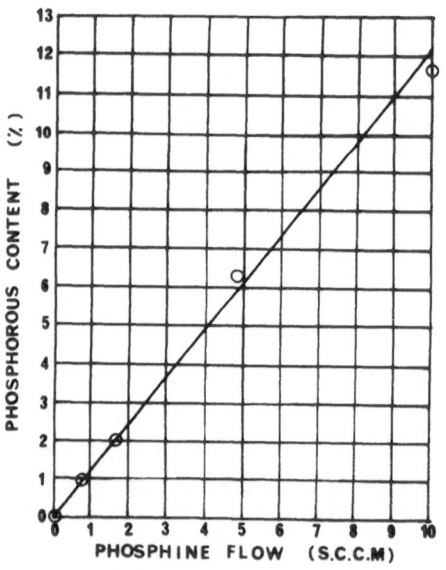

Fig.4 : Linear variation with
phosphine flow of the
elemental phosphorous
percentage in n-doped
films, measured by E.D.X.S.

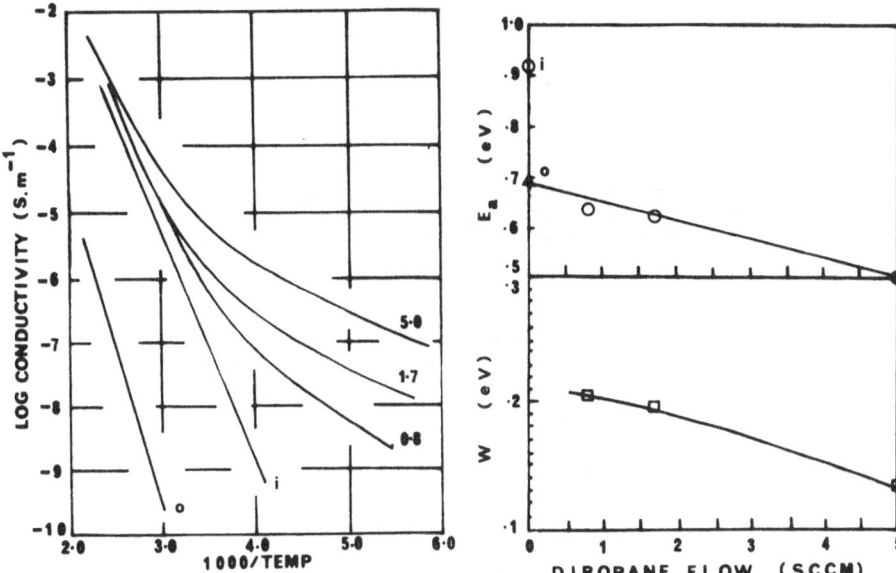

Fig.5 : Temperature dependence of conductivity for p-doped films
prepared with diborane flows
0-5 S.C.C.M., also showing the
change in intrinsic material
before (i) and after (o) the
doped series.

Fig.6 : Activation energies
for extended state and
hopping conduction in
p-doped films prep-
ared with diborane
flows 0-5 S.C.C.M.

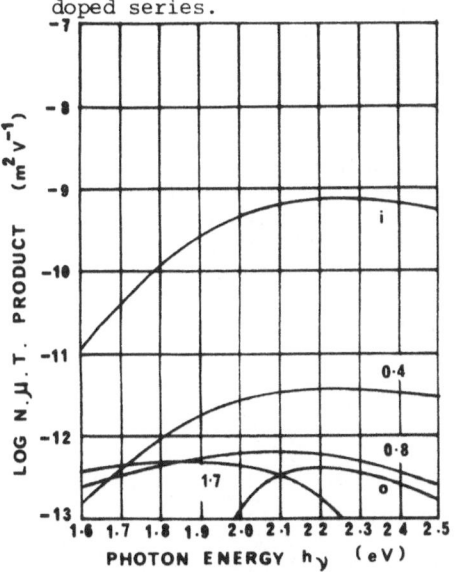

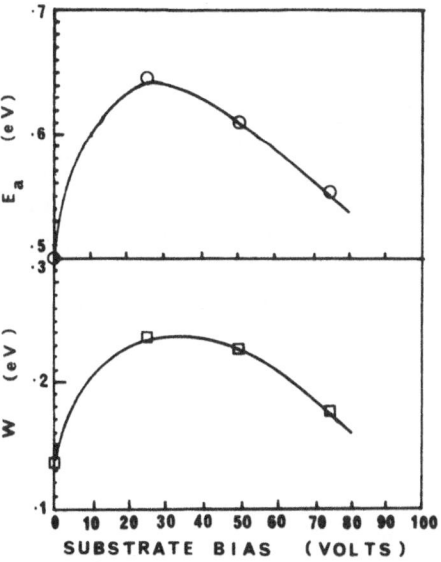

Fig.7 : Normalised photoconductivity
spectra for p-doped films at
various diborane flowrates.
The sample prepared at 5.0
S.C.C.M. shows no photo-
response.

Fig.8 : Variation of the hopping and extended state
activation energies
with substrate bias in
p-type films doped by
5.0 S.C.C.M. diborane.

PREPARATION, STUDY AND CHARACTERIZATION OF HYDROGENATED AMORPHOUS SILICON FOR PHOTOVOLTAIC CELLS

Authors   : C. COLUZZA, F. EVANGELISTI, P. FIORINI,
           G. FORTUNATO, A. FROVA, C. GIOVANNELLA,
           P. MIGLIORATO.

Contract number   : ESC-R-029-I (S)

Duration   : 36 months 1 July 1980 - 30 June 1983.

Head of project   : Prof. A. Frova, Istituto di Fisica,
           Universita' di Roma.

Contractor   : University of Rome, Istituto di Fisica.

Address   : Istituto di Fisica "G. Marconi",
           P.le Aldo Moro, 2 - 00185 ROMA, Italy

Summary

A novel technique for the deposition of a-Si:H by a double ion-gun sputtering apparatus is described. Properties of some preliminary samples prepared by this method are discussed. A procedure for gap-state spectroscopy by photoconductivity measurements has been developed and tested on glow-discharge samples. It is shown that there is a close correspondence between gap-state density and re-combination kinetics. Some preliminary work on Schottky-barrier cells is reported.

1. Introduction

Hydrogen passivation of defects in amorphous silicon films has proven to be a powerful tool for the improvement of phototransport properties towards higher efficiency solar cells. Research in this and related areas has received therefore a continuously increasing attention in recent years. Several deposition techniques have been adopted. Some of them, i.e. RF glow-discharge in silane and RF sputtering, have been thoroughly investigated and are now widely used, others, e.g. CVD or laser-induced pyrolysis of silane, are still at a rather developmental stage.

The present report describes the preliminary results obtained with a novel technique, which is based on sputtering silicon off a target by an argon ion-gun and separately implanting hydrogen during film growth by means of an auxiliary gun. In order to compare the properties of the films obtained by this method with those produced by glow-discharge, a capacitively coupled RF glow-discharge reactor has been assembled and optimized for lowest density of gap states. This has implied a systematic characterization of the films, in particular gap-state spectroscopy, as a function of deposition parameters.

In the next chapter, the operating conditions of the double ion-gun sputtering apparatus are described. Infrared and transport properties of the early samples produced by this technique are being presented and discussed. The following chapter will describe the experimental techniques that have been set up for the characterization of the material. Results correlating the density of gap states and recombination kinetics, as obtained on glow-discharge samples grown for different deposition conditions, will be illustrated and discussed. Preliminary results on the behavior of Schottky-barrier cell structures will also be presented.

2. a-Si:H produced by double ion-beam sputtering

2.1 Double ion-beam sputtering apparatus

The system, which is now assembled, offers a variety of alternative methods for the incorporation of hydrogen and different dopants, capable of providing accurate control of particles density and energy. A basic scheme of the apparatus is illustrated in Fig. 1. It comprises a stainless steel high-vacuum chamber and two high current ion-guns. The beam from gun 1 after going through a neutralizing filament, aims at an appropriate target support, which has various degrees of freedom so as to allow silicon or other materials to be exposed to bombardment. Transition from one target to the next (e.g. degenerate n-doped silicon to degenerate p-doped silicon) can be achieved in a continous manner. The other gun is directly oriented

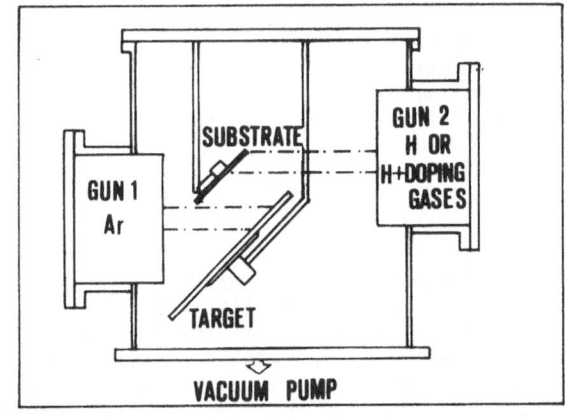

Fig.1 - Scheme of the double ion-beam sputtering apparatus.

toward the substrate, the latter being also rotable and temperature stabilized.

The deposition procedure goes through the following steps. The
chamber is first evacuated to 10^{-7} mbar by means of a high-speed turbomo-
lecular pump. Argon is then admitted into gun 1 at a flow rate of 2.5-3.5
s.c.c./min. so as to raise the pressure locally to about 10^{-4} mbar as
needed to start the discharge. The flow of neutral silicon atoms reaching
the substrate is controlled by the current level and the accelerating voltage of the argon beam, while their energy is restricted to a range of
10-20 eV which insures good adherence to the substrate and hopefully limits the formation of extended microvoids in the amorphous matrix. Gun 2
is used to shoot hydrogen ions into the growing film, with careful and independent control of their flow and energy. Depending on the latter, they
can be implanted in the amorphous matrix and/or react with the growing
surface of the film (1). Uniform film thickness and hydrogenation are
achieved by rotation of the substrate. As the substrate is floating, it
can be polarized for the purpose of studying the effect of an electric
field on growth kinetics. Moreover, the substrate, under a negative bias
of ~25 V, allows separate measurement of the ion current density from either gun 1 (after reflection at the target) or from gun 2; which corresponds to the amount of hydrogen impinging on the sample per second.

Typical values of the parameters used in the growth of a-Si:H are
listed in Table I. It is important to note that the discharge voltage in
gun 2 was maintained at ~100 V to increase the H^+ to H_2^+ ratio in the
beam (2).

TABLE I

	Gun 1(Ar)	Gun 2(H_2)
Beam diameter	2.5 cm	2.5 cm
Beam current	10-27 mA	5-13 mA
Gas flow rate	2-4 s.c.c./min.	35-65 s.c.c./min.
Acceleration potential	200-1000 eV	150-400 eV
Grid-sample distance		~13 cm
Grid-target distance	10-15 cm	
Chamber operating pressure	$\sim10^{-4}$ mbar	
Target temperature	Water cooled	
Sample temperature	220 ^{0}C	
Growth rate	1-3 Å/sec	

2.2 Materials characterization

Assembling and mastering of the sputtering equipment required a
longer time than expected. For instance, it was difficult to obtain, from
a Kaufman source like ours, a stable beam of hydrogen ions with low kinet-
ic energy. Moreover it was necessary to control the spatial distribution
of the ions from the two guns, in order to minimize contamination from the
chamber walls and from other metallic parts. Therefore the production of
the films has started only recently, so that characterization measurements
are at a preliminary stage. X-ray diffraction, IR and near-gap absorption, dark and photoconductivity measurements are presently being carried

out. We have obtained the following indications. The X-ray analysis
shows that the films are completely amorphous in all deposition conditions explored (see table I). Infrared absorption data exhibit a continuous shift of the 2000 cm^{-1} peak (due to SiH) to 2090 cm^{-1} (due to SiH_2) when the accelerating potential of hydrogen ions is increased from 150 eV to 400 eV, without changing the other growth parameters (Fig. 2). The effect is thouroughly reproducible. This confirms that the method allows careful control of the process of hydrogen incorporation. A correlation of this behavior with transport and phototransport properties,

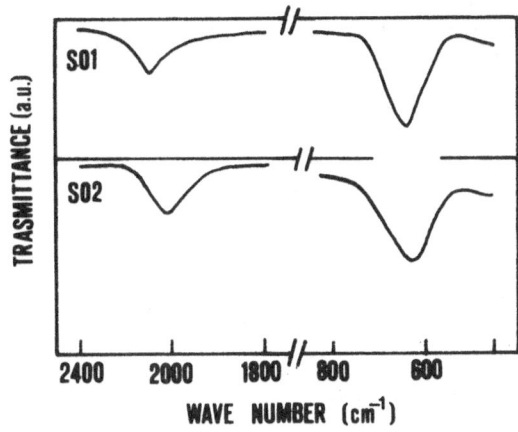

Fig.2 - IR transmission spectra of a-Si:H films produced by double ion-beam sputtering. Sample thicknesses are 0.9 and 1 µm for S01 and S02 sample, respectively.

as well as with optical properties, is in progress. The dark conductivity is high, typically 10^{-3} $\Omega^{-1}cm^{-1}$ and does not show thermally activated behavior at high temperature. The photoconductivity signal is two orders of magnitude lower than that of the best glow-discharge material. These results suggest that a large density of states is present in the pseudo-gap. However, a reduction of the conductivity as well as an increase in the photoconductive response have been obtained upon film annealing at 300 °C for 2 hours. We expect, then, that a systematic investigation of the influence of growth parameters, mainly the temperature, will result in an appreciable improvement of the photoconductive properties of the films.

3. a-Si:H produced by glow-discharge in SiH

3.1 Gap-state spectroscopy

Information on the density of gap-states has been obtained by a technique which combines photoconductivity and transmission measurements, allowing determination of absorption coefficient values as low as 1 cm^{-1}. We have found (3) that the results are critically dependent on the exponent β that relates photoconductivity and generation rate, so that its determination as a function of photon energy and excitation level is a basic prerequisite to this type of experiments.

To explore this technique, we have applied it to a series of glow-discharge samples with medium to low density of gap-states (i.e. optical absorption coefficient due to gap-states below 10^3 cm^{-1}). Typical behavior of the absorption coefficient α is shown in Fig. 3 for samples deposited under different conditions. Two separate regions deserve attention. In the interval 1.5-1.8 eV an Urbach tail is present whose slope is almost constant from sample to sample and independent of hydrogen content. This is attributed to transitions involving tails of intrinsic localized states. Below 1.5 eV the extension of the Urbach tail is limited by the taking over of the low-energy absorption plateau, which varies considerably from sample to sample and is attributed to extrinsic states. Their density Ng can be estimated from the corresponding integrated area of the absorption coefficient. We have found that these states affect considerably the photoconductivity response. The effect is shown in Fig. 4, where we have plotted the product mobility-lifetime $\mu\tau$ (for $h\nu=1.8$ eV)as a function of Ng. A clear inverse correspondence between the two quantities is found. Finally, by comparison of the Urbach-tail slope and the values of the exponent β, we have been able (4) to rule out the current interpretation of recombination kinetics intermediate between 0.5 and 1 found in a-Si:H in terms of the movement of the quasi-Fermi level in an exponential tail of trapping centers. Quite to the contrary, we have shown that a simple model, which takes into account the hole localization, can explain the kinetics between 0.5 and 1.

The above technique and analysis will be soon extended to ion-gun produced samples.

3.2 Schottky-barrier cells

We have made Schottky barrier devices using a Pd/a-Si:H(i)/a-SiH:P/Cr/glass structure, in order to further test the quality of our glow-discharge material for applications.

The preliminary results are satisfactory, We obtain values for I_{sc} (~ 6 mA/cm^2) and V_{oc} (~ 0.4-0.5 V), that are similar to the ones reported in the literature for the above structure. However FF's are low (~ 0.4) and some effort is now

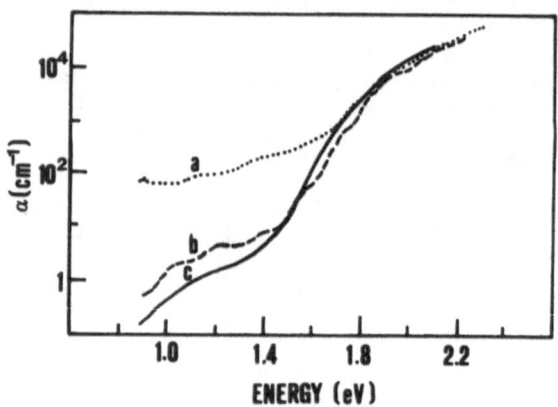

Fig.3 - Absorption coefficient as a function of photon energy for three typical samples.

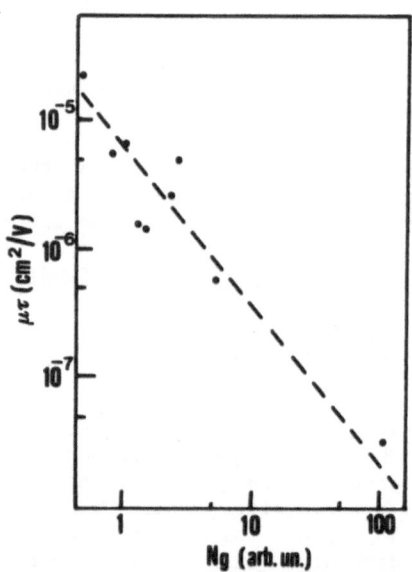

Fig. 4 - Mobility-lifetime product as a function of the number of gap states Ng.

being devoted to minimizing the back contact resistance and investigating the effects of surface treatments.

References

1) C. Weissmantel, Thin Solid Films 32, 11 (1976); C. Weissmantel, Le Vide 183, 107 (1976).
2) P.J. Wilburn and H.R. Kaufman, AIAA Paper n.78, 667 (1978).
3) F. Evangelisti, P. Fiorini, G. Fortunato, A. Frova, C. Giovannella and R. Peruzzi, J. Non-Cryst. Solids (submitted).
4) F. Evangelisti, P. Fiorini, G. Fortunato and C. Giovannella (to be published).

STUDIES TO IMPROVE THE EFFICIENCY OF Cu$_2$S-CdS SPRAY SOLAR CELLS

Authors   : M. SAVELLI, S. MARTINUZZI, J. VEDEL, J. LAHAYE

Contract number : ESC - R - 030 F

Duration   : 12 months   -   1 July 1980 - 30 June 1981

Head of project : Dr. M. SAVELLI, Centre d'Electronique de Montpellier

Contractor   : Université des Sciences et Techniques du Languedoc

Address   : Université des Sciences et Techniques du Languedoc,
                    Place Eugène Bataillon
                    F - 34060 MONTPELLIER cédex

Summary

The research on Cu$_2$S-CdS sprayed solar cells were dispatched between the following different French laboratories:

-L.E.A.A.: Laboratoire d'Electrochimie Analytique et Appliquée, ENSCP (Paris, J. Vedel)

-CRPCSS : Centre de Recherches sur la Physico-Chimie des Surfaces Solides (Mulhouse, J. Lahaye)

-L. P.   : Laboratoire de Photoélectricité (Marseille, S. Martinuzzi)

-C. E. M.: Centre d'Electronique de Montpellier (Montpellier, M. Savelli).

Within a pilot plant framework, 25 Cu$_2$S-CdS cells of back-wall structure were made in each run by the spray technique. The substrate of each cell is 2cm x 2cm x 0.2cm. We first describe the fabrication process of each compound thin film; a review is then given of their chief optical and electrical properties. On unencapsulated cells, for 1cm^2 area and under 100 mW cm^{-2} illumination we obtained a statistical mean efficiency of 6%, the best result being 8%. After that, in the monoglass encapsulation we present the results obtained, then the first stability tests are given.

1. Cu₂S-CdS unencapsulated solar cells

a- Fabrication of back-wall Cu₂S-CdS "sprayed" cells.

The method used to fabricate the ITO and CdS thin film was the chemical spray deposition process. We could obtain a homogeneous deposit on a 100 cm² surface (25 cells on a 2cm x 2cm Pyrex substrate). The CdS was deposited by the spray technique onto an ITO conductive transparent thin film, the Cu_2S was obtained by dipping the CdS in a hot solution of CuCl. To prevent the diffusion of cuprous ions along the grain boundaries toward the ITO during dipping, we used a $CdS(Al_2O_3)$ film between the pure CdS and ITO. The stoichiometry of the Cu_xS is of great importance to cell efficiency. The best efficiency is obtained with perfect stoichiometry. Therefore copper refilling was needed to optimize the photovoltaic parameters, and this was carried out by heating the cell which had been previously covered with a 500 Å evaporated copper thin film. The heat treatments for the copper refilling process and the junction formation were performed in the same step. Generally, annealing in air at a temperature of about 150°C for 2 h is used. The cross-section of the cell and the thicknesses of the differents layers are shown in Fig.1.

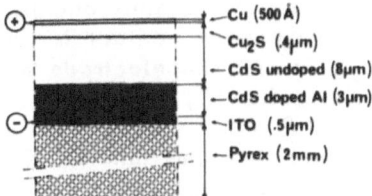

Fig. 1. Cu₂S/CdS cross section.

The physical properties of the different thin films, the standard cell is setted with, are as following:

-ITO : The transparency of the transparent electrode in the visible region is up to 90% when the sheet resistance had been optimized to 7 ohms.

-CdS-Al₂O₃ is a mixed film which had a high resistivity of about 10^3 - $10^4 \, \Omega$ cm.

-CdS : X-ray studies showed the columnar structure of the film with the c axis perpendicular to the substrate. The structure is hexagonal with grain dimensions around 2 microns. Typical results obtained from electrical measurements are 20 ohms cm, 3 cm²/V.S, 10^{17} cm⁻³ respectively to the resistivity, mobility and carrier density.

-Cu₂S : By dipping we obtained a Cu_xS film with x greater than 1.99 for a layer thickness of 0.3 micron. The resistivity of the layer is of about 3 10^{-3} ohm.cm, and the diffusion lenght estimed around 0.4 micron. We experimentally determined the absorption coefficient α (cm⁻¹) of Cu_xS for wavelengths between 0.5 and 1.2 um : α is in the range of 10^5 to 3 10^4 cm⁻¹.

b- Photovoltaic performance

With the fabrication process described and by optimization of the junction formation, the highest efficiency obtained was 8% on an 1 cm^2 area unencapsulated cell under a power illumination of 100 mW. Figure 2 shows the I-V characteristics in darkness and under air masse one (AM 1) illumination. The statistical photovoltaic performances obtained on a batch of cells are plotted in figure 3. The mean values of short circuit current, open-circuit voltage and efficiency are 19 mA, 443 mV and 5.2% respectively. Figure 4 presents typical spectral response of our cells, we observe the two front edges at 0.52 and 1 µm owing to the gap of the two CdS and Cu$_2$S materials; the maximum is situated at about 0.55 µm because of the window effect of CdS.

A mercury ohmic back contact was used on the Cu$_2$S and enabled us to control the evolution of the parameters at each step of the optimization process.

In conclusion, as our results obtained on unencapsulated cells was promising, a transfer of technology is beginning between our laboratory and Saint Gobain Recherche.

2. Cu$_2$S-CdS encapsulated solar cells

In order to realize the solar cells encapsulation we have chosen monoglass encapsulation process, the biglass one having to be performed by Saint Gobain Recherche; the back contact is taken by setting a conductive epoxy on the Cu$_2$S layer. The electrode and encapsulant were selected taking into account following qualities:
- small resistivity component $\rho \sim 10^3 - 10^4 \,\Omega$ cm
- drying at temperatures lower than the junction formation temperature
- unchemical resistivity with the Cu$_2$S layer
- easy using and low cost

After several tests with different component the HD 31 silver conductive epoxy was selected. From figures 3 and 5 we can observe that the photovoltaic results obtained with the epoxy contact are similar to the mercury contact one's. Otherwise these performances do not decrease in a few days. With this contact the first reliability tests were setted.

3. Stability

These tests were performed on standard cells batches. Two different tests were carried out:
- Darkness test: the cell is maintained at a temperature between 60°C and 90°C.
- Illumination test: the cell is illuminated (100 mW/cm^2) under open circuit conditions.

a- Darkness results:

The figure 6 shows mean values variations of the photovoltaic parameters as a function of the test duration:
- During the first six hours, a sharp decrease of the efficiency is noted, the short circuit current and the open circuit voltage staying unchanged.
- Then, the open circuit voltage slowly increases, the short circuit current and fill factor decrease as shown Table I.

d(days)	V_{oc}(mV)	I_{sc}(mA)	η %
0	448	18	4.8
75	461	14.5	2.8
120	474	12.1	2.1

- We think that humidity reaching the junction induces the decrease of I_{sc}: biglass encapsulation will be prefered in future works.
- The decrease of the efficiency is connected, in addition to the I_{sc} lowering, to the fill factor reduction. This last one is due, on the one hand, to a contact degradation (increasing the serie resistance) and on the other hand to the interface transport mecanisms evolution.

b- Illumination results:

During illumination test, the cells present a similar behaviour (figure 7); yet on several cells a great decrease of the shunt resistance is noted. This failure may be imputed to macroscopic impurities being introduced during the fabrication process (dust ....)

4. Conclusion

We now have a good control over the pilot plant cells fabrication. Homogeneous cells having a mean reproductible efficiency of 5% are fabricated.

The stability study of the monoglass encapsulated cells has displayed some degradation process; the first goal is the improvement of the back contact-encapsulation to prevent air humidity intake: this will be done by the use of biglass encapsulation.

Though it would not exactely be measured, a copper diffusion may occur during the tests; so we have started to study the introduction, in the CdS layer, of impurities such as copper that provided the evaporated cells stability.

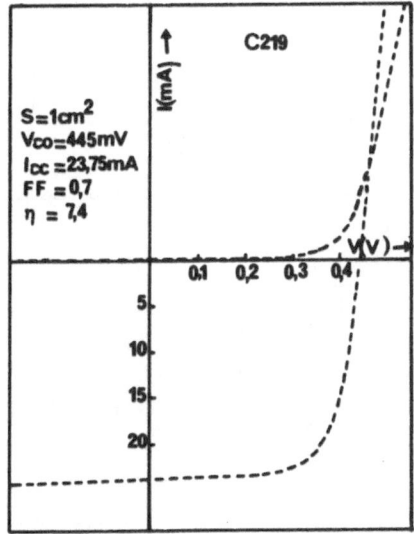

Fig. 2

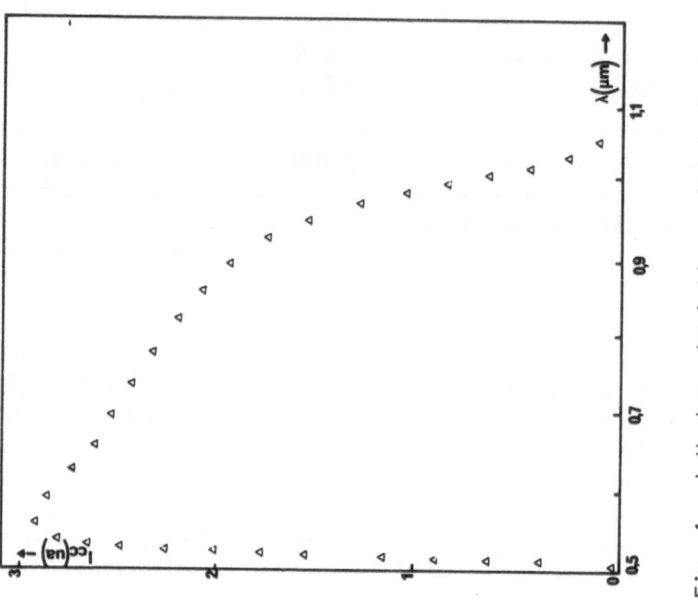

Fig. 4 – I-V characteristics – Spectral response

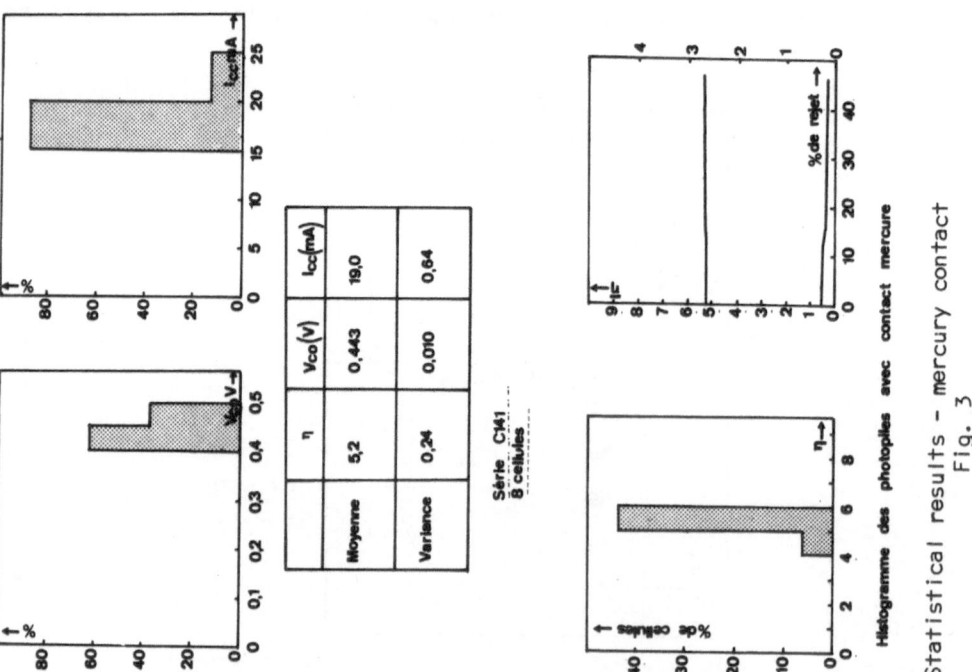

Statistical results – mercury contact
Fig. 3

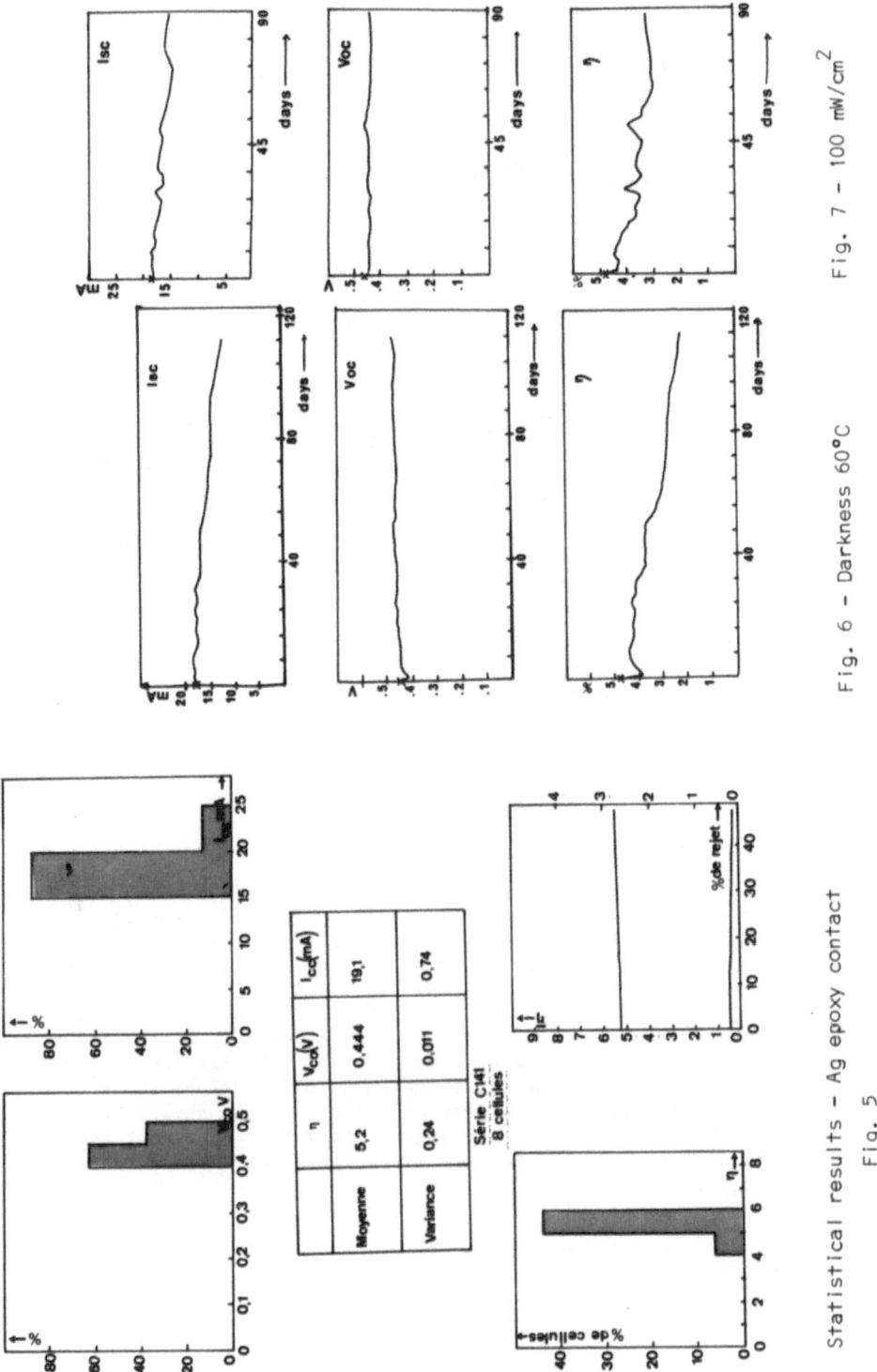

Fig. 7 – 100 mW/cm²

Fig. 6 – Darkness 60°C

Statistical results – Ag epoxy contact

Fig. 5

ELECTROCHEMICAL PREPARATION OF CUPROUS SULFIDE

Authors :   : J. VEDEL, P. COWACHE, D. LINCOT

Contract number   : ESC-R-031-F

Duration   : 24 months 1 July 1980 - 30 June 1982

Head of project   : Dr. J. Vedel, Centre National de la Recherche
                         Scientifique

Contractor   : Ecole Nationale Supérieure de Chimie de Paris

Address   : Laboratoire d'Electrochimie Analytique et
                         Appliquée
                         Ecole Nationale Supérieure de Chimie
                         11, rue Pierre et Marie Curie
                         75231 PARIS Cedex 05

Summary

The objective of this work is to show the possibility of forming the
cuprous sulfide of CdS-Cu_2S solar cells by cathodic reduction.
Cuprous sulfide of stoichiometric ratio greater than 1,99 can be
prepared in this way owing to the use of an electrolyte containing
a convenient masking agent (TRIEN). The reaction time (10 to 20 mn)
allows an easy control of the fabrication. A co-deposit of copper
is observed. The collection grid is also electrodeposited, following
an original process. Cells of 2 cm^2, having 7 % efficiency were
realized. The analysis of the junction by C^{-2} -V plots shows that
there is no significative migration of the copper during the electrodeposition. The compensated region after heat treatment seems to
be less disturbed by the texture effect than for cells obtained by
ion exchange.

1. Introduction

The purpose of the present contract is to establish an alternative way to the classical method of making CdS-Cu$_2$S solar cells with vacuum evaporated CdS layers. This new way is based on the use of electrochemical techniques in which the ion-exchange (dipping) reaction :

$$CdS + 2\ CuCl \rightarrow Cu_2S + CdCl_2 \qquad (1)$$

is replaced by

$$CdS + 2\ Cu^{2+} + 2e \rightarrow Cu_2S + Cd^{2+} \qquad (2)$$

and the post treatment (usually vacuum deposition of copper or reduction in a plasma of hydrogen) by :

$$Cu^{2+} + 2e \rightarrow Cu \qquad (3)$$

Both cathodic reactions (2) and (3) are controlled by the electrode (i.e. CdS) potential and allow reaction times high enough for an easy process. In addition, we have added an electrochemical technique for the current collection, that is the electrodeposition of a collecting grid.

At the previous coordination meeting, (Sorrento, 1981), it was demonstrated that good cuprous sulfide can be formed on large areas, provided one uses a convenient electroplating solution containing a complexing agent (TRIEN) ; co-deposition of copper was evidenced and cells of 4.8 % efficiency were obtained. This paper is devoted to the continuation of the work (grid electroplating, co-deposition of copper, physical caracterization of the cells) which led us to cells having efficiencies up to 7 %.

2. Experimental

The CdS plates were given by the Institut für Physikalische Electronik (Stuttgart). After cutting in smaller plates of $\sim$ 4 cm^2, a contact was taken on the silver electrode by soldering. The substrates were cleaned and etched using either the "cyanide procedure" or the "acidic procedure". In the first case, a deposit of Cu$_2$S (thickness 0.6 μm) is dissolved in a cyanide solution. In the second case, the layer is immersed for 10 s in a diluted HCl solution (1/3) at 60°C. The texturization of the surface increases from the lack of etching ("NE") to the cyanide etching ("CE") to the acidic etching ("AE").

The composition of the electroplating solution is :
- Sodium acetate   0.1 $\underline{M}$
- Copper (II) perchlorate 2.10^{-3}M
- Triethylenetetramine (TRIEN) 7.10^{-2}M

The solution is purified by anodic oxidation on Pt electrode at - 0.1 V vs Ag/AgCl. Temperature : 80°C.

The electrochemical apparatus is essentially a three electrode potentiostatic system, allowing the integration of the current and its use in the galvanostatic mode as necessary for the analysis operations. Analysis, which is a solid state coulometry (SSC) gives the mean thickness of cuprous sulfide, its deviation from stoichiometry as well as the amount of cupric oxide, if present [1].

I V characteristics were obtained using a halogen lamp, standardized by means of a silicium photocell (SAT). Spectral responses under monochromatic light were obtained using a JOBIN-YVON H 20 monochromator (0.1 to 3.2 μm). The source is a 70 W iodine lamp. The spectrum was obtained with a KEITHLEY microammeter. The C V characteristics were obtained using a potentiostatic system (PAR 173 potentiostat + GSTP 3 Tacussel ramp generator with a WAVETEK model 188 function generator allowing the addition of an alternating signal to the previous DC signal. An ORTHOLOC SC 9505 lock in amplifier led to the quadratic component of the signal. The doping level of CdS was determined by studying the capacity of the CdS/NaOH 1M interface in the Mott-Schottky range.

3. Co-deposition of copper

It was previously shown that the best working temperature was 80°C. For lower values, the reaction time is too long. For higher ones, the co-deposition of copper is too fast. The deposition potential was chosen equal to - 0.2 V, for which the best stoechiometric ratio was obtained.

In these conditions, the co-deposition of copper was demonstrated by comparing the quantity of copper present as copper sulfide to the total quantity of copper. The first one was given by the SSC curves, the last one by chemical analysis. The mean values of several determinations are given in Table I, for an apparent thickness of 0.6 μm.

Nature of the etching	Stoichiometric ratio	$Cu(Cu_2S)/\mu g$	$Cu(total)/\mu g$
acidic	1.980	32 ± 2	44 ± 4
cyanide	1.985	33 ± 2	40 ± 4
no etching	1.993	29 ± 2	40 ± 4

Table I. Comparison of the total quantity of copper to the quantity present as cuprous sulfide.

The behaviour of the co-deposited copper was studied by SSC. On figure 1 are represented several titration curves of a standard Cu_2S, on non etched CdS (0.6 μm, x = 1.995). Curve 1 corresponds to the titration of Cu_xS just after formation. The observed wave is due to the non-stoichiometry of the copper sulfide (NB : the reference potential $-E = 0-$ corresponds to that of the plateau of the $Cu_2S \rightarrow Cu$ reduction). The lack of wave at +0.85 V indicates the absence of cupric oxide. Curves 2 and 3 are obtained after thermal treatment in air, at 180°C, for 10 and 60 mn, respectively. Curve 2 shows the compensation of Cu_2S following

$$Cu_xS + (2-x) Cu \rightarrow Cu_2S \qquad (1)$$

and the formation of a small amount of Cu_2O, characterized on the curve by the preceding overshoot. For longer time (curve 3), a wave corresponding to the formation of CuO begins to appear as well as increase of Cu_2O. For comparison, curves 4 and 5 were obtained with Cu_xS covered by a vacuum deposited layer of copper of 130 A°. Thermal treatment were also of 10 and 60 mn, resp. There is a protection against oxidation by air but the compensation seems to be not so efficient as previously, probably because a slower compensation reaction takes place at the interface Cu/Cu_2S.

In conclusion, the co-deposition of copper leads to a well dispersed metallic phase, allowing a fast compensation reaction and giving a good protection against the oxidation by air during the thermal treatment.

4. Current collection

The fabrication of photocells having good efficiency was necessary to demonstrate the feasability of the electrolytic way for Cu_xS preparation. The improving of current collection was done by developing an electrodeposition technique of the collection grid. The main feature of the technique is that gold is electrodeposited on cuprous sulfide formed upon non-etched CdS. This diminishes the tendancy to make short-circuits. For that, the photoresist is settled down the non etched CdS surface, then the grid is exposed and developed and cuprous sulfide is electroplated in the grid design. On the plain surface, a bright layer of gold can then be deposited, using a cyanide - free bath [2]. After removal of the remaining photoresist, the CdS is etched and the photovoltaic Cu_2S is electroplated.(Fig. 2)

5. Réalisation and characterization of photocells

Photocells were realized using the following conditions : area 2 cm^2 ; electrodeposited grid ; deposition potential of PV Cu$_2$S : - 0.2 V (vs Cu) ; mean thickness : 0.6 μm ; thermal treatment : 10 mn, 180°C, air.

Immediately after formation, the cells exhibit high photocurrent (20 mAcm^{-2} at AM 1), but a low shunt resistance limits the cell properties. The thermal treatment improves them. Dark and light curves intersect due to the formation of a compensated layer [3]. This effect also appears on the I - λ curves (fig. 3). Before treatment (curve A), a good quantum efficiency is observed (0.7 between 0.7 and 1 μm). After treatment, a decrease in the quantum efficiency is observed (curve B), but the spectral response depends on the illumination state before determination (figure 4), as already observed [3].

On figure 5 are plotted the C^{-2}/V variations for non etched devices, as well as the same plot corresponding to the junction non etched CdS/NaOH. The slope corresponding to the liquid diode (curve A) is about equal to that corresponding to the Cu$_2$S/CdS diode just after formation. This indicates that there is non diffusion of copper in CdS during the reaction, and no preferential penetration of Cu$^+$ along the grain boundaries (which would cause an increase of the apparent area). The barrier height ($\sim$ 1.1eV) shows the good quality of cuprous sulfide. After thermal treatment, the C^{-2} vs V plot shows two ranges of variation. For V < 0.5 volt, the space charge regions lies in the non-compensated CdS. For V > 0.5 volt, the linearity of the plots supports the uniform doping of the compensated layer. (N$_D'$ = 10^{15} cm^{-3}) as well as the lowering of the barrier height.

For etched substrates, the observed phenomena is qualitatively the same. The junction area is increased. For instance, at zero bias, the capacity is $\sim$ 83 nF cm^{-2} instead of 13 nF cm^{-2} for non etched ones. This gives an area factor equal to 6.4 . A difference between electroplated and dipped Cu$_2$S can be observed on figure 6. For electroplated Cu$_2$S, the plot remains practically constant when V < 0.5 V, contrary to the pronounced variation observed with the dipped Cu$_2$S [4]. This could be due to a more regular growth of cuprous sulfide.

6. Conclusion

This work shows that the electrochemical way is an interesting alternative for the fabrication of CdS/Cu$_2$S solar cells. However they are some remaining questions, requiring subsequent work. These questions are :
- the stability of the grid against the etching agents,
- the copper deposit on the grid, during the electroplating of Cu$_2$S,
- the post-deposition of copper.

This needs a special study of the etch process and the investigation of other deposition conditions, for instance by using deposition at constant current.

7. References

1. J.Vedel and M.Soubeyrand, J. Electrochem. Soc., 127, 1730, (1980).
2. F. Mornheim, Plating, (1961), 1104.
3. W.D. Gill and R.H. Bube, J. Appl. Phys., 41, 3731, (1970).
4. F. Pfisterer, Workshop on the II-VI solar cells and similar compounds, Montpellier (1979).

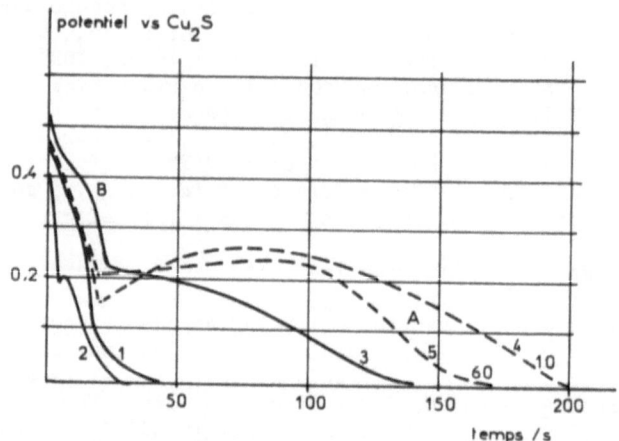

Fig. 1 - Titration curves of electrodeposited Cu_2S.
1. Just after formation; 2 and 3, after thermal treatment, in air, at 180°C. (2) : 10 mn; (3) : 60 mn; 4 and 5 Cu_2S covered with copper by vacuum deposition. Reference potential : $\frac{}{2}S/Cu$ plateau.

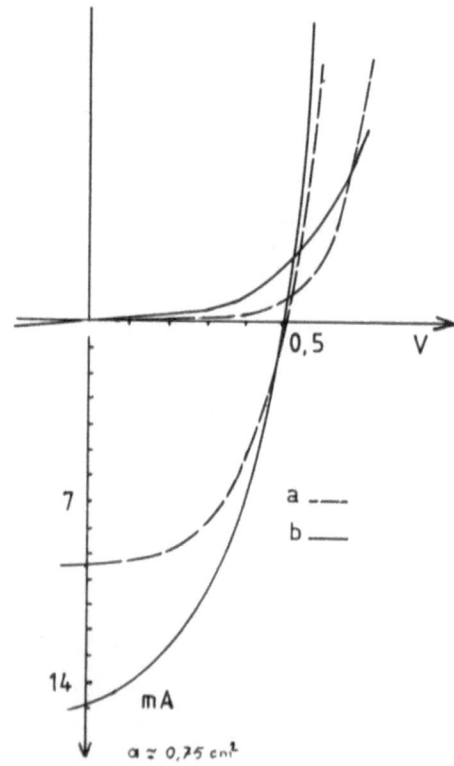

Fig. 2 - Comparison of IV characteristics obtained with pressed contacts (a) and electrodeposited contact (b).

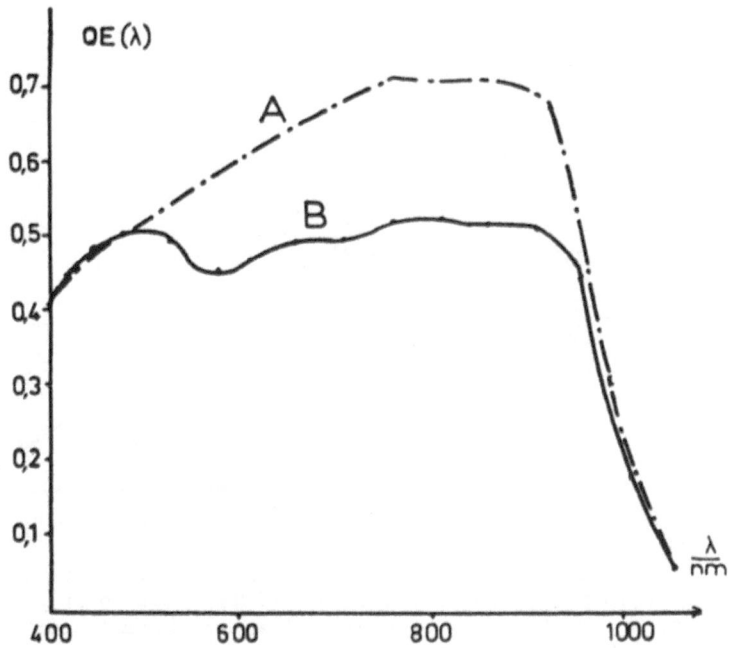

Fig. 3 - Influence of thermal treatment on spectral responses.
A) after electroplating - B) after annealing.

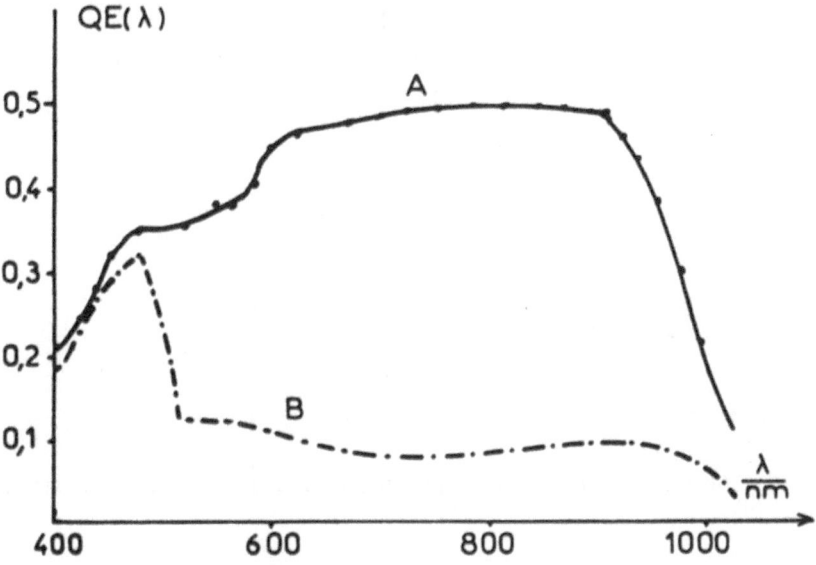

Fig. 4 - Spectral response just after illumination at 0,5 μm(A) and after a long stay in the dark (B).

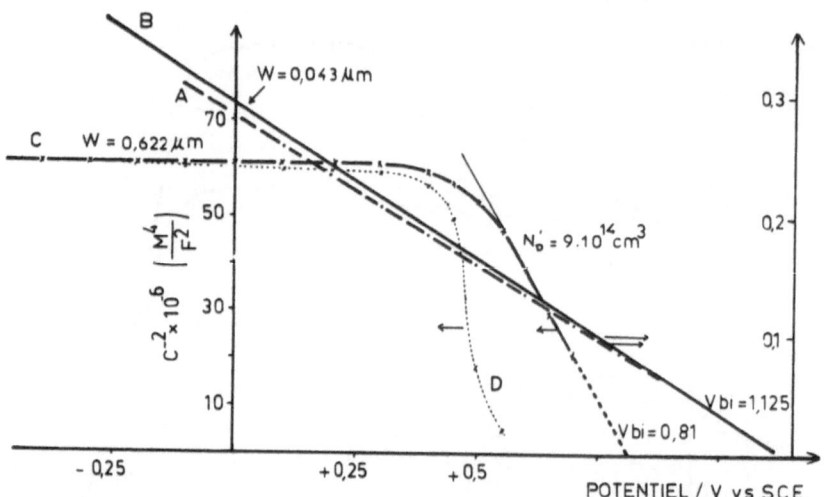

Fig. 5 - C^{-2}vsV plots (non etched CdS). A : MottSchottky plot of non-plated CdS in a 1M NaOH solution; B : just after electroplating; C : after thermal treatment; D : etched CdS after thermal treatment (an area factor $\approx$ 6 is taken into account).

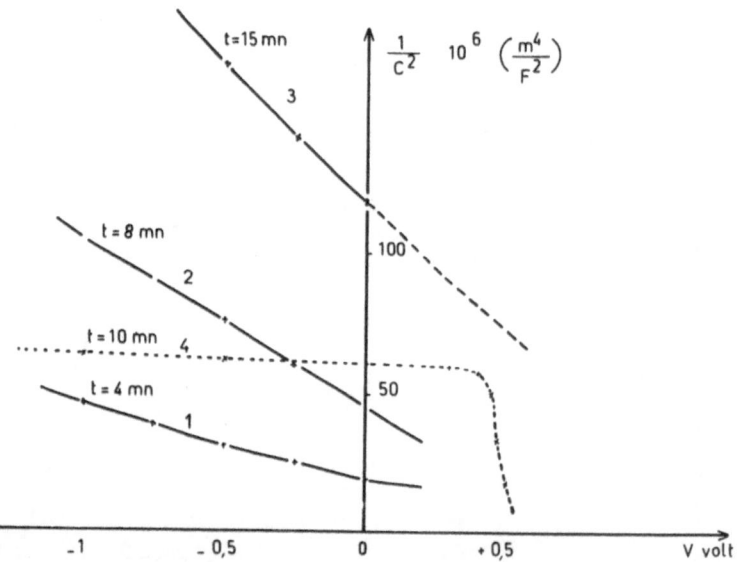

Fig. 6 - Comparison between dipped Cu_2S (1-2-3) and electroplated Cu_2S. (4).

ELECTROPHORETICALLY DEPOSITED THIN FILMS FOR

LOW COST SOLAR CELLS

Authors   : T.J. CUMBERBATCH, I.D. McINALLY

Contract number   : ESC-R-032-UK

Duration   : 36 months   1 July 1980 - 30 June 1983

Budget   : Total: £650,000   CEC Contribution: £165,000

Other sponsors   : UK DoI £160,000 (2 October 1981 - 1 October 1983)

Head of project   : Dr. L. Holt

Contractor   : THORN EMI plc, Central Research Laboratories

Address   : Trevor Road
                      GB - HAYES, Middlesex UB3 1HH

Subcontractors   : Newcastle Polytechnic
                      Southampton University

Summary

The aim of this work is to employ electrophoresis for the deposition
of thin semiconductor films suitable for photovoltaic cells. Layers
of CdS, CdZnS and Cu_xS have been produced with heterojunctions successfully fabricated.

Progress of the work has been hampered by two major problems. In
practice very thin layers prepared by electrophoresis are pinhole free;
however, in the course of this work it has been found that small
projections on the surface of the deposition electrode lead to microcracks on film drying - polishing the substrates has been shown to
overcome this. Secondly, the recrystallization and phase transformation of a thin powder film is a non-trivial problem and one that was
seriously underestimated at the initiation of this contract. However,
after investigations involving a variety of techniques it has been
found that pulsed and CW laser radiation offer the means to achieve
this.

Heterojunctions have been prepared in several ways, the best results
so far being provided by the dry process. In addition a number of
novel approaches using laser heating have been investigated and shown
to be feasible.

The results obtained to date provide a basis for the production of
thin film solar cells at a very low cost with very efficient material
usage employing simple technology.

1. Electrophoretic Deposition and Substrates

Electrophoretic deposition involves the movement of colloidal particles of CdS (or ZnCdS or Cu_xS) by the application of an electric field and their subsequent deposition onto a suitable conductive substrate acting as an electrode (1)(2).

The hydrophobic colloidal suspensions are prepared by using hydrogen sulphide to precipitate CdS from aqueous solutions of cadmium salts (or mixtures of zinc and cadmium salts to produce ZnCdS). Each colloidal particle typically 20 - 40 nm in diameter has a surrounding sheath of S^{2-} ions which impart an overall -ve charge to the particle. Deposition therefore takes place at the +ve electrode. Typically the substrate used as the +ve electrode is either stainless steel or tin oxide coated glass.

Although electrophoretic deposition produces pinhole free films, imperfections on the surface of the substrate (fig. 1a), particularly with tin oxide, can result in small cracks in the CdS films (fig. 1b) as the deposited film drys. This particular problem, with the consequent problem of producing short circuited junctions, can be overcome if smooth substrates are used e.g. polished stainless steel. However cracking of the film on all substrates remains a problem for thicknesses above 2µm.

2. Recrystallization

Recrystallization is required to transform the as deposited CdS layer, whose powder nature is shown in fig. 2a and whose crystalline structure is indicated in fig. 5a, into a polycrystalline thin film with hexagonal structure and c-axis perpendicular to the substrate.

It has been discovered that for thermal annealing a temperature >∿520C is required for phase conversion (fig. 5b) and that little grain growth occurs below ∿700C (fig. 2b). Unfortunately at these temperatures instability of the substrates occurs, with glass softening and CdS reacting with stainless steel above 600C. To overcome these problems transient heating (using a graphite filament) has been investigated as a method of reducing the substrate temperature with encouraging results. Also, standard fluxes have been studied in attempts to lower the recrystallisation/phase transition temperatures. Promising results have been obtained using cadmium chloride but in the final analysis it seems certain that thermal heating alone cannot produce the desired results.

More encouraging behaviour has been observed using both pulsed and CW laser radiation. However, the interaction of a laser beam with a powder layer is complex making it difficult to predict the required power and energy levels. Results so far have yielded a phenomenological understanding allowing the definition of an experimental 'window'. For pulsed treatment, a degree of thermal prebaking (400C ∿30 mins or transient) is required to improve the adhesion of the particles both to each other and the back contact so preventing ablation. The reaction of a prebaked CdS film to a 300 ns pulse is shown in fig. 2c and that of an as deposited film to a 3µs pulse in fig. 2d. Typical RHEED patterns are shown in figs. 5c and d where conversion to the hexagonal phase is evident as is the 'spottiness' associated with the melted surface produced by a short pulse.

Very recent experiments with a scanned CW Ar ion laser have revealed that no prebaking is required and that prospects for columnar grain growth are good (figs. 3a and b) as is the degree of phase change (3W produces a RHEED pattern similar to fig. 4d). For high powers/long dwell times the surface topography is seriously affected (fig. 4c) as for the case of high energy pulses.

At this stage, cathodoluminescence is being used to monitor the electrical activity of the CdS during laser recrystallization[3]. Increases in the

Fig. 1   Microcrack Generation

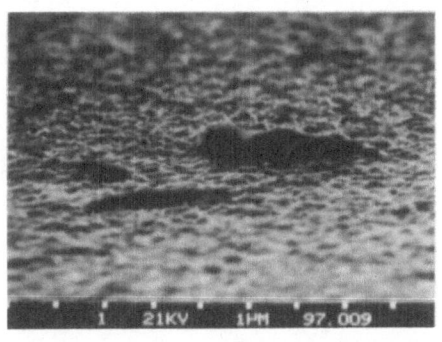

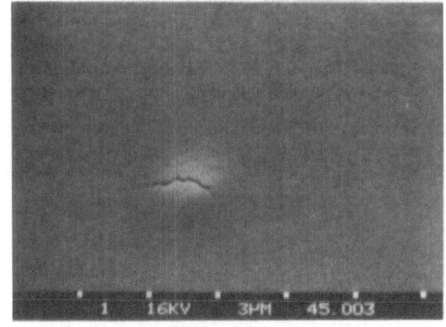

a)   Feature in SnO$_x$ coating

b)   Microcrack in CdS layer

Fig. 2   CdS Layer After Thermal and Pulsed Laser Heating

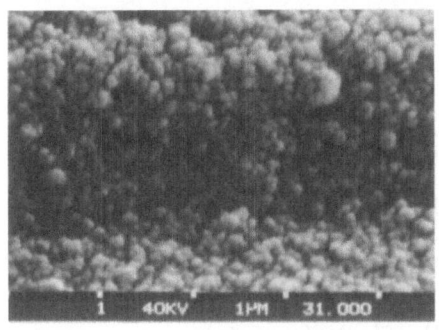

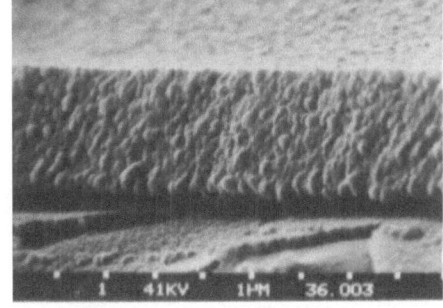

a)   As deposited

b)   After thermal heating: 700C
     10 mins N$_2$

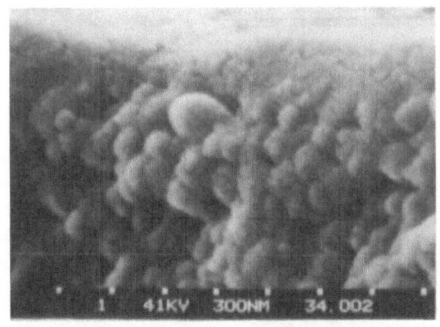

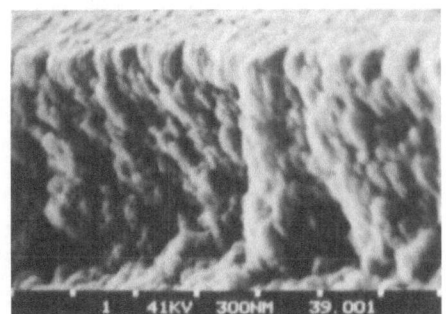

c)   After 300 ns Pulse
     (650C 10 min prebake)

d)   After 3μs Pulse

Laser Recrystallization (λ = 505 nm, E $\sim$200 mJcm^{-2})

Fig. 3   CdS Layer After Scanned CW Laser Heating

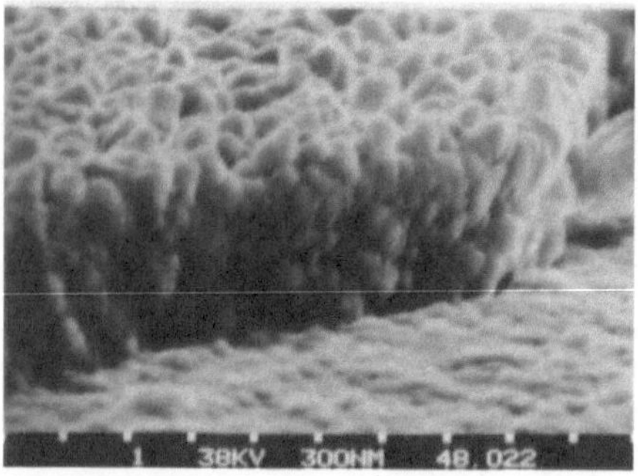

a)   2W

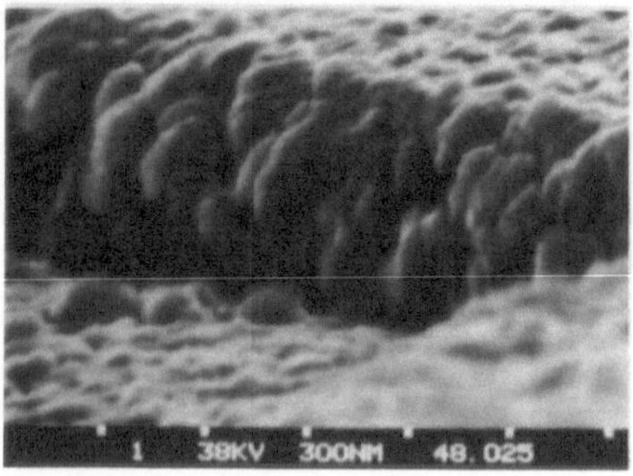

b)   3W

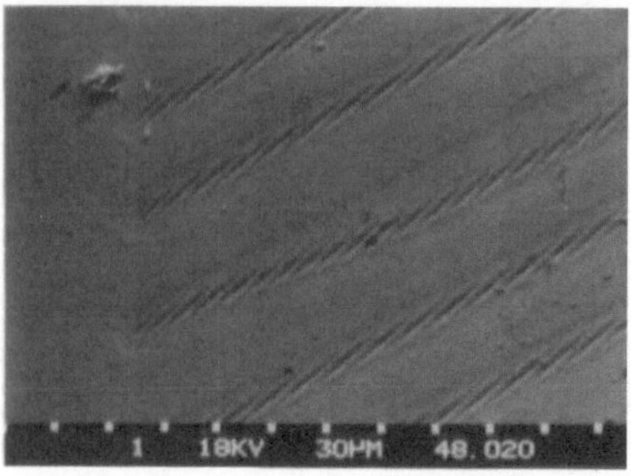

c)   4W - Emissive

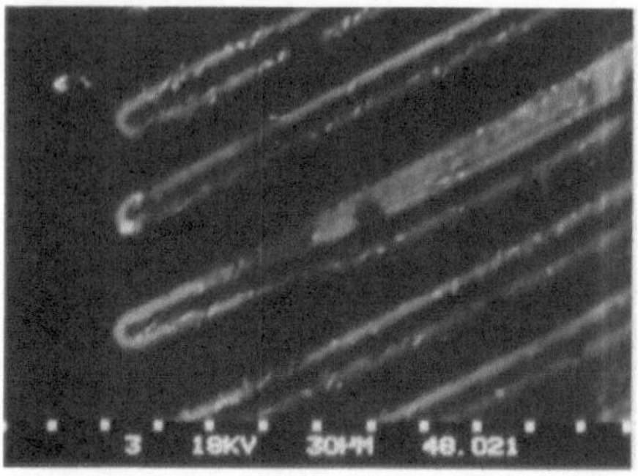

d)   CL Image of (c)

Fig. 4   CdS Layer After 3µs Laser Pulse

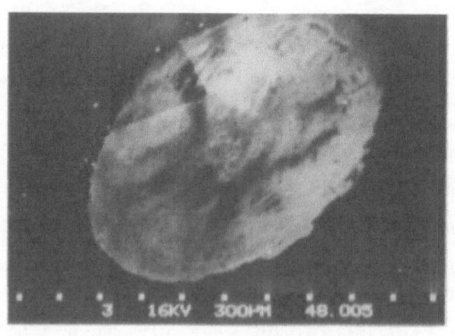

CL Image of Exposed Region λ_{pk} ~550 nm

Fig. 5   Crystal Structure
(Arrows Indicate Conversion to Hexagonal Structure)

a)   As Deposited CdS (cubic)

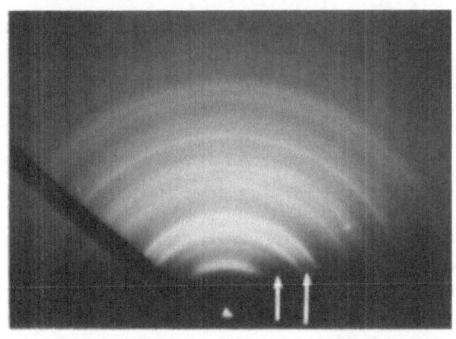

b)   After Thermal Heating: Evaporate
      $CdCl_2$ 350C 90 mins 530C 10 mins

c)   After 300 ns Laser Pulse
      (λ = 505 nm, E = 200 $mJcm^{-2}$)

d)   After 3μs Laser Pulse
      (λ = 505nm, E = 220 $mJcm^{-2}$)

Fig. 6   Laser Produced Heterojunction

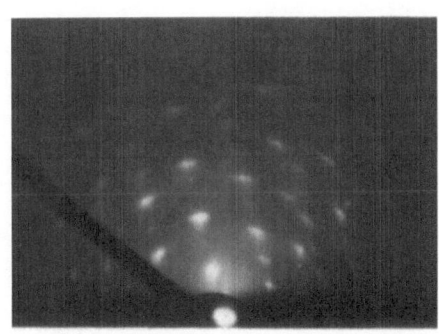

a)   Evaporated CdS
      (IPE Stuttgart)

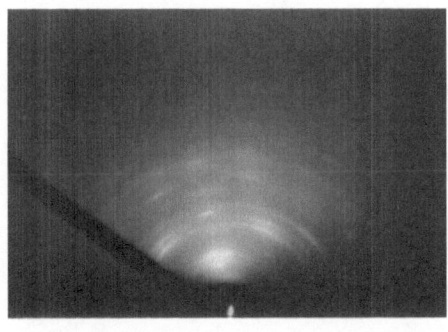

b)   Evaporated CdS after formation of
      $Cu_{1.96}S$ layer using CW laser

luminescence intensity correlate with grain growth and phase transformation (figs. 4d and 5); the variation in intensity for the former attributed to the beam shape and for the latter both the collection efficiency of the CL optics and the state of the CdS. More comprehensive electrical measurements are in progress.

In conclusion, the use of scanned or pulsed optical energy is felt to be the solution to the recrystallization problem; the possible advantages of a $CdCl_2$ flux have yet to be determined.

3. Heterojunctions

The dry process has produced the best junctions to date on CdS layers onto which $100\text{\AA}$ of $CdCl_2$ has been evaporated followed by thermal anneals of 350C for 90 mins and 530C for 10 mins. This is followed by the evaporation of $250\text{\AA}$ of CuCl and a bake at 150 - 180C for 1 - 2 mins producing heterojunctions whose characteristics are shown in figs. 7a and b. The exact role of the flux is not currently understood but its use appears to seal the microcracks. From the spectral response it is immediately apparent that the phase of the Cu_xS is not chalcocite; this and optical absorption measurements suggest a mixture of diginite and djurelite. However, the CdS film is still essentially a powder layer giving rise to a complicated junction geometry and precluding the fabrication of thicker Cu_xS layers.

Clearly, the advantage of using electrophoretic deposition of CdS is lost if it is necessary to use vacuum processes for junction formation. Also, our layers must be considered too thin for wet dip junction formation. Therefore, several novel alternative approaches have been attempted.

Electrophoretic deposition of Cu_xS layers has been achieved by a similar method to that used for producing CdS layers. Analysis indicates that the phase is a mixture djurelite ($Cu_{1.96}S$) and diginite ($Cu_{1.8}S$). Future work will include attempts to form heterojunctions from layers of electrophoreted Cu_xS deposited onto layers of electrophoreted CdS by using a laser.

Lasers have also been used as local heat sources to initiate topotaxial exchange reactions between copper ions and CdS layers. A number of sources of copper ions have been used, most importantly,

1) Cu^{2+} ions electroplated from solution.
2) Solutions of copper compounds sprayed onto the surface of the CdS layer and allowed to dry (notably $CdCl_2$ in methanol and $CuSO_4$ in water).
3) CdS layer hit by the laser beam while suspended under the surface of a solution of a copper compound.

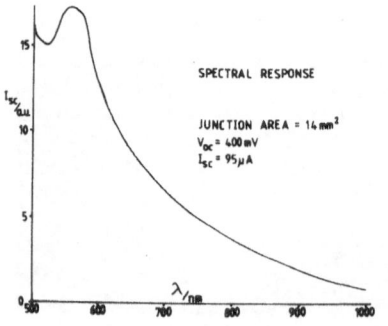

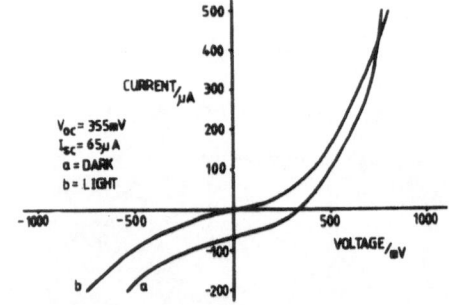

Fig. 7a Spectral Response   Fig. 7b Current Voltage

Each technique has proved successful both with CW and pulsed lasers. The result of a recent experiment with evaporated CdS is illustrated in figs. 6a and b with a probable djurelite layer well defined.

4. Future Work

The future programme will concentrate on the two major problems that of microcracks and the recrystallization of the CdS layers with the following topics covered.

Microcracks:- Optimise substrate polishing; determine the role of $CdCl_2$.

Electrophoretic Deposition:- Optimise deposition of CdZnS; investigate pulsed deposition.

Recrystallization:- Determine the role of $CdCl_2$; continue thermal heating experiments; optimise the wavelength, prebake time and ascertain the total energy required for pulsed laser recrystallization; optimise dwell time/power level for CW laser recrystallization.

Alternative heterojunctions:- Optimise electrophoretic deposition of Cu_xS; investigate other non-vacuum exchange reaction systems; investigate Cu_2O formation on resultant Cu_xS surfaces.

Theory:- Model Cu_xS-CdS solar cell without heterojunction interface states; explain behaviour and degradation solely in terms of Çu distribution i.e. mixed phases in Cu_xS layer, electronic activity of Cu in CdS layer.

5. References

1. E.W. Williams et al, Solar Cells 1 357-366 1971/80.
2. T.J. Cumberbatch et al, 732, Fourth E.C. Photovoltaic Solar Energy Conference, Stresa, Italy, 10-14 May 1982.
3. T.J. Cumberbatch et al, 551, Fourth E.C. Photovoltaic Solar Energy Conference, Stresa, Italy, 10-14 May 1982.

DEVELOPMENT OF A CdSe THIN FILM SOLAR CELL

Author   : E. RICKUS

Contract number   : ESC-043-D (B)

Duration   : 30 months 1 January 1980 - 30 June 1982

Head of project   : Dr. E. Rickus

Contractor   : Battelle-Institut e.V.

Address   : Battelle-Institut e.V.
                      Am Roemerhof 35
                      D-6000 Frankfurt 90

Summary

The application of the inherently simple MIS-technology to polycrystalline CdSe films leads to thin film solar cells which can be fabricated solely by standard vacuum evaporation processes.

Different investigations show that the "I-layer" plays a dominant role with respect to the photovoltaic properties of the cells. Specific steps to influence the electronic properties of this layer resulted in AM 1 efficiencies up to 7 percent for cells of 0.6 cm^2 active area.

Stability tests have established that the structures are very insensitive to humidity. Storage on the shelf for more than one year and even immersion into water does not seriously affect the unsealed cells.

Further optimisation can enhance the efficiency of this cost-effective, stable solar cell to about 10 percent, which is the final goal of this work.

1. Introduction

Besides the efficiency of solar cells, their cost will determine the future extent of their terrestrial utilisation for electricity generation. The PV-module cost is mainly determined by material consumption, production cost and by the cost of encapsulation. The application of the inherently simple MIS technology to thin, polycrystalline semiconductor films fulfils all major conditions for the production of low-cost devices. Simple and cheap encapsulations make high demands on the stability of the cells.

The present state of work indicates that the CdSe thin film solar cell can meet all these requirements. Its overall material consumption of about 10 g/m^2 (corresponding to about 10 DM/m^2) shows that the material cost is negligible compared to the production cost. Standard vacuum deposition techniques generally are considered to be sufficiently cost effective for the production of solar cells. Beyond this, alternative deposition techniques such as spray pyrolysis are applicable to CdSe.

2. Preparation of the Cells

The CdSe cells (Fig. 1) presently are produced solely by standard vacuum evaporation techniques. The deposition conditions of the CdSe film, the starting material and the substrate material influence the crystallographic and photovoltaic properties of the solar cells /1,2/.

A thin chromium layer on standard glass substrates serves as back contact. At substrate temperatures around 400°C the CdSe layer, which presently is about 2 μm thick, grows with good orientation of crystallites which have their c-axis perpendicular to the substrate /1/, so that light-generated carriers need not cross grain boundaries. Hall effect measurements show that our CdSe films have a native doping level of 10^{13} to 10^{14} electrons per cm^3, which is enhanced to 10^{15} to 10^{16} cm^{-3} by illuminating the films with white light of 100 mW/cm^2. The carrier mobilities are in the range of 10 to 30 cm^2/Vs.

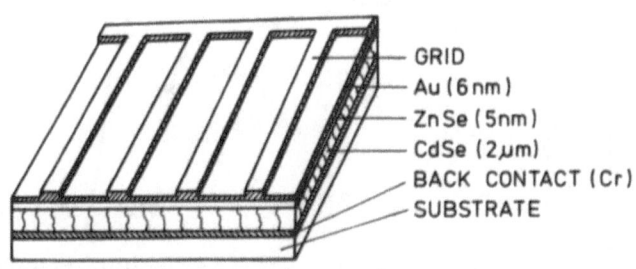

GRID
Au (6 nm)
ZnSe (5 nm)
CdSe (2 μm)
BACK CONTACT (Cr)
SUBSTRATE

Fig. 1: Structure of the CdSe thin film solar cell.

As has been reported, previously /3/, the best I-layer found so far is a ZnSe film of about 5 nm thickness (Fig. 1).

The current collection grid which has not yet been optimised and the 6 nm thick Schottky contact are made from gold (Fig. 1). In order to reduce the reflection from the gold film, a simple antireflection coating is evaporated, which consists of 55 nm ZnS on top of the structure. The AR coating produces a bluish hue.

After deposition of the various layers, the structures have to be annealed in a nitrogen atmosphere for periods of 10 to 30 min at about 200°C to improve the photovoltaic properties /1/.

3. Properties of the Cells

3.1 Spectral Response

The quantum efficiency of a standard cell with antireflection coating (Fig. 2) indicates high sensitivity over the whole spectral region. Taking into account that there still is a certain amount of reflection from the cell surface, the measured high external quantum efficiency shows that the internal collection efficiency has to be near unity, leading to relatively high short-circuit current densities. This positive behaviour results from the high absorption coefficient of CdSe (10^4 to 10^5 cm^{-1}) leading to field-assisted collection of the light-generated carriers because nearly all of the light is absorbed within the interface region.

As can be seen from spectral response measurements made on structures with platinum electrodes which are known to have a uniform transmission coefficient over the visible part of the spectrum /4/, the decrease in yield at short wavelengths (Fig. 2) is not essentially an internal effect that is due to a high density of recombination centres at the CdSe/ZnSe interface or to absorption in the ZnSe layer, but rather results from enhanced reflection at short wavelengths (bluish hue of the cells). Optimising the antireflection coating, therefore, can definitely enhance the response in the blue.

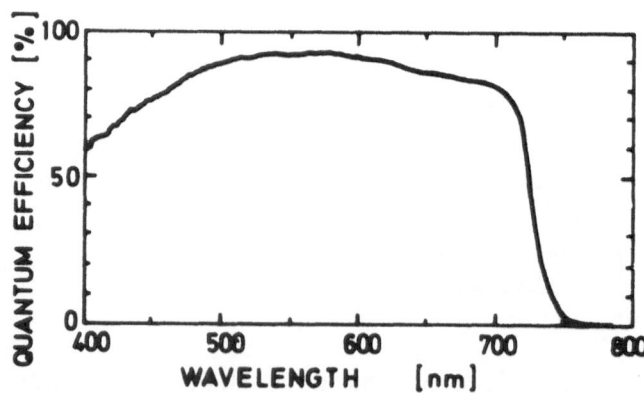

Fig. 2: External quantum efficiency of the CdSe cells.

3.2 Influence of the Front Contact

To study the influence of the work function of the front-contact metal on the open circuit voltage, we deposited various metals onto the I-layer. The resulting open-circuit voltage (U_{oc}) of the cells can be seen from Fig. 3. The standard annealing step is omitted in order to show the direct influence of the metals. As predicted by the Schottky theory, the open-circuit voltage varies significantly with the work function of the contact material. This behaviour is in contrast to studies on MS diodes on single-crystalline CdSe, which resulted in surface-state-controlled barriers that are practically independent of the metal /5, 6/. The different behaviour of our structures (Fig. 3) are obviously due to the ZnSe film. With respect to pure MS diodes, this layer drastically reduces the number of surface states so that the Fermi level no longer is pinned at a fixed position.

Because of interactions between contact material and I-layer which can change the electronic properties of ZnSe, annealing of the cells causes deviations from the behaviour shown in Fig. 3 and opens up possibilities to influence the photovoltaic properties of the cells. The greatest increase in U_{oc} after annealing occurs with gold contacts. Open-circuit voltages up to 850 mV have been observed.

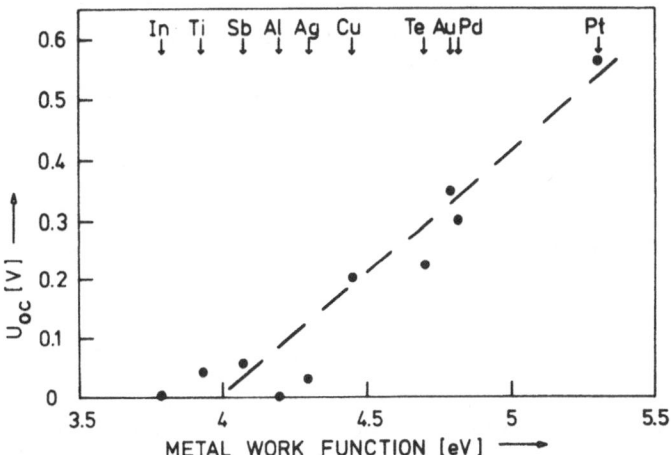

Fig. 3: Influence of the contact material on the open circuit voltage of unannealed cells

3.3 Effect of UV Irradiation

Irradiating the active area of our solar cells with light of 250 nm to 350 nm wavelength positively influences the internal electric field. This procedure probably creates negatively charged centres in the ZnSe film which enhance the potential barrier in CdSe. The improvement of the photovoltaic parameters of the CdSe cells induced by this process saturates after an irradiation dose of about 1.5 Ws/cm^2. This results in a final enhancement of the open-circuit voltage up to 15 percent and in a slight improvement of the fill factor and the short-circuit current density (Fig. 4). These electronic centres in the ZnSe layer are stable up to temperatures around 150°C. Under normal operation conditions they therefore cause no stability problems, so that UV irradiation represents a very simple method to improve the efficiency of this CdSe solar cell.

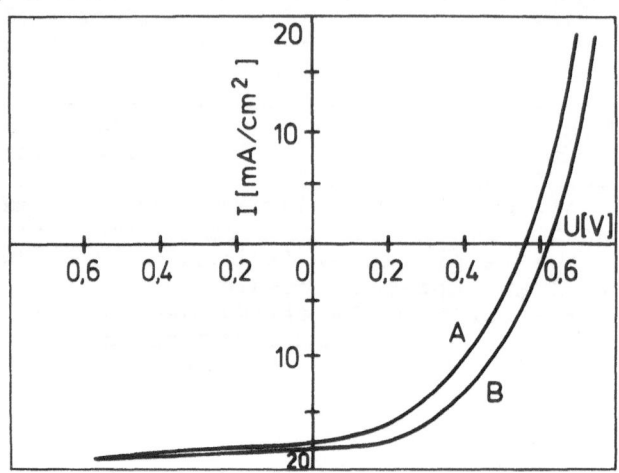

Fig. 4: Changes in the photovoltaic properties of CdSe cells induced by UV irradiation
A: Before irradiation, B: After irradiation

3.4 Stability

Stability tests on our CdSe cells have established that these devices are indeed very stable. Storage of the cells without any encapsulation on the shelf for more than one year does not seriously affect their photovoltaic parameters. Furthermore, humidity and even immersion into water causes nearly no degradation of the cells.

This behaviour was demonstrated by storing cells at elevated temperatures of 75°C in moist air of 100 percent humidity. Even after 70 days under these extreme conditions, the degradation is lower than 10 percent (Fig. 5). Whereas the short-circuit current density and the open-circuit voltage are unaffected within the experimental error, the main parameter that determines the cell's stability is the fill factor (Fig. 5). It is supposed that its slight reduction is caused rather by an increase in series resistance due to oxidation of the back contact (Cr) than by changes in the MIS structure itself.

Similar results confirming the high stability of our cells were achieved by outside storage of some small cells in a simple plastic box for eight months. Though even the plastic box was deformed the efficiency of these cells was lowered only by about 10 percent.

The high stability of our cells is probably due to the inherent stability of CdSe which is a material that is not seriously corroded even by common acids. In addition, the use of ZnSe as "I-layer", i.e. an oxygen-free material, has the advantage that there is no oxygen that can diffuse into the photoactive semiconductor causing changes in its electronic properties. Besides this, oxygen cannot cause this "I-film" to grow further with time, which represents a pronounced degradation mechanism in native oxide MIS devices.

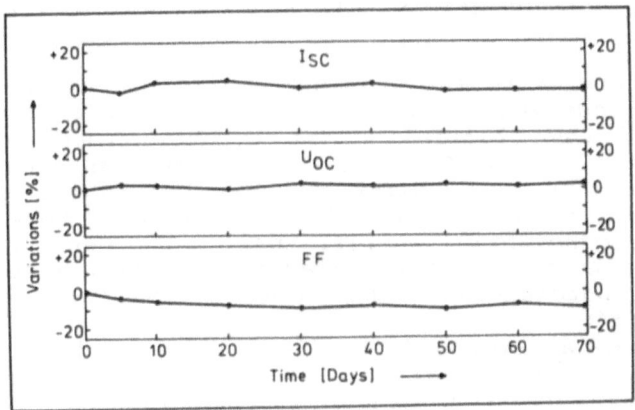

Fig. 5: Stability of <u>unsealed</u> CdSe-cells in hot, moist air (75°C, 100 % humidity)

4. Discussion

It is well known that the I-layer strongly influences the photovoltaic parameters of MIS solar cells /7/. Its main task is the reduction of the reverse current by generating an additional barrier that reduces thermionic emission of majority carriers into the metal. On the other hand, the minority carriers must be able to pass this barrier by tunnelling, i.e. the I-film generally has to be very thin. Usually the I-layer in MIS cells is formed by producing an approximately 2 nm thick native, insulating oxide on the photoactive semiconductor.

As has been shown on our Au/ZnSe/CdSe cells, the insulator in MIS cells can be replaced by a wide-bandgap semiconductor. Compared to common insulators such as SiO or SiO_2, the thickness of these semiconductor "I-layers" can be enhanced because the additional barrier introduced by these layers is lower than that of usual insulators. Thus the same tunnelling probability, which is proportional to the barrier height and width, can be achieved with thicker layers.

In the case of our CdSe cells we can use ZnSe layers up to 3.5 nm thickness without decreasing the short-circuit current density (I_{sc}) or the fill factor (FF). Even 5 nm thick "I-films" do not affect the I_{sc} but reduce the fill factor to values between 40 and 50 percent.

On the other hand, the open-circuit voltage (U_{OC}) of the CdSe structures is enhanced with increasing ZnSe film thickness /8/. This probably results from the reduction of inhomogeneities and pinholes in this layer, which are due to the surface roughness of the underlying CdSe /1/.

Our results show that in order to improve the efficiency we have to strive for high fill factors at simultaneously high "I-layer" thickness. At first glance this seems to be impossible, but recent experiments have shown that doping of the ZnSe layer by indiffusion of copper results in the desired effect. Fill factors around 60 percent have been achieved by applying this technique to 5 nm thick ZnSe films, resulting in cell efficiencies around 7 percent on 0.6 cm^2 active area (Fig. 6).

There are different explanations for this behaviour. Most likely the observed effect is due to a shift of the energetic position of the "I-film" caused by Cu doping. Copper forms deep acceptor levels (0.6 eV) in ZnSe. From an energetic point of view it therefore shifts the barrier caused by ZnSe upwards. With respect to the different carriers this results in an asymmetric potential barrier that approaches the ideal case of blocking only the majority carrier emission into the metal, while the photogenerated minority carriers can easily cross the "I-layer".

These Metal/Semiconductor/Semiconductor (MSS) structures probably represent a transition between MIS diodes and heterodiodes. They even show higher fabrication-related advantages than MIS diodes because thickness variations in the "I-film" play a minor role compared to native-oxide devices. These films can simply be physically deposited even on relatively rough surfaces of as-grown polycrystalline semiconductor films.

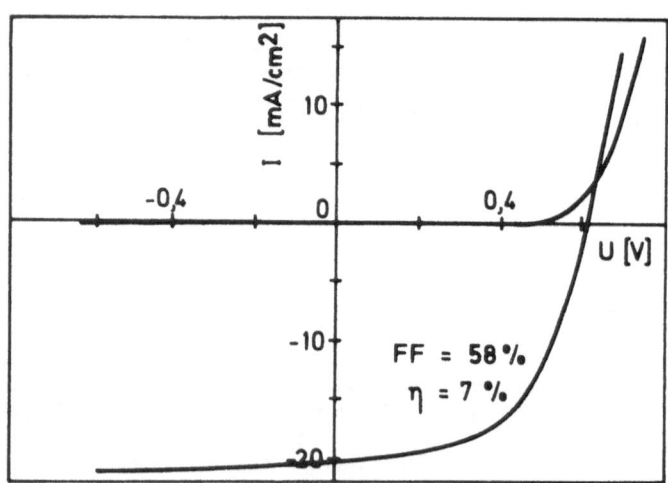

Fig. 6: I-U curve of a CdSe cell with 5 nm Cu-doped ZnSe as "I-layer"

5. Conclusion

Investigations on the CdSe cell have demonstrated that these structures may have a significant potential for terrestrial use as stable, low-cost thin-film solar cells. Efficiencies around 7 percent on an active area of 0.6 cm^2 achieved with relatively little effort compared to other solar cell developments give reasons to expect further progress in reaching the goal of developing a cost-effective solar cell of 10 percent efficiency.

6. References

/1/   D. Bonnet, E. Rickus: Proc. 14th Photovolt. Specialists
      Conf., San Diego, 1980, p. 629

/2/   E. Rickus, D. Bonnet: Proc. 3rd E.C. Photovolt. Solar
      Energy Conf., Cannes 1980, p. 871

/3/   D. Bonnet: Proc. 2nd E.C. Photovolt. Solar Energy Conf.,
      Berlin 1979, p. 387

/4/   C.R. Wronski, B. Abeles, G.D. Cody: Solar Cells, 2 (1980)
      p. 245

/5/   R.L. Consigny III, J.R. Madigan: Solid State
      Communications, Vol. 7 (1969) p. 189

/6/   C.A. Mead: Appl. Phys. Lett., Vol. 6, No. 6 (1965) p. 103

/7/   R.Singh, M.A. Green, K. Rajkanan: Solar Cells, 3 (1981)
      p. 95

/8/   E. Rickus: Proc. 16th Photovolt. Specialists Conf.,
      San Diego, 1982

SYSTEM STUDIES

Optimization research into a complete photovoltaic generator/consumer appliance system for small independent electricity supply

Development of a microcomputer for the control of a photovoltaic power station

Feasibility study for small solar cell operated units

Photovoltaic system for a solar house

Hybrid thermal and photovoltaic concentration collector

Optimization research into a complete photovoltaic generator/consumer appliance system for small independent electricity supply

Authors   : R. BURGEL, H. WELSCHEN, R. SONNEVILLE

Contract number   : ESC-R-047-NL

Duration   : 18 months   1 July 1980 - 31 October 1982

Contractor   : Holec N.V.

Address   : Holec N.V.
                     Stationsplein 93
                     Utrecht
                     The Netherlands

Summary

This study is directed at the design of an appropriate photovoltaic system for the purpose of Rural Electrification. It is divided into 2 phases:
1. System-analysis
2. Experimental phase
Main attention during the first phase has been focussed on the adaption of the photovoltaic generator and the battery-storage system including consumer-appliances. Besides the specification of this power matching device (MPPT), a rather unique inverter design will be specified considering specific parameters for this specific application. During the experimental phase working models of both devices will be realised (not finished yet). Final testing during at least 5 years will be carried out as a part of the CEC awarded pilot project Terschelling.

1. Conventional electric energy versus photovoltaic electric energy generation

Within the context of this contract the system has been concentrated on the application of photovoltaics in rural areas. Still one should consider at least other appropriate conventional systems such as dieselgenerators for the same purpose.

Figure 1 illustrates the difference between both systems. As Energy Consumption Density will be rather low (100-500 W per dwelling and 20 ÷ 50 dwellings per square kilometer) also compared with other "renewables" a photovoltaic system offers good possibilities to match installed power and load in an efficient way.

Another important argument in favour of photovoltaics is the modular approach of the solar-generator, the inverter and the storage system which means that investments per kWpeak, to a large extent, are independent of the dimension of installed power.

If these considerations lead to the decision to accept the photovoltaic solution one must consider:
- restrictions for Pmax
- average Pmax rather low due to the small Pmax number consumers per system.

This means thorough investigations of the daily load profile have to be executed in order to make possible the dimensioning of the power-conditioning equipment.

From this load profile the relation between Paverage and Pmax can be learned and from that it can be decided whether or not power conditioning equipment (inverter) should be split-up in smaller units (master-slave configuration).

Another important consideration should be the lay-out of the array-field. Large solar generators have to be split-up into smaller arrays for several reasons:
- flexibility
- serviceability
- extension possibilities
- tolerances in module characteristics (intrinsic and due to different insolation and temperature)
- availability dc-switchgear.

2. System description

A photovoltaic (p.v.) system for application in rural electrification is characterized by:
- the generator is composed of a large number of small individual sources
- a few of the (relatively) small number of consumers can consume a major part of the power
- the individual source shows a current source[1] character up to a limited voltage, then it changes to a voltage source[2].

So a p.v. generator differs markedly from conventional electro-mechanical generators which is a single source generator with a voltage source character. Therefore in a p.v. generator additional requirements have to be met, to obtain a maximum total power which is close to the sum of the maximum individual power of the sources. The p.v. generator is supplied by a varying solar energy that cannot be stored, therefore, a storage of electrical energy is necessary to meet the load demand that varies also but in an unpredictable manner.

The combination of a p.v. generator plus a battery storage together provides the virtually constant power source (within certain limits) which is necessary to meet the load demand.

However in order to be compatible with current practice in

electricity supply and with availability of components and apparatus, it is strongly recommended to provide the user with an a.c., constant voltage grid (3-phases).

The inverter, therefore, has not only to transform the d.c. power into the a.c. one, but also to provide the a.c. grid with sufficient reactive power for inductive user (as a.c. motors, transformers, chokes).

Moreover, peak currents up to 10 x Inom are necessary to blow (safely) fuses in case of short circuiting and starting currents of a.c. motors (highly inductive) have to be provided by the inverter as well.

These transient power and current cannot be delivered by a high-efficient semiconductor inverter, so an additional element in the inverter is required. A free running synchronous generator in parallel with the a.c. grid provided the required transient energy from its own mechanical moment and can generate the required reactive load by adjusting its d.c. supply in the field winding.

So far, the basic system lay-out except the organisation of the numerous small sources in the p.v. generator are considered. For this organisation the laws of the electrical network theory concerning power sources, have to be followed i.e. without power loss and without equilizing currents.
* multiple d.c. current sources can be connected in parallel with only one voltage source forming a high power voltage source
* multiple d.c. voltage sources can be connected in series with only one current source forming a high power current source,
* series connection of current sources requires equal current of each source.

Therefore, considering the nature of p.v. cells as power sources, series connection of cells in a module (and stacks of modules) require current match over the whole operating voltage rate. If this requirement is met for all the p.v. cells, and the insolation is the same for each cell, and the degradation is the same too, then only then one may connect the whole p.v. generator to the voltage source (battery). This, however is a purely hypothetical case, in current practice one has to recognize the scattering in properties of the cells and modules, the inhomogeneous insulation and temperature distribution in larger generators and the degradation (including hot-spots and cracked cells) of the modules. So, in order to maintain the maximum power extraction of the generator (array), suitably sized sub-arrays have to be defined each of which to be provided with an electronic converter giving them the required current source character over the operating voltage range.

An up-converter with a feedback maximizing the output power by increasing the output voltage is a good answer to this requirement.

In the present state of the art a suitable size of such a sub-array, giving sufficient flexibility in replacing modules and future extensions is between 0,5 and 2 kW.

In figure 2 the different functions are indicated together with suitable voltages.

3. Conventional power control versus Maximum Power Point Tracking

This comparison has been made with the Terschelling-project as an example. Besides the purely technical argument in favour of maximum power point tracking for larger systems also financial factors had to be considered. The system was therefore also calculated on the basis of a straight forward control realizing switching on and off sub-arrays by means of electronic devices and appropriate d.c. switchgear.

With respect to the consumer load (40.000 kWh on annual basis) and an average power-transfer efficiency of 70 per cent a 50 kWp solar generator should cover 80 per cent of the total energy consumption.

Assumptions:
- Sub-arrays to be connected directly to battery bank
- Un battery is 400 Vdc (varying between 346 Vdc and 432 Vdc)
- Tolerances in maximum module output $\pm$ 10%.

To assure the current source behaviour of all parallel connected strings the number of modules in one string is estimated by the formulae:

Ubmax = N x Vmpm - 10%
Ubmax = Battery voltage max.
Umpm = Module voltage at MPP (mean value)
n   = modules series connected.

In the actual case:

$$n = \frac{Ubmax + 10\%}{Umpm} = \frac{432 + 10\%}{8,72} = 55 \text{ modules per string}$$

The mean system voltage (Ums) is 389 V which means a mean system current (Ims) of 128,5 Amps should be reached in order to get a system power to the battery of 50 kW (without wiring and diode losses).

With a mean module current at MPP (Impm) of 2,2 Amps a total number of $\frac{128,5}{2,2}$ = 58 strings has to be installed.

or 58 x 55 = 3190 modules or 3190 x 2,2 x 8,72 = 61,25 kWp.

The alternative is to install per 3 strings of 30 modules a max. power point tracker. With 29 sub-arrays each consisting of a MPPT and 90 modules the total installed peak power is 29 x 3 x 30 x 19,2 = 50,112 kWp (2610 modules). Supposing losses of the MPPT's of 10% 3 additional subarrays of 3 x 30 modules each has to be installed in order to reach a net peak power output of 50 kWp to the battery-bank.

This means 2880 modules plus 32 Max. Power Point Trackers deliver the same mean system power as 3190 modules plus appropriate switching devices.

Breakdown of prices (3)

US dollars/kWp	With MPPT	Without MPPT	
modules	8,5	9,5(4)	
control equipment	0,62	0,6	Fig. 3
mech. structures incl. wiring	0,9	1,0(5)	
	]0,--	11,--	

4. Specification Max. Power Point Tracker

The maximum power point tracker (MPPT) as described above is a microprocessor controlled up-chopper. It consists of three main parts as there is the power circuit, the control circuit and the status guard circuit. Apart from these three parts there are the power supply, the voltage and current probes and the signal conditioning. A schematic diagram has been drawn in fig. 4.

The specifications for the power circuit are:

input   Uin = 210 - 273 V   output   Uon = 324 - 432 V
        Uioc = 370 V   Ion = 0-5 A
        Iin = 0-7 A   Iosc = -3 kA
        Iisc = 7,5 A

4.1 Power Circuit

With the switch S closed energy from the subarray is transferred into the self induction. This energy can be taken from the self induction at any voltage level.

This will happen when S opens, energy forms the self induction and the subarray will flow into the battery which will have a voltage level $U_o > U_i$.

The condensators Ci and Co merely act as filters for the ripple currents, which are in the order of 1,5 App and 5 App respectively. The voltage ripples are in the order of 0,5 Vpp at a frequency of 20 kHz. This frequency has been chosen to minimize the volume of the self induction and condensors.

4.2 Control circuit

To extract the maximum power from the subarray the input current and voltage are measured, filtered and multiplied. The multiplication takes place in the microprocessor which also contains an algorithm to seek the maximum power point. At start-up the duty-cycle of the switch S is varied from minimum to maximum.

The duty cycle at which the maximum power is extracted is registered and adjusted. Due to variances in irradiation and temperature the maximum power point will vary. Therefore, the microprocessor will constantly seek for improvement of extracted power by varying the duty-cycle. To check whether the MPPT has not been locked at a local extreme, the duty is varied at regular intervals from mimimum to maximum. The extra power loss by this behaviour is less than 1 %.

4.3 Status guard circuit

The protection circuit has the following functions:
- signalling of working conditions
  . normal: MPPT is in working order
  . off: MPPT has been shut off by hand or EMS (energy management system)
  . power low: irradiation is too low
  . voltage limit: output voltage is at maximum due to interrupted output connection to battery.
- protection of switching element. At low power conditions a switching behaviour coule arise, damaging for the switching element. In this case the pulses to the switch are blocked.

5. Specification Inverter

Although the main target was to develop an inverter suitable for the stand-alone mode, for the purpose of application in the Terschelling project this inverter had to be designed equally for grid-connected systems.

5.1 Working principle

The main items of the system are
- 60 kVA load commuteated current source
- 40 kVA brushless synchronous motor.
(Fig. 5, principle diagram of the inverter system).

The latter provides the Electro-magnetic Force, needed for natural commutation in the stand-alone mode. The inverter operates in the minimum extinction angle mode in order to optimize its power factor. Thus a DC-chopper is needed for adaption of the battery-voltage.

The chopper also provides additional protection in case of commutation failures.

A 40 kVA harmonics filter, consisting of three tuned branches, ensures low voltage distortion.

At the same time it compensates lagging power factor of the inverter. In this way the motor rating and so its power loss, can be kept to a minimum.

The system has two control inputs. The chopper duty cycle controls the transfer of active power to either the consumer or the mains.

In stand-alone mode the frequency is a measure for the proper active power balance.

The motor-excitation determines the reactive power balance within the system. If the motor is kept running in the mains mode it may be used for power factor correction, the system offers then no-break capability at the expense of decreased efficiency.

5.2 Main parameters

- DC input voltage	324 - 432 V
- nominal output voltage	220/380 V
- nominal output current	90 A
- maximum output power	48 kW/60 kVA
- short circuit current	720 A - 10 msec.
- stationary voltage stability (within specified range of input voltage, output current and ambient temperature)	+/- 2%
- stationary frequency stability (same conditions as above)	+/- 0.1%
- total harmonic distortion	better than 5%
- initial voltage drop at 100% load variation (pf = 1)	40 V
- recovery time	0,2 S
- operating temperature range	-5 to + 35 °C
- power losses at 400 DC input voltage:	
. no load	2,7 kW
. full load, pf = 1	5,1 kW
. full load, pf = 0,8	6,3 kW
. full load, excl. motor	3,1 kW

Figure 6 shows a complete survey of all different losses at full load.

5.3 Features

The complete inverter-system consists of 2 sub-systems which both can be characterized as a well-proven design.

Taking into account the severe circumstances for systems to be applied for Rural Electrification it is of utmost importancy that the system offers high quality, is simple to service. Its behaviour should be as close as possible in accordance with the facilities e.g. short circuit current, response to surge currents etc. as they can be offered by conventional electricity generation.

Summarized the inverter offers for R.E. systems
- simple, proven design
- ability to support surge currents
- inherent power factor control
- high reliability
- high efficiency
- low distortion.

6. Notes

(1) current source: the current is almost constant and the terminal volage depends on the load.

(2) voltage source: the voltage is almost constant and the output current depends on the load.

(3) Solar modules incl. supporting hardware and control equipment based on 1982-prices.

(4) taking into account mismatch between battery-voltage and system voltage in MPP.

(5) as a consequence of additional modules (10% extra).

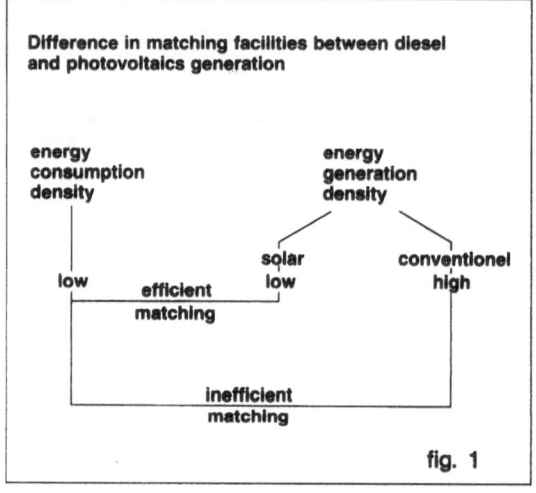

fig. 1

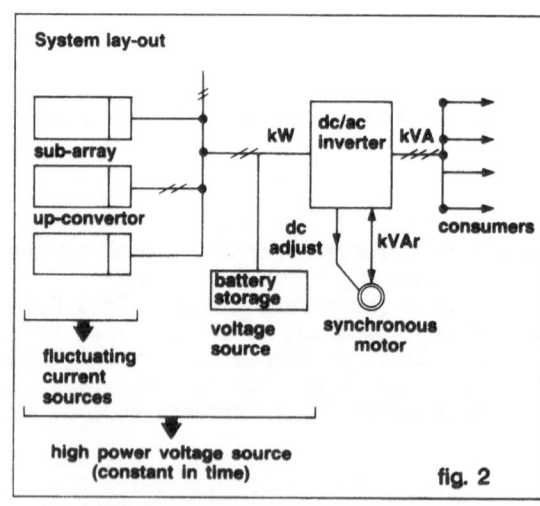

fig. 2

Breakdown of prices (US dollars per kWp)	with M.P.P.T.	without M.P.P.T.
modules	8.5	9.4
control equipment	0.6	0.6
mech. structures incl. wiring	0.9	1.1
total	10.0	11.1

fig. 3

Schematic diagramm of MPPT

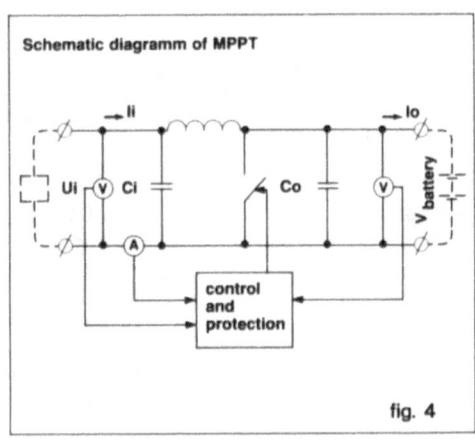

fig. 4

Principle diagram of the inverter system

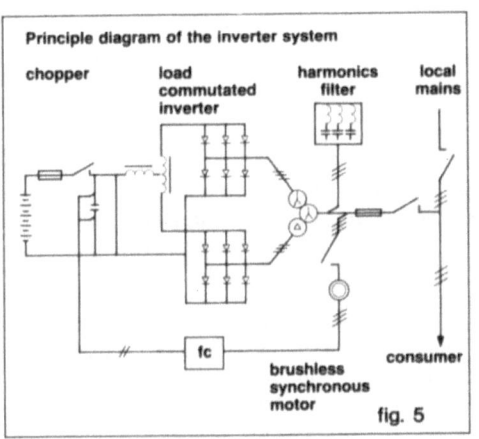

fig. 5

Power losses [kW]	no load	full load pf = 1	full load pf = 8
chopper (300v, 175 A)	.2	.9	.9
inverter (60 kVA)	.2	1.6	1.6
filter (40 kVA)	.3	.6	.6
motor (40 kVA)	2.0	2.0	3.2
total	2.7	5.1	6.3

fig. 6

DEVELOPMENT OF A MICROCOMPUTER FOR THE CONTROL
OF A PHOTOVOLTAIC POWER STATION

Authors   : S. McCARTHY, G.T. WRIXON

Contract number   : ESC-R-081-EIR

Duration   : 1.6.81 - 1.6.82

Head of project   : Prof. G.T. Wrixon, National Microelectronics
                      Research Centre

Contractor   : National Microelectronics Research Centre

Address   : University College Cork
                      Ireland

Summary

The 50KWp photovoltaic power station on Fota Island in Cork Harbour
is designed to operate in a stand-alone mode, i.e. to supply
electrical energy to the dairy farm independent of the utility grid.
As well as operating independent of the utility grid the station
must operate at its highest efficiency while ensuring that each unit
in the station is protected from faults. To meet these requirements
a microcomputer is used to control the station's operations. The
microcomputer consists of a MOTOROLA 6800 microprocessor system,
computer terminal, printer and a magnetic tape recorder. As well as
controlling the photovoltaic power station the microcomputer monitors
and displays data on the printer, allows changes of operating
parameters via computer terminal, stores desired data on magnetic
tape and performs tests on hardware interface to the microprocessor.

The 24K of software for the microcomputer is stored in EPROMs and
8K of RAM is required for the system's operations.

The microcomputer is an intelligent and flexible system. It has a
quick fault debugging facility and keeps a permanent record of the
photovoltaic power station's operations. Such characteristics are
required to determine the most efficient operation of a large
photovoltaic power station.

1. Introduction

The Fotavoltaic project is part of a photovoltaic pilot program set up by the Commission of the European Communities to investigate the suitability of photovoltaics to Europe. The power station is located on Fota Island in Cork Harbour and is designed to supply electrical energy to the 250 herd dairy farm independent of the utility grid.

The system comprises of the following hardware units as shown in Figure 1. A 50KWp solar array located on the roof of a south facing

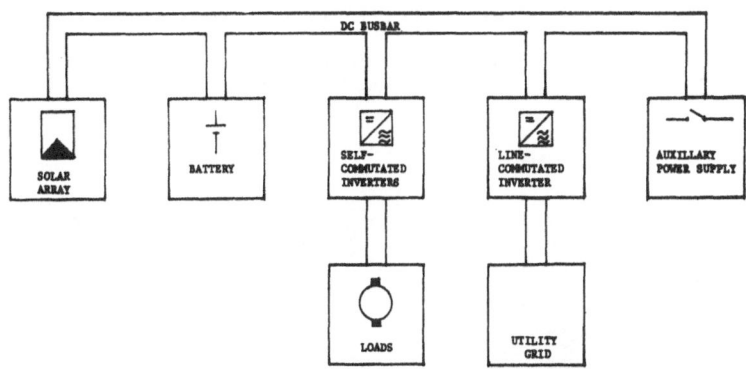

Figure 1. Layout of Photovoltaic System

building at an angle of 45° to the horizontal, two battery groups each operating at 268V and each having a storage capacity of 300Ah, three 10KVA self-commutated inverters which interface the photovoltaic system to the 3-phase AC loads on the dairy farm and a 50KVA line-commutated inverter used to supply excess energy from the photovoltaic system to the utility grid, and an auxilliary power supply is used to supply power to the microcomputer and switches in the control room. The 50KVA line-commutated inverter also acts as a rectifier and is used to bring the battery groups to full charge if both battery groups become deep-discharged. Switches are used to interconnect the hardware units and thus regulate the energy flow in the system.

As well as operating in the stand-alone mode, i.e. independent of the utility grid, the power station must operate at its highest efficiency while ensuring that each hardware unit is protected from faults. To meet these requirements a microcomputer is used to control the complete operation of the power station.

2. Microcomputer Hardware

The microcomputer is located in the control room and consists of a central microprocessor system, computer terminal, printer and magnetic tape recorder. The central microprocessor system known as GEAMIC 10 was developed by AEG-Telefunken. It consists of a number of standardised plug-in cards as shown in Figure 2. The central operating card (ZK101) contains the MOTOROLA 6800 microprocessor and this operates two buses, i.e. storage bus and input/output bus. The storage bus selects the Programmed Read Only Memory (PROM) and Random Access Memory (RAM). For

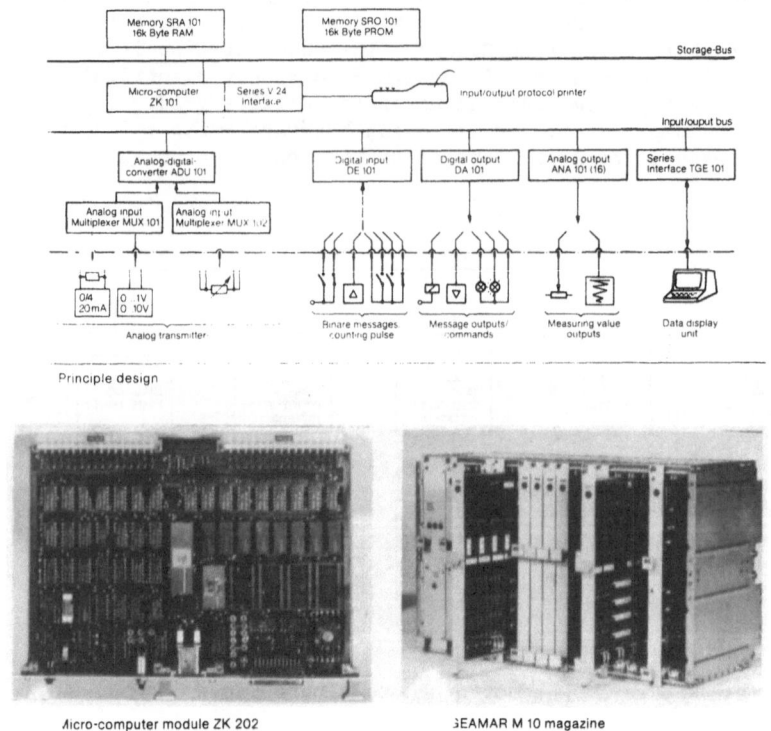

Micro-computer module ZK 202   3EAMAR M 10 magazine

Figure 2.   Microprocessor System

the control of the Fotavoltaic project 24K of software is stored in PROMs
and 8K of RAM is required. The PROMs and RAM are stored on different
cards. Also, the RAM card contains a separate battery backup which
prevents the loss of information or data during a power failure.

The input/output bus selects the cards which interface the microprocessor to the photovoltaic system as shown in Figure 1.

(i)   Analog data is interfaced using an analog to digital converter
and a multiplexer. The multiplexer card MUX 101 accepts data as 0 - 10V
or 0 - 20mA. Therefore, current, voltage and power measurements in the
photovoltaic system are converted to these ranges using transducers. The
card MUX 102 is designed to measure temperature directly using platinum
resistance thermometer (PT100).

(ii) Digital inputs representing the status of switches or the status
of hardware units are interfaced using the card DE102.

(iii) Digital outputs which are used to operate switches and operated
display signals are outputed via DA102.

(iv) Analog outputs with a range 0 - 10V are outputed via ANA 101.

(v)   A serial interface card, TGE 102, is used as an interface for
the computer terminal, printer and magnetic tape recorder.

As well as controlling the storage bus and input/output bus the
central card, ZK 101, contains a hard wired circuit which tests the
microprocessor by generating interrupts at regular intervals. If the
microprocessor fails to respond the microprocessor is shut-down and a
relay is activated which also shuts-down the photovoltaic system.

3. Microcomputer Control

The microcomputer has complete control over each hardware unit and each interconnecting switch in the photovoltaic system. By defining the system's operations in terms of switch configurations the correct energy flow in the system is controlled. The following sections describe the main control operations in terms of switching configurations.

The main control operations performed by the microcomputer are:
- (i)   Battery charging using solar array
- (ii) Load supply using inverters
- (iii) Delivery of excess energy to the utility grid
- (iv) Battery charging using the utility grid

3.1 Battery Charging Using Solar Array

The solar array is divided into twelve sub-arrays for monitoring and control purposes. Each sub-array can be switched to battery 1 or battery 2. Battery charging involves supplying both battery groups with a proper charging current (50 - 65A) while ensuring that both the solar array and the battery are protected from faults. The number of sub-arrays connected to battery 1 depends on the insolation (and thus the output of each sub-array) and load demand. The remaining sub-arrays are used to supply electrical energy to charge battery 2. As well as controlling the charging current the microcomputer performs a daily start-up and daily shut-down of the solar array. This prevents the flow of a discharge current in the solar array during the night.

3.2 Load Supply Using Inverters

Under normal operating conditions the inverters are connected to battery 1 and the loads are connected to the inverters. However, when battery 1 becomes deep-discharged the inverters are disconnected from battery 1 and connected to battery 2. Battery 1 is then recharged using the solar array. During system shut-down the loads are disconnected from the inverters and connected to the utility grid. This ensures a continuous energy supply to the dairy farm.

3.3 Delivery of Excess Energy to Utility Grid

When excess energy exists in the system the line-commutated inverter is connected to the utility grid. In the absence of the utility grid the line-commutated inverter does not operate. The current flow through the line commutated inverter is controlled by a 0- 10V signal from the microprocessor interface.

3.4 Battery Charging Using Utility Grid

When both battery groups require recharging from the utility grid the polarity of the firing pulse to the line-commutated inverter is reversed. The unit then acts as a rectifier for charging. The polarity is reversed using a digital signal from the microcomputer interface.

The above operations are described in terms of switch configurations. The microcomputer then controls the system by scanning the measuring points every 500ms and when measured data exceeds predefined limits a new switching configuration is adopted, e.g. when battery 1 charging current is less than 50A an extra sub-array is switched to battery 1.

For each of the thirty four interconnecting switches in the system a logic equation describes its operation. The operands of the logic equations are data bits which are set (0, 1) when predefined limits on measured data are exceeded, e.g. sub-array 1 is connected to battery 1 when: battery 1 charging current is less than 50A AND battery 1 temperature is less than 55°C AND solar array voltage is greater than battery 1 voltage. This logic is processed by the microprocessor and the result (0, 1) determines the operation of the switch.

Therefore, the microcomputer control of the interconnecting switches is the basis for controlling the system's operations.

4. Monitoring System

As well as controlling the power station the microcomputer monitors and displays data. The microcomputer collects analog data from the forty six measuring points in the system every 500ms. This data is converted to a form suitable for display on the printer or computer terminal. The microcomputer also performs calculations on the measured data, e.g. it calculates average values, deviation from the average, maximum values and minimum values. Digital data is also collected and displayed, e.g. the status of the interconnecting switches is indicated as being 'ON' or 'OFF'. This data is stored in the microcomputer in thirteen data groups. At any time during the operation of the system any or all of the data groups are available on printer by request via computer terminal. Figure 3 is an example of a printout. Each measuring point has a text indicating its name, the units in which it is measured and its instantaneous measured value. Also included are the time and date at which the printout occured.

```
SYSTEM JOURNAL          FOTA-VOLTAIC PROJECT          11 DEC '81  18.16  LT

02       SOLAR ARRAY OUTPUTS..I,V.

0400:    ARRAY VOLTAGE..U.GEN.                         251.       V.DC .....
0401:    OUTPUT CURRENT OF SUB-ARRAY  1                9.56       A.DC .....
0402:    OUTPUT CURRENT OF SUB-ARRAY  2                9.02       A.DC .....
0403:    OUTPUT CURRENT OF SUB-ARRAY  3                9.56       A.DC .....
0404:    OUTPUT CURRENT OF SUB-ARRAY  4                9.50       A.DC .....
0405:    OUTPUT CURRENT OF SUB-ARRAY  5                9.56       A.DC .....
0406:    OUTPUT CURRENT OF SUB-ARRAY  6                9.16       A.DC .....
0407:    OUTPUT CURRENT OF SUB-ARRAY  7                9.53       A.DC .....
0408:    OUTPUT CURRENT OF SUB-ARRAY  8                9.51       A.DC .....
0409:    OUTPUT CURRENT OF SUB-ARRAY  9                9.03       A.DC .....
0410:    OUTPUT CURRENT OF SUB-ARRAY 10                9.20       A.DC .....
0411:    OUTPUT CURRENT OF SUB-ARRAY 11                9.29       A.DC .....
0412:    OUTPUT CURRENT OF SUB-ARRAY 12                9.42       A.DC .....
```

Figure 3. Printout of Data Group

A further use of the monitoring system is that of fault analysis. Figure 4 shows an example of a printout occuring when a fault condition exists. The printout includes a text indicating the measuring point, the

```
                    FOTA-VOLTAIC PROJECT          16 December '81 LT

0553   BATTERY 1   TEMPERATURE 1   56.0   "C   **B   TOO HIGH  08.01.23

0553   BATTERY 1   TEMPERATURE 1   55.0   "C   **E   TOO HIGH  08.30.56
```

Figure 4. Printout of Fault Message

instantaneous measured value, a comment (TOO HIGH) indicating the fault, the time it began (**B) and the time it ended (**E). A similar fault detection is performed on the computer terminal whereby the measuring point on which a fault exists appears in a flashing mode. The option to obtain a printout or flashing when a limit is exceeded is defined in the initial software. Both options allow a quick fault detection facility.

5. Changes Via Computer Terminal

The software for the control of the system is stored in the PROMs. However, using the computer terminal changes in the operating parameters may be performed, e.g. the nominal charging current for a battery group is 60A. This may be changed simply by request via the computer terminal. The options to remove measuring points, include delays, manually simulate inputs, shut-down or start-up the system may also be performed via computer terminal. Any changes which are made via the computer terminal are recorder on the printer. The message indicates the measuring point affected, the change made and the time it occured. As a safety measure a key operated switch inhibits any changes via the computer terminal.

6. Data Recording

The principal aim of the photovoltaic pilot program is to collect information on the performance, efficiency and reliability of the photovoltaic systems. Data recording for the Fotavoltaic project is performed using the same microcomputer which controls the power station. The microcomputer collects data from the power station every 500ms. This data is then processed and stored on the magnetic tape recorder in a format selected by the CEC, e.g. the energy input to an inverter is calculated as follows: The unit for recording is KWh (Kilowatt hours); the DC input current to the inverter and the DC voltage is measured; the power is calculated (V x I) and this value is sampled every minute to calculate the Wh (Watt hours). This is then converted to KWh for storage on magnetic tape recorder. The data is stored once every hour. A standard recording format has been selected by the CEC to allow a comparison with other projects in the pilot program. Figure 5 shows an example of data stored in this format.

% FOTA % 04-07-82, 21.15, FOTA-VOLTAIC PROJECT

```
1,  300.,  300.,  300.,  300.,  300.,  300.,  300.,  300.,  300.,  300.,
2,  311.,  311.,  311.,  311.,  311.,  311.,  311.,  311.,  311.,  311.,
3,  320.,  320.,  320.,  320.,  320.,  320.,  320.,  320.,  320.,  320.,
4,  25.0,  25.0,  300.,  311.,  25.0,  300.,  311.,  25.0,  320.,
5,  300.,  300.,  300.,
```

Figure 5. Data Recording Format

It shows the power station (% FOTA %), the date and the time at which the recording occured. Six data groups are shown. Data groups 1 and 2 contain data requested by the CEC and data group 3, 4, 5 and 6 contain data for analysis in University College Cork.

7. Conclusion

The microcomputer developed for the Fotavoltaic project is an intelligent and flexible system. It has a quick fault debugging facility and has a permanent record of the photovoltaic power station's operations.

Such characteristics are required to determine the most efficient mode of operation of a large photovoltaic power station.

FEASIBILITY STUDY FOR SMALL SOLAR CELL
OPERATED UNITS

Authors   : K.F.SORAS AND V.MAKIOS

Contract number   : ESC-R-077-GR

Duration   : 18 months 1 January 1982 - 30 June 1983

Head of project   : Professor V.Makios, Head Electromagnetics Laboratory,
                     University of Patras

Contractor   : University of Patras (School of Engineering)
                     Laboratory of Electromagnetics

Address   : University of Patras
                     PATRAS-GREECE

Summary
    This feasibility study aims in identifying all the possible present
and future applications of small stand-alone photovoltaic systems in Greece.
For the purpose of this study three major steps should be undertaken. The
first step, which has been performed already, was to interview major public
organizations who seem to be potential customers and others involved in
photovoltaics. The second step is to adopt the best methodology to analyze
potential markets. For this purpose it is necessary to estimate the size of
the photovoltaic system components before proceeding to an economic
analysis. A computer program has been written for the above and first results
are presented. The third step will be to go back to the customers with the
complete analysis and estimate the future markets.

1. Introduction

The purpose of this study is to identify all possible applications of flat plate stand-alone small solar cell operated units, up to 5 KWp, which can operate in the Greek environment. Emphasis will be given to systems which can find immediate application and can compete with existing solutions. Applications which although today seem to be very expensive and not feasible, will be also studied, in terms of becoming attractive in the near and long range future.

In the process to implement our goals a number of steps have to be undertaken. At first potential customers in Greece were interviewed to find the extend of their present needs, the rate at which these needs are growing and their willingness and ability to buy photovoltaic systems. The interviews resulted to the conclusion that it will be profitable to segment the market into groups of similar applications. The market analysis for each application requires as a first step an estimation of the system component size i.e. array size - battery storage. For this purpose a computer program has been formulated, which takes into account systems design characteristics and economic factors. The results have to be systematically analyzed for each system in order to prove its viability. The final step is to discuss the results with the potential customers and sense the future markets for such systems.

In this paper will be presented the results of the initial stage of this study. They include interviews with potential customers and the approach adopted to calculate the systems size.

2. Identification of Possible Project Applications

For this purpose contacts were undertaken with several local organizations which seem to be potential applicants and also others interested in promoting photovoltaic systems. The results of the first round discussions can be summarized as follows:

a) Public Power Corporation (ΔEH)

Although the Greek Public Power Corporation is mainly interested in large systems, they have been looking into applications of smaller systems, i.e. remote island summer houses, remote villages etc. Since they are very well informed, they can be of help in gathering various important figures, e.g. estimation of the true cost of produced KWh using diesel-generators on several Greek islands.

b) Railroad Organization of Greece (OΣE)

It seems that a very strong potential candidate for PV applications are railroad crossing signals,since Greece is in the process of railroad electrification and extension of the existing system. Discussions are underway for the use of PV systems in certain locations. Our discussions included such data as number of railroad crossing signals estimated to come into operation in the near future, description of installations currently in use, their energy requirements and cost to connect them to the utility grid via transformers.

c) Lighthouses

Greece along with Norway use the greatest number of lighthouses in Europe. Therefore PV operated lighthouses are very strong candidates for applications. Our discussions with the lighthouse authorities have shown that in many locations in Greece still antiquated and expensive methods are being used. They have decided to replace many of these by photovoltaic systems and have already replaced some of them.

d) Public Telecommunication Organization (OTE) and Greek Radio and Television CO. (EPT)

Due to the mountainous formation of Greece many radio, TV and telephone repeaters are still needed. The interest of the above organizations and the market potential is estimated to be high.

  e) Armed Forces
  We think that there could be possible applications in this area, but
we found that the people involved were not at all informed about PV Systems.
This is an area where strong public relations are still needed.

  Also contacts were made with local representatives of PV manufacturers
in order to identify their problems of selling such systems in the area and
also about promoting ideas of PV products. We have discussed the current
prices of PV components with them. They reported that due to high cost of
PV systems only public sector can still afford to buy and use such systems.
The level of awareness is generally very high among the technical staff of
public organizations and therefore strong propaganda is still needed in the
private sector.

3. Size Selection of the Photovoltaic System for a Given Application
  In this section a description will be given for the procedure adopted
in this study, for the selection of the optimum size of a small stand alone
photovoltaic system for a given application. Sizing of a system mainly means
the selection of the array size and the storage capacity of the battery,
since the above parameters of the components contribute the largest portion
in the system's cost. For this purpose a computer program is developed based on simplified design methods, using monthly average daily parameter
values, rather than detailed simulations using hourly parameter values. The
above selection was made since we want to estimate the size of the System
rather than computing its performance.

3.1. Calculation of the Monthly Average Daily Electrical Output of the Array
3.1.1. Solar Radiation Data
  There are 34 stations around Greece, which collect meteorological data.
As far as the solar radiation data on a horizontal surface is concerned we
found that only three stations use pyranometers while the others measure
hours of sunshine. A number of investigators (1)(2)(3) have applied an
Ångström-type regression equation between the duration of sunshine and the
total radiation for these stations and reported estimates of the mean
daily values for each month. The best data were available from the National
Observatory of Athens including maximum values of insolation, probability
to occur a number of consecutive days with low insolation, etc. This type
of data for solar radiation is the best available today in Greece.

  Flocas (1) has identified three distinct types of solar radiation
areas according to the values of the annual total solar radiation over
Greece and the values of the Ångström-type regression coefficients a and b:

  Type I, with three sub-types Ia, Ib and Ic with values of total global
solar radiation of 5,440-6,100 $MJm^{-2}yr^{-1}$ and values for a=0.20 and b=0.51;
Type II, with two sub-types (IIa, IIb) with values of a=0.19 and b=0.53 and
global solar radiation ranging between 5,020 and 5,439 $MJm^{-2}yr^{-1}$; and Type
III, with total amounts of global solar radiation <5,020 $MJm^{-2}yr^{-1}$ and
a=0.19 and b=0.54. The distribution of all these types and their sub-types
over Greece is shown in Fig.1.

  In Fig.2 (1) the distribution of monthly mean amounts of global radiation throughout the year is shown for 6 stations each of them representing
each type and sub-type mentioned above.

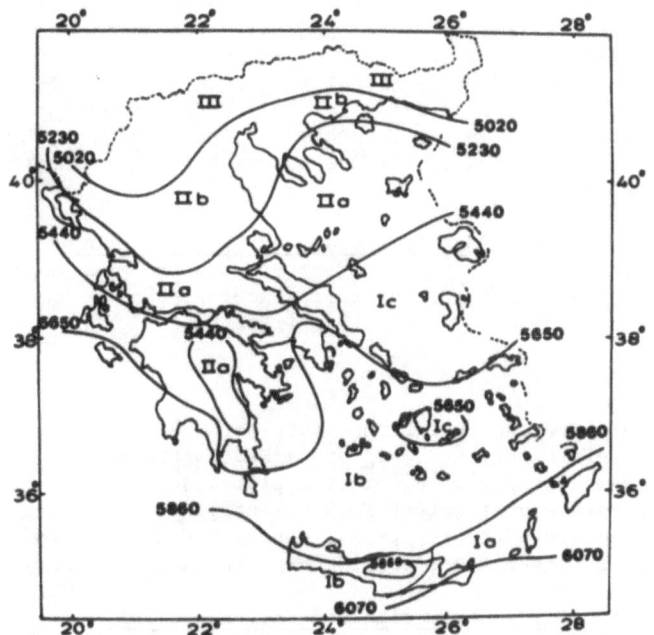

Fig.1. Estimates of global solar radiation in Greece, annual totals
(MJ m^{-2}) after Flocas (1).

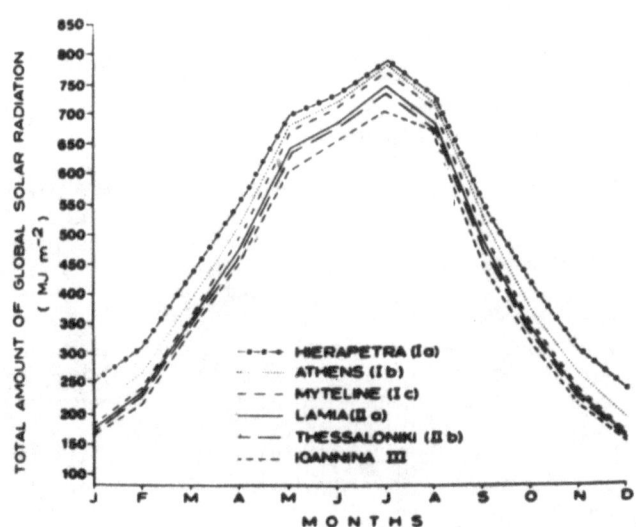

Fig.2 Seasonal distribution of solar radiation in Greece of each
representative station for each type and sub-type of global radiation,
after Flocas (1).

3.1.2. Radiation on a Tilted Surface

The next step was to convert the above radiation data to those on a tilted surface facing towards south for different angles of tilt. The method adopted was that of Liu and Jordan (4) for long-term monthly averages of the daily values of solar radiation on a horizontal surface. This method has been verified by actual long-term data world-wide and the results of Liu-Jordan's method agrees closely with those of modern, more sophisticated methods (5). Several doubts (6), published, concerning the Liu and Jordan's correlation of diffuse to total radiation are not important for flat plate PV systems.

3.1.3. Calculation of the Monthly Average Array Efficiency

Using typical values of array parameters and the mean monthly ambient temperature, the mean monthly array efficiency, $\bar{\eta}$, can be computed through a simplified method, given by Evans (7), from the following equation:

$$\bar{\eta} = \eta_r \left[1 - \beta (T_c - T_r) \right]$$

where T_c = monthly average cell temperature
  η_r = array efficiency at $T_c = T_r$ and 1 KWm^{-2} insolation
  T_r = reference temperature for cell efficiency
and β = temperature coefficient for cell efficiency.

Thus we can compute the electrical energy at the output of the array which is available to the power conditioning unit and the load.

3.2. Selection of optimum array size and storage capacity

Having calculated the monthly average daily radiation on a tilted surface, $\bar{H}_T$, and the monthly average array efficiency, $\bar{\eta}$, the monthly average daily electrical energy at the output of the array, $\bar{E}$, is given by the following formula (8) in KWh/day:

$$\bar{E} = A \, \bar{\tau} \, \bar{H}_T \, \bar{\eta}$$

where A is the array area
  and $\bar{\tau}$ monthly average transmittance of the array cover.

On hand of Figure 3 the selection procedure of sizing the array and storage capacity will be discussed.

Depending on the specific application different components either separately or in combination can be used, i.e. DC-DC converter, DC-AC inverter, voltage regulator. The DC-DC converter can actually power track on the array side and float with the battery on the battery side. The DC-AC inverter is needed when we have an AC load and the voltage regulator protects the battery from overcharging and excess discharging. In the example, which will follow for a specific system, say a microwave repeater with constant load during daylight hours, only a DC-DC converter will be considered to show the sizing procedure.

The energy at the output of the DC-DC converter is $\eta_c \bar{E}$, where η_c is the efficiency of the converter. From Siegel et al.(8) a critical radiation level, G_c, is defined at which the rate of electrical energy production is equal to the load, which along with K_T (ratio of monthly average total to extraterrestrial radiation on a horizontal surface), latitude and orientation computes (9) the monthly average daily utilizability, $\bar{\varphi}$. Then the monthly average daily energy available to charge the battery equals $\eta_c \bar{E} \cdot \bar{\varphi}$ while the rest $\eta_c \bar{E}(1-\bar{\varphi})$ supplies directly the load.

On the average each month the load accepts $\eta_c \bar{E}(1-\bar{\varphi})$ directly from the array plus $\eta_B \eta_c \bar{E} \bar{\varphi}$ from the battery. It is assumed that the effective

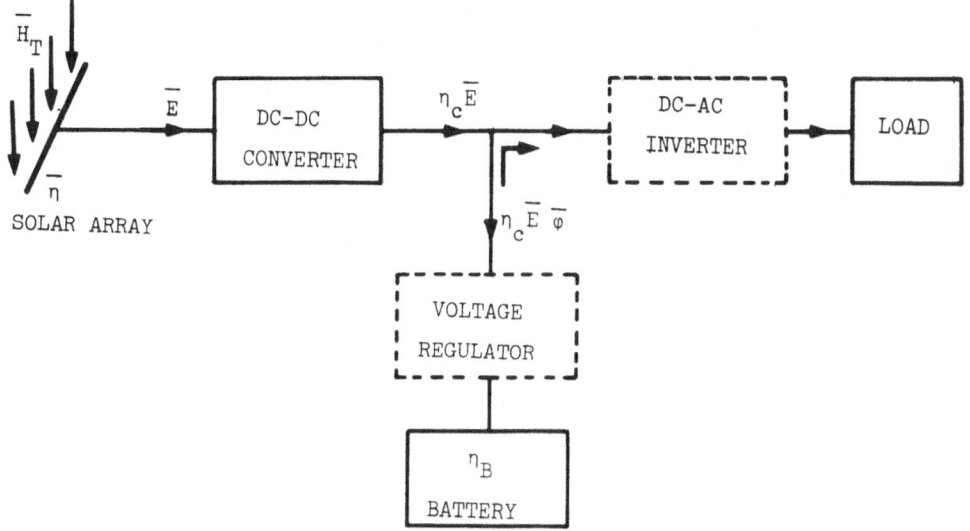

Fig.3. Energy flow diagram.

battery capacity consists of the sum of the long and the short term storage capacity, which account for the transfer of energy to the winter months and daily variations plus night supply respectively. For a given array size the long term storage capacity is given by the sum of monthly energy deficits where for a specific month the deficit is:

$$\left\{L - \left[\eta_c E(1-\overline{\varphi}) + \eta_B \eta_c \overline{E}\,\overline{\varphi}\right]\right\}^+ \times D\,P\,M$$

where DPM is the number of days per month
and "+" means that only positive values are encountered.

The short term storage capacity has to be selected initially as the number of consecutive days for which the battery is able to supply the load. The probability of occurence of such low insolation periods gives the probability of systems failure and is taken from statistical meteorological data (2).

Up to this point we are able to compute the battery size, B, for a given array size, A. However, there are two array limits, A_{min} and A_{max}, between which the array size should lie (10). A_{min} is defined as the minimum area able to produce the yearly load. A_{max} is the upper limit at which the array is capable to cover the load even at the month with the lowest insolation.

A computer program has been formulated to find array size-battery storage pairs capable to cover the load year round and select the most economical pair among these. This program has two loops: The inner loop is concerned with values of A (between A_{min} and A_{max}) while in the outer loop tilt angle is varied from latitude $-20°$ to latitude $+20°$. Each pair corresponds to a candidate system, for which the life-cycle cost is calculated in order to find the optimum selection.

3.3. A Sizing Example

Following, we select an example of a system in the Patras area and we present initial results from our computer program. Because the parameters

that vary are many, i.e. efficiency of the array, battery performance
characteristics, economic factors etc., the results are only preliminary
since we have chosen only typical values.
     The input data for our computer program are the following:
     a) Site characteristics: Latitude = 38°N

M	J	F	M	A	M	J	J	A	S	O	N	D
$\overline{H_i}$	6897	10241	13376	16720	22990	24035	25916	24035	18392	12540	7524	6897
T_M	10	10	12	15	20	24	26	27	23	18	14	11

$\overline{H_i}$ = monthly average insolationon horizontal surface (KJ/m² day)
T_M^i = monthly average ambient temperature (°C).

     b) array characteristics: 20 years lifetime
          In this example calculations were performed for two array
          efficiencies η=8% and η=10% at 25°C and 1 KW/m² insolation
          and two array costs 10 $/$W_p$ and 15 $/$W_p$.
     c) Battery characteristics: Battery efficiency $η_B$=85%
                                 Depth of discharge DOD=80%
                                 Life of the Battery 5 years
                                 Battery cost 100 $/KWh.
     d) Load characteristcs: Total daily consumption 2.4 KWh/day
                                 of which 35% during daylight; continuous
                                 power load 100 W.
     e) Economic factors: Systems lifetime 20 yrs
                                 Inflation rate   20%
                                 Discount rate   25%
                                 Array and battery maintainance cost 1%
                                 of initial.
     In Figure 4 life cycle costs vs. array area are presented with para-
meter values η=8% and η=10%, tilt angles 38° and 58° due south, and array
cost 10 $/$W_p$ and 15 $/$W_p$ with a constant 100 $/KWh battery cost. In Figure
5 the life cycle and initial cost vs. array area are shown for η=10%.
     From Figure 4 it is apparent that the optimum solution, for the above
system and the parameters chosen, is:

                η = 10%; tilt angle 58°; array area 9.6 m²;
                        storage capacity 16 KWh
          Initial cost $ 19,250 and Life Cycle Cost $ 23,900

     The above results are only a first attempt torwards the solution of
the problem and a more systematic approach is still needed to draw final
conclusions.

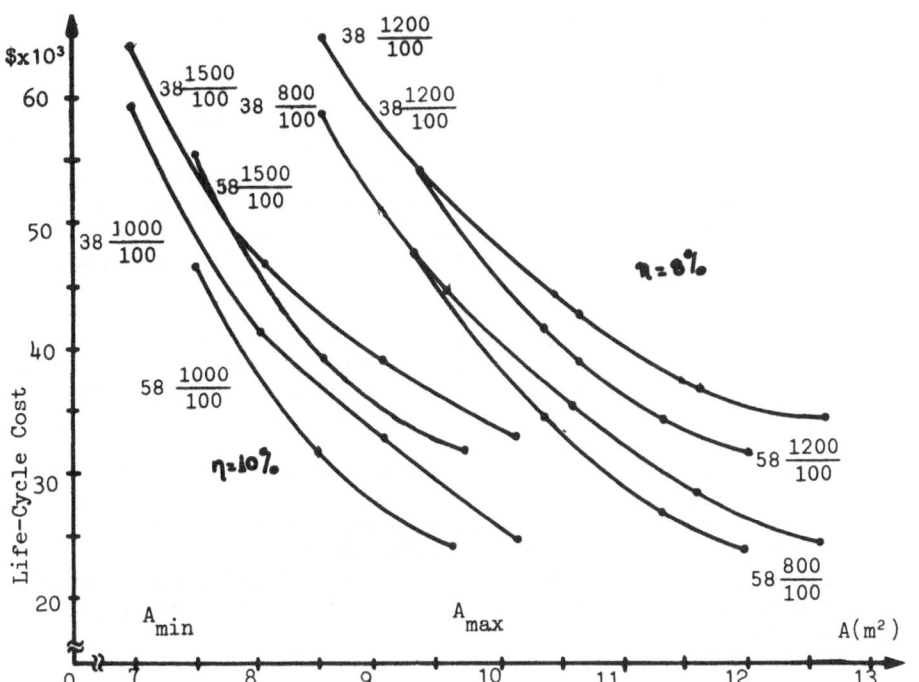

Fig.4. Life-Cycle costs vs array area for η=8% and η=10% tilt angles 38° and 58°.

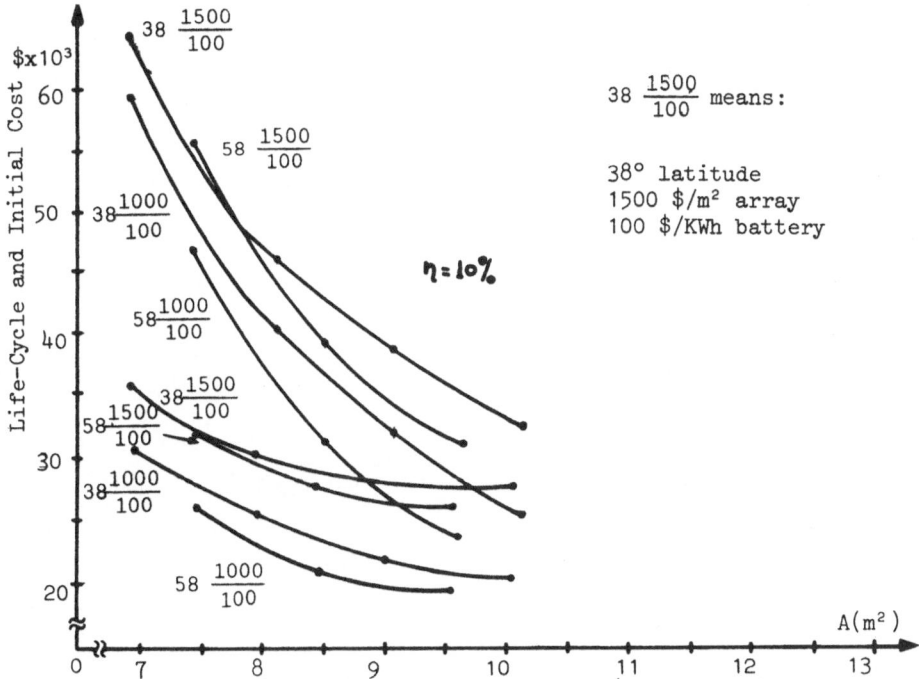

Fig.5. Life-Cycle and Initial costs vs array area for η=10% and tilt angles 38° and 58°.

ACKNOWLEDGEMENT
    The authors wish to thank Mr. Ch.Bouras for his valuable help in the
computer program.

REFERENCES
(1) A.A.Flocas, "Estimation and prediction of solar global radiation over
     Greece", Solar Energy, Vol.24, pp.63-70 (1980).
(2) B.Katsoulis and C.Papachristopoulos, "Analysis of Solar Radiation
     Measurements at Athens Observatory and Estimates of Solar Radiation in
     Greece", Solar Energy, Vol.21, pp.217-226 (1978).
(3) G.Makris, "On the Distribution of Solar Energy in Greece", Memoirs of
     the National Observatory in Athens, Ser.II, Meteorology, No.43 (1976).
(4) S.A.Klein "Calculation of monthly average insolation on tilted sur-
     faces", Solar Energy, Vol.19, pp.325-329 (1977).
(5) Manuel Collares - Pereira and Ari Rabl "The average distribution of
     solar radiation - correlations between diffuse and hemispherical and
     between daily and hourly insolation values", Solar Energy, Vol.22,
     pp.155-164 (1979).
(6) D.W.Ruth and R.C.Chant, "The relationship of diffuse radiation to
     total radiation in Canada" Solar Energy, Vol.18, pp.153-154 (1976).
(7) D.L.Evans, "Simplified method for predicting photovoltaic array out-
     put", Solar Energy, Vol.27, pp.555-560, (1981).
(8) M.D.Siegel et. al. "A simplified method for estimating the monthly-
     average performance of photovoltaic systems" Solar Energy Vol.26
     pp.413-418 (1981).
(9) S.A.Klein "Calculation of flat-plate collector utilizability"
     Solar Energy, Vol.21, pp.393-402 (1978).
(10) A.Kipperman "The photovoltaic solar system, analysis and basic design
     rules" Proceedings of the fourth E.C. Photovoltaic Solar Energy
     Conference p.280 (1982).

PHOTOVOLTAIC SYSTEM FOR A SOLAR HOUSE

Author   : J. SCHMID

Contract number   : ESC-R-083 D

Duration   : 18 months
                        1 January 1982 - 30 June 1983

Head of project   : Dr. J. Schmid

Contractor   : Fraunhofer-Institut für Solare Energiesysteme

Address   : Fraunhofer-Institut für Solare Energiesysteme
                        Oltmannsstr. 22
                        D-7800 Freiburg

Summary

The first German utility-interactive photovoltaic residential system
is under construction in Munich. The project is managed by the
Fraunhofer Institute for Solar Energy Systems (ISE) in Freiburg and
supported by German Industry and the Commission of the European Communities in Brussels as well. The building, which is designed for
optimal passive heat gain will be equipped with a 60 m² solar cell
area which will provide about 5 kW_p. For grid-conncection, a highly
efficient inverter is used, which was developed by the ISE especially
for solar energy purposes. Its main features are: low harmonic distortion, efficiency well above 90 % over full power range and inherent
safety functions for grid-connection. Completion of the whole system
is expected in mid '83. A five year test period is planned to obtain
all necessary information concerning efficiency, reliability and grid
interference.

1. General

Electric power generation with small distributed photovoltaic power systems can be economically attractive, when a connection to the utility is possible and when solar cell costs decrease further /1/. Because power is produced at the same place as it is used, distribution losses in the grid do not occur. Since private households consume more than one third of the total produced electric energy /2/, private houses are well suited for photovoltaic power systems. Its roof areas are satisfactory for the needed power production. Since there is no such grid-connected photovoltaic system in Germany, the Fraunhofer Institute for Solar Energy Systems has decided to erect such a plant. The main target of the experiment is to test all necessary components under realistic conditions, to evaluate the construction costs and to find good methods for a harmonic integration of the solar cell-areas into the roofs. Solar cells and batteries for the experiment are financed by German industries and the activities of our institute are supported by the Commission of the EEC.

2. List of Project Participants

Fraunhofer Institute for Solar Energy Systems	- Project management
	- Integration of the system
	- Inverter
	- Conduction of measurements
Prof. Dr. T. Herzog	- Building's architect
Siemens AG, München	- 30 m² solar cells
AEG, Hamburg	- 30 m² solar cells
Varta, Kelkheim	- 10 kWh storage batteries
City of Munich -Utility Department-	- Measurements on the grid

3. Technical Details of the Photovoltaic Plant

Figure 1 shows the principle of interaction between the photovoltaic generator, the electricity consumers and the utility grid, as it is planned for the experiment: During high solar radiation, the solar cells can provide the electric power need of the consumers. If there is no sun, the power is supplied by the grid in a conventional manner. During solar radiation, when most consumers are switched off, the excess power is fed into the grid.

Short term storage of electrical energy is provided by a set of 10 kWh storage batteries. In principle storage is not necessary, but during the tests we want to test the impact of such damping elements on the grid's voltage stability. For the interface between the photovoltaic generators and the grid an inverter is used which converts the solar DC to synchronized AC. This part is a vital element of the whole system. It must have a high conversion efficiency over a large power range, it must supply AC with low harmonic power distortion, it must withstand lightnings and other overvoltages in the grid and it must be cheap. Further it has to fulfil the safety conditions for grid connection the main requirement being that it stops power feeding into the grid immediately when the grid is switched off by the utility.

Because there was no such device on the market, we tried to develop an inverter which could meet all the mentioned requirements. The result was a unit which consists only of electronic elements without any parts such as

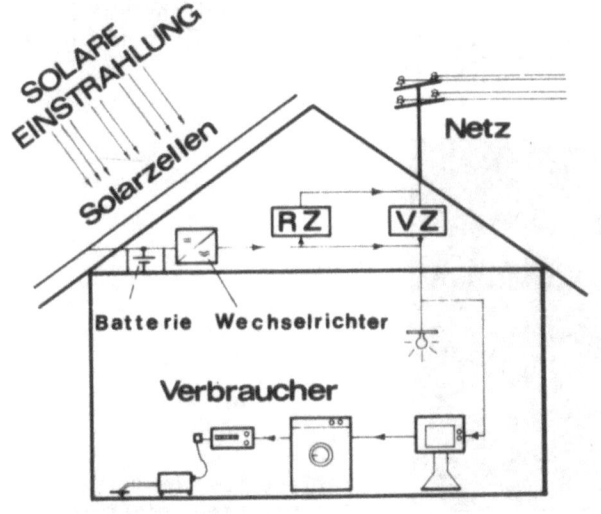

Fig. 1

Utility interactive photovoltaic residential system

transformers or coils. Its basic principle is to switch as many solar cells in series as necessary to get the same voltage as the grid at each moment. In order to save switching stages we did it in a binary mode so we could get 16 voltage levels from 4 stages. Figure 2 shows the simplified circuit diagram of the inverter.

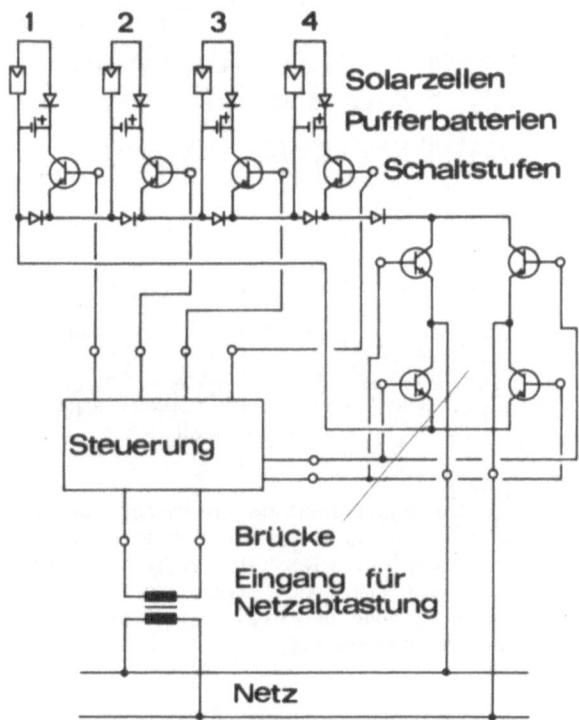

Fig. 2

Simplified flow-sheet of the FhG-inverter and its connection to the grid

The solar cell areas and its transistor switches are numbered by 1 to 4. The binary control of these transistors is done in an extremely simple manner: The grid's voltage is sensed by a fast ADC with binary output. These output signals can be directly used to control the transistors. So by definition the inverter's output voltage is proportional to the grid's voltage in a 16 step-approach (see Fig. 3).

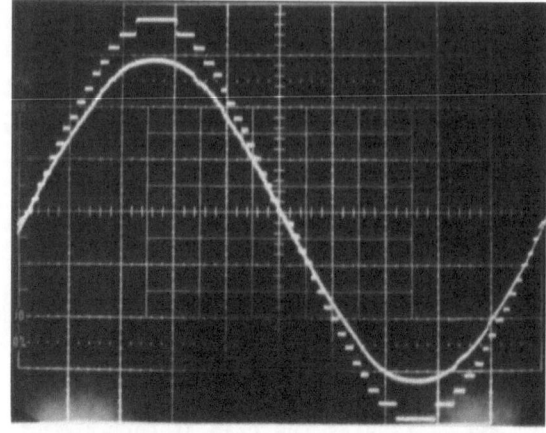

Fig. 3

Perfect synchronization of inverter output to the grid

The following features are maintained by this concept:
a) The synchronization is given without any time delay and so the inverter can be coupled to grids of different frequencies without any change (we tested it between DC and 200 Hz).
b) The inverter's output does not only follow all given frequencies but all given voltages as well. Because of that there is no excess power to the grid, when its voltage is decreasing.
c) When the grid is switched off, the sensing element will sense zero voltage and force the output to stay at zero. So no power is fed into the grid.
d) Absolutely silent. This is an important fact because many inverters make such a noise that they must be separated from the living area in dwellings.
All necessary safety functions are therefore inherent and no additional unit has to be used.

In our laboratory we have built two prototypes. Figure 4 shows a 1 kW inverter. Most of the volume is used for the cooling ribs of the transistors. The transistors are of the power MOS-type and need very low power control signals so that no power interface between the control device and the transistors is used. The "on"-resistance of the transistors is extremely low so that the whole device has the excellent efficiency which is shown in Fig. 5. Even at 10 percent load, it is well above 90 %.

4. Integration of Solar Cells into the Roof
The building we have choosen for the experiment is optimized for passive solar heat gain. The total roof area of the house, which is shown in Fig. 6 is glazed. The basic idea was to replace glass sheets by frameless solar cell modules. In that case we can get full roof credit for the solar panel and its mounting costs. The costs for the photovoltaic plant are reduced to the wiring and the rest of the components.

Fig. 4   1 kW FhG-inverter during its first tests

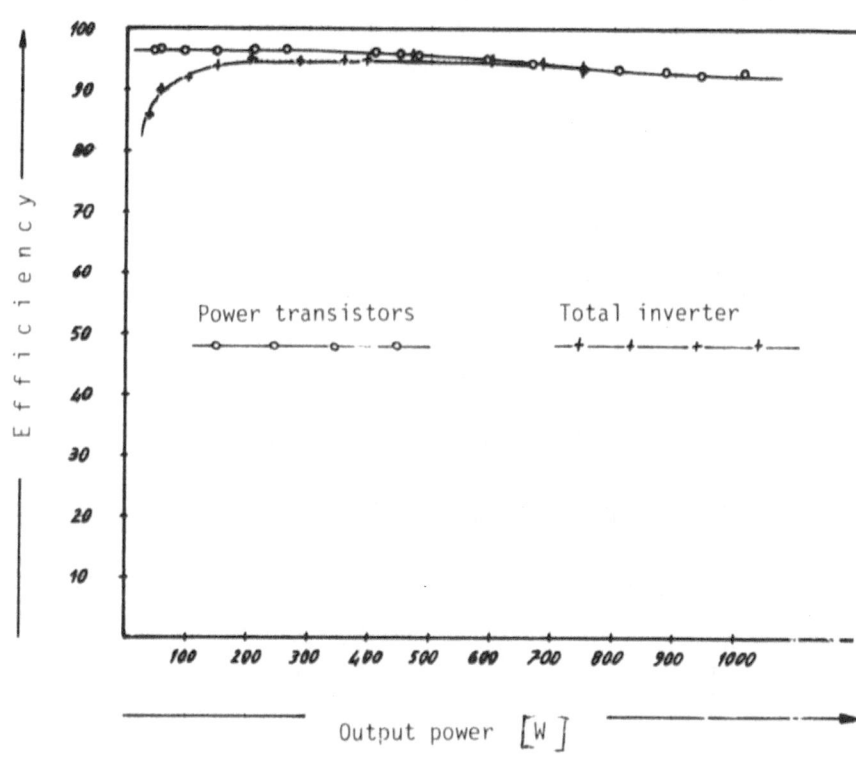

Fig. 5   Efficiency of the inverter between 5 and 100 % of rated output

Fig. 6   Photovoltaic residence

5. Status of the Work

15 m² of solar cells are already integrated into the roof. Further 15 m² are to be installed in the near future. The inverter has performed successful long term tests of 4000 hours without any failure. Preparation of the 30 m² experiment area inside the photovoltaic house is in progress. The final system is expected to be finished in mid' 83. A five year test period is planned after completion.

6. References

/1/ P.L. Temple, D.E. Mahone
     Photovoltaic Systems, Retrofitting to Residences
     Photovoltaics, Premiere Issue, June/July 1982
/2/ P. Schnell, M. Delhi
     Die Einsparung von Energie - ein Schwerpunkt
     energiewirtschaftlicher Aufgaben
     Energiewirtschaftliche Tagesfragen 9, 1979

HYBRID THERMAL AND PHOTOVOLTAIC
CONCENTRATION COLLECTOR

Anthor	:	Ph. BUFFET
Contract number	:	ESC-R-034-F
Duration	:	24 months
Head of project	:	Ph. BUFFET
Contractor	:	SEP (Société Européenne de Propulsion)
Adress	:	B.P. 802 - 27207 VERNON (France)

SUMMARY

SEP is developping a hybrid Solar Energy Concentration Collector in which either Electrical and Thermal energies are collected. Such a concept is expected to save capital costs by sharing the most part of it between both thermal and eletrical users and by reducing the photovoltaic cells area.

After a short presentation of the principle, a progress status is given in which an important delay is due to the cancellation of the concentration cells production by the accepted producer, leading to the choice of a second manufacturer.A prototype built to represent as closely as possible the definitive industrial product is described and, because of cells shortage, only thermal performances tests are reported.

A provisionnal energy production estimate made with some assumption on electrical yield is compared to a cost analysis in order to give figures for an economical viability assessment.

1. PRINCIPLE

The principle of hybrid Solar Energy Collection consists in recovering the thermal Energy which has not been transformed in Electricity by photovoltaïc cells in concentration collectors. Such a concept allows to share most of the collector capital costs (concentrator, sun tracker , supports and set up) between both thermal and electrical users and to reduce consequently the cost of Energy.

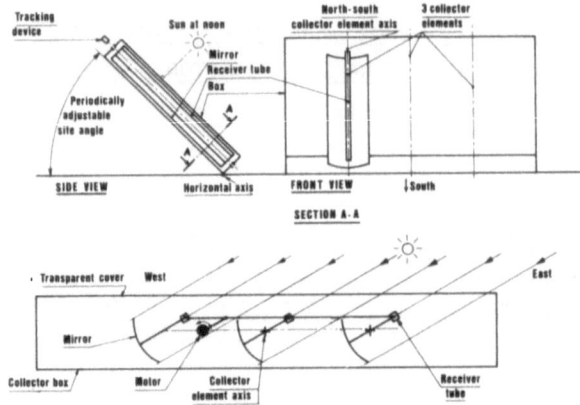

Fig. 1

In the SEP development, as shown in fig.1 above, the collector is made of a battery of cylindro-parabolic collector elements, North South Oriented, assembled on a single frame and rotated together to track the sun. The frame either can be incorporated in a box with a top glass cover or, as an alternative, mounted under a glass roof. This design is interesting because the collection function is separated from the protection function which allows to use light, accurate and economic mirrors.

Each collector element, is a thin reflecting foil trough mirror with an associated receiver tube. Two shaped supports are used to maintain the shape of the mirror and as a frame to link it to the receiver. This last is a tube onto which a row of concentration solar cells is stuck and in which water or any cooling fluid flows.

2. PROGRAM

The aim of the contract was the definition of a hybrid collector able to be manufactured industrially at low costs and the assessment of the economical viability. Consequently, the work was divided in three tasks, as shown on the planning chart below.

		1980	1981	1982	
SPECS	MODEL CHOICE OF APPLICATION SPECIFICATIONS		[SORINTO MEETING]		
DESIGN CONSTRUCTION	DESIGN CALL FOR TENDER CONSTRUCTION	(*)			
PROTOTYPE CONSTRUCTION	CALL FOR TENDER DESIGN WITH MANUFACTURER A CHOICE OF MANUFACTURER B PROCUREMENT PRELIMINARY STICKING TESTS 1st RECEIVER EQUIPMENT	CELLS	CANCELLATION	1st BATCH EXPERIMENTAL	
TESTS	ADJUSTMENTS THERMAL PERFORMANCES ELECTRICAL PERFORMANCES		(*) PROTOTYPE CONSTRUCTION		

Task 1: Specifications
This work was reported at the Sorinto Meeting. It consisted in detailed Specifications establishment on the basis of a theoretical performances analysis and of a specific application choice.
Task 2: Prototype design and construction
The progress of this work has been marked by the fact that, for industrial policy reasons, the cell manufacturer who was choosen decided to give up his concentration cell production. This main event resulted in :
- deletion of the projected cooperation in sticking and encapsulation process
- delays for a new negociation with a second manufacturer
- delays again for cells delivery, production problems beeing not all solved

The expected delivery date for the whole cells is end of december
Task 3: tests
This task is not fully carried out because of the lack of cells explained before. However general operation and thermal performances tests have been performed.

3. SPECIFICATIONS (as a reminder)
    The secluded individual dwelling (with or withount air cooling) was found to be the major application theme and the following specs were issued:

3.1 Performances
    Instantaneous efficiency given by

    Thermal $\eta_{th} = 0.72 - 1.36 \dfrac{Tc - Ta}{Wi}$

    Electrical $\eta_{el} = 0.123 - 0.00046\ Tc$

    Fluid temperature range : $20°C < Tc < 100°C$

3.2 Maintenance & Reliability
    Conformity of casing to building weathering standards
    Casing tightness to dust
    Upper glazing washability
    Life time : 20 years except mirrors, seals, hoses : 5 years
    Utilisation factor: 95 %

3.3 Main design guideline
    6 mirrors per casing
    Cells : 2 cm x 2 m; 57 cells/row ; 45 Wp/row
    Glazing : White glass $\tau = 0.915$
    Mirrors : aluminium $\rho = 0.86$
    Effective concentration C = 20

4. PROTOTYPE DESCRIPTION
    The prototype had to be designed in a way to represent as closely as possible the definitive industrial product. it is shown on the next page picture.

The main frame and the casing are made of brake formed sheet. The glazing is made from normal glass instead of better quality white glass which was not found commercially available. Glass is fasten to casing with the help of standard extruded profiles.

The six collector elements are attached on the frame via plastic plain bearings which would normally be injection moulded. The elements are linked together and to the driving tracking unit by a cog-belt cascade. The maximun tracking error is 15'.

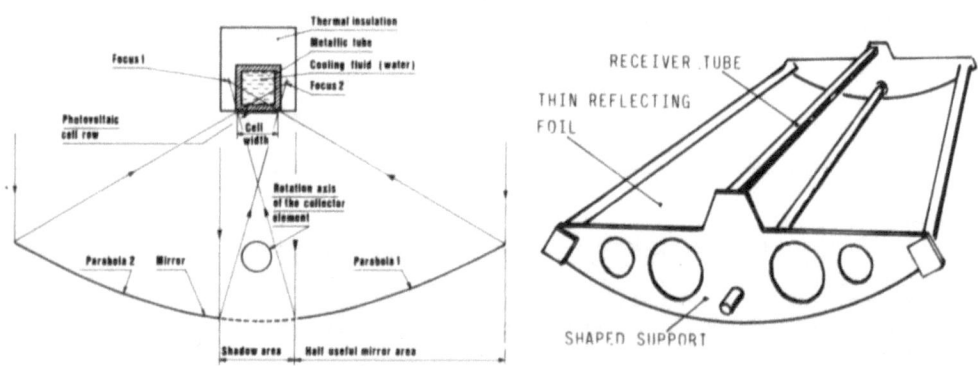

Fig. 2   Fig. 3

The cross section of each collector element is given in fig. 2 which shows the mirror profile composed of two arcs of distinct paraboloe arranged in such a way that the two convergent beams reflected by each of them cross exactly on the whole width of the cells leading to a nearly uniform resulting illumination.

As presented in fig. 3 the mirror is a thin reflecting aluminium sheet bent onto two shaped moulded supports linked by two longitudinal square rods on which the sheet edges are riveted. Aluminium sheet (ρ = 0.86) was Preferred to thin flexible tempered glass (ρ = 0.90) because bending creates parasitic stresses modifying the geometrical profile ; those stresses can be avoided by a sheet prerolling with aluminium but not with glass. On the other hand aluminized plastic films were discarded due to their rather low spectral reflectivity (ρ = 0.77) and no evidence of good aging.

The main caracteristics of a collector element are the follwing :

length	overall	:	1370 mm
	cells	:	1200 mm
width	overall	:	583 mm
	useful aperture:		550 mm
Pitch between two elements:			730 mm
Effective concentration			
ratio		:	20
(nominal, with glazing)		:	
cells	width	:	20 mm
	length	:	30 mm
	number	:	40 cells per row
	efficiency	:	14 %
peak power	electrical	:	50 W
(800 W/m2)	thermal	:	500 W

5. RESULTS

This work has been carried out partly as explained in paragraph 2.
Only general operation and thermal performances tests were performed.
The former were found satisfactory while the later were conducted
by replacing the unavailable solar cells by a nearly thermally equivalent
black coating.

The experimental law for THERMAL EFFICIENCY was found to be :

$$\eta th = 0.63 - 3.66 \frac{Tc - Ta}{Wi}$$

The rather low optical factor (0.63 instead of theoretical 0.72) is explained mainly by the use of normal glass, the transmission coefficient of
which is only 0.85 instead of 0.92 and also by possible reduction of aluminium reflectivity (during preroll process ?) Anayways , it seems rather
uneasy to improve the opticol factor.

On the other hand, the very bad heat l ss coefficient (3.66 instead
of 1.36) has been found obviously due to a bad insulation design an will
be easy to reduce down to 2.

ELECTRICAL tests were possible with only one experimental cell batch
equipping one tube receiver. But, infortunately some cells break down appeared a few days after sticking process, in spite of satisfactory sticking
preliminary tests. Afterwards, the bad weather conditions did not make possible any valuable test.

6. YEARLY ENERGY PRODUCTION ASSESSMENT

In spite of the lack of definitive experimental results, we tried to
have provisional effciency estimates. For thermal efficiency, as staded
before, the following formula seems likely to be satisfactory :

$$\eta th = 0.63 - 2 \frac{Tc - Ta}{Wi}$$

For eledrical efficiency, a plausible estimate can be found from the theoritical equation by correcting it with the ratio of experimental to theoritical optical factor. It gives :

$$\eta el = 0.108 - 0.0004 Tc$$

As an examples, calculations of yearly production were made for Marseille,
in south of France, for fluid at 50°C (sanitary hot water) and at 90°C
(air cooling). Results are the following :

	Fluid at 50°C		Fluid at 90°C	
	Elect.	Heat	Elect.	Heat
Production KWh/m2	150	830	125	665
Average Yield	8,5 %	47 %	7 %	38 %

7. COST BREAKDOWN

A cost analysis was performed on a 500 collector/year basis (2000 m2/year). This figure may seem rather small but the production rate will probably be low the first years and an extrapolation to bigger series would have seemed questionable. The cost break down is the following :

TROUGH

Reflecting sheet	0.9 %	
Frame	7.9 %	35 %
Cells (encapsulated)	25.2 %	
Assembly	1.0 %	

TRACKING

Sensor & Electronics	4.8 %	
Motor & gear box	8.0 %	16 %
Belts & Pulleys	3.2 %	

CASING

Box & Trough supports	31.7 %	
Glazing cover	3.0 %	36 %
Hoses & piping	1.3 %	

SUPPORTS   6.5 %

ASSEMBLY & INSPECTION   6.5 %

                                                      100 %

And the cost related to the useful area (i.e useful aperture X cell length) is 7500 F/m2.

Assuming that the price of energy is 2 FF/KWh electrical and 0.5 FF/KWh thermal, one founds, on the previous paragraph calculation basis, that each m2 brings in a return of 715 FF/year for fluid at 50°C and 583 FF for fluid at 90°C. This leads to a pay back time of respectively 10.5 and 12.8 years.

CONCENTRATION

Test and demonstration of concentrating photovoltaic generators
SOPHOCLE under mediterranean climatic conditions

Conversion of solar energy using fluorescent collectors : installation of a test collector to deliver several watts power

Holographic thin film system for multijunction solar cells

High concentration P.V. 100 Wp module making use of spectrum
splitting and Si-GaAlAs coupled cells

TEST AND DEMONSTRATION OF CONCENTRATING PHOTOVOLTAIC GENERATORS SOPHOCLE UNDER MEDITERRANEAN CLIMATIC CONDITIONS

Authors   : A. DUPAS, D. ESTEVE, B. LAURENT, G. VIALARET

Contact number : ESC-R-035 F(G)

Duration   : 36 months   1 July 1980 - 30 June 1983

Head of project : A. DUPAS, D. ESTEVE

Contractor   : C. N. R. S.

Address   : P. I. R. S. E. M.
                       282, Boulevard Saint-Germain
                       75007 PARIS

                       L. A. A. S.
                       7, Avenue du Colonel Roche
                       31400 TOULOUSE

Summary

The SOPHOCLE program (Solar Photovoltaïc with Limited Concentration of Energy) has for its purpose the development and the experimentation of a family of photovoltaïc generators using concentration of Solar rays (C=45). These generators make use of an altazimut mounting with two axis tracking. The conception of the system is modular with Silicon Solars Cells and Concentrating Fresnel lenses integrated six by six in passivly cooled modules.

After an international program on tests of reliability with low power SOPHOCLE systems, the L.A.A.S. has undertaken an experiment on 500 W_p - generator provided with a DC - motor-pump in parallel with a similar experience at the University of Athens. The aim of this experience, supported by EEC, is in the one hand to analyse the individual behaviour of the modules, concerning watertightness, insulation and disparity of electric characteristics on the other hand to evaluate the energy production line generator-converter - motor-pump. In this periodic report, we are presenting the means and the first results of this experience.

1.1. Introduction

This program research, supported by EEC (Contract Nr ESC-R-035 F(G)), comes within SOPHOCLE program of C.N.R.S., which began by the demonstration of the feasibility for a photovoltaïc concentrating system with Fresnel Lenses. Then eight 100 Wp-prototypes have been erected in various climatic environments to be tested under difficult conditions. To complete these informations, it was necessary to study a more powerful system with an utilization; so a 500 W generator for pumping was set up in L.A.A.S. (Toulouse). At the same time, another system is working in Greece, at the Physical Electronics Laboratory of the University of Athens directed by Professor Caroubalos, and will supply a microwave telecommunication relay.

The aim of this study is to precise :
* the efficiencies of the different elements generator-converter-motor-pump and the role of MPPT-DC/DC Converter ;
* the behaviour of the concentrator modules, concerning the both important problems of watertightness and insulation, which appeared on the prototypes (Results of international Sophocle 100 program) ;
* the effective ageing of physical and electrical characteristics ;
* the role of Cells - or modules - Characteristics disparity on the system efficiency ;
* the total and direct irradiation on various surfaces (horizontal, tilted or oriented 2-Axis).
  This periodic report describes the data acquisition system and the first results ; the data processing of records will allow a characterization of ageing.

1.2. Some conclusions of SOPHOCLE 100 Program (1980-1982)

Thanks to 100 Wp-generators with seven track magnetic cassette recorder, we have characterized particularly
- the solar irradiation, especially in LIBREVILLE and ALGER ;
- the SOPHOCLE'S behaviour (dependance of the efficiency up on temperature, irradiation and time.)
  Reliability was not assured on two points :
- watertightness (modules outof service because leaks) ;
- insulation (between cell and heat sink).
  It is necessary to find solutions to these both problems. 500 Wp-prototypes make clearer the second default, because its output voltage is higher (current leakage more important).

Except for these problems, this program has shown the good behaviour of heliostat, Fresnel lenses, M.P.P.T. and passive cooling, hence the interest of this type of concentration.

2. SOPHOCLE 500 Program (1981-1983)

Complementary to Sophocle 100 program and in the aim to solve these problems, a 500 Wp-prototype has been erected at L.A.A.S.(Toulouse) in 1982.
A lot of steps were necessary to see this experience through :

2.1. Preparation of the ground

A 6 meter-deep well and a basin to keep water have been bored.

2.2. Setting up of the generator

A 12 m²-heliostat with 40 modules (from which 12 special modules with electric output between each cell) is fixed to a concrete base.

2.3. Choice of the motor-pump

We have associate a DC-motor (with permanent magnets) with a volumetric pump. To charge this group, a gate give a drop in pressure, which increases with the flow from 0 to 6 bar. No storage is required ; a M.P.P.T. allows adaptation between generator and pumping.

2.4. Definition of measure system in two parts

- permanent record by a "Schlumberger Solartron 3430" ;
- automatic characterization of cells or modules with a calculator X1.

2.5. Using of computer for data processing

3. Measures'organisation

3.1. Permanent measures

Table I precises the different recorded parameters, we can classify in 3 groups :
* meteorological measures ;
* electrical measures ;
* hydraulic measures.

All these measures are recorded on cassettes, then processed on calcula-tor "SEL" to be changed into daily tables or curves.

3.2. Automatic characterization

The described system (figure 1) allows to study in the same conditions cells or modules characteristics. A storage on discs makes easier a data accumulation, which will give the evolution of Cell'performance and the in-fluence of diverse parameters.

The principle consists to use a variable power supply to make the cur-rent vary in a circuit with a resistance R, a solar cell under irradiation and a shunt to measure current intensity.

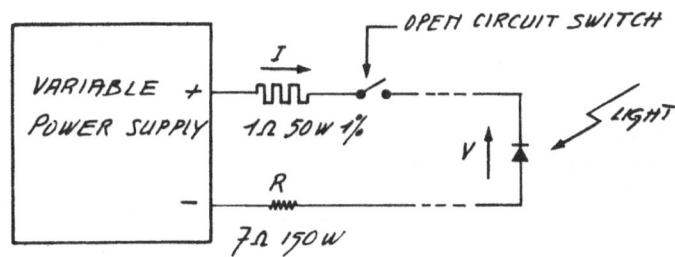

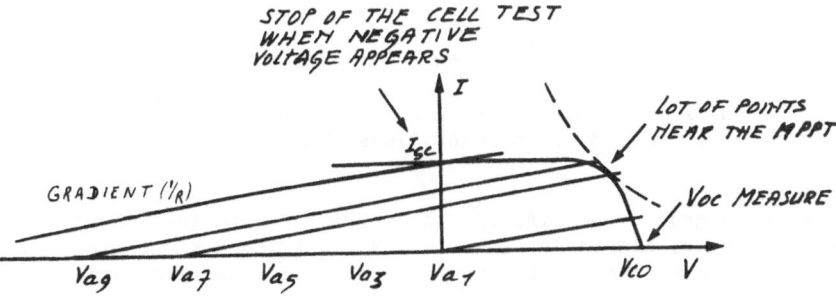

For each value V_{ps} from the power supply, a couple $(V,I)_i$ is stored. The data processing gives V_{oc}, I_{sc} and P_{max}.

For the extension to group of cells, it is necessary to use a protection diod for each cell, which avoids to be limited by current I_{cc} of the worse cell.

4. First results

4.1. Role of module's disparity on system efficiency

* Fig. 3 shows the efficiency loss due to two bad modules in a serie of 14 modules. Without diod in parallel an every module, current couldn't be greater than short circuit current of worse cell. Diod limits the loss of voltage to 3.6 V;
* two different series of 14 modules connected in parallel have an efficiency more or less equal to the average of efficiencies (figure 4) ;
* Efficiency of a module is near of the worse cell efficiency.

4.2. Efficiency of the production line (figure 5)

The average efficiencies of the different elements are :

GENERATOR	CONVERTER	LINE	MOTOR	PUMP	TRANSMISSION
9,5 %	90 %	92 %	95 %	80 %	75 %
			MOTOR-PUMP (p $\simeq$ 5 bar)		
			56 %		

The role of M.P.P.T. is very important with a volumetric pump when hydraulic head is high. Without it, the system doesn't run.

4.3. Insulation

This problem has been solved with insulation of the modules from the heliostat structure by new fastenings. This attitude obliges to keep an output voltage under 48 V for security.

S.N.I.A.S. has used a 3M-film, which ensures a good insulation, but increases cell temperature of ten C-degrees (Mali-2kW experience).

4.4. Watertightness

The pressure variation inside the module caused expansion and leakages apparition. To prevent that, we have used a membrane, which ensures a constant pressure by its volume variation (photo 3). This system is until today satisfying.

The ageing takes a longer time and will be characterized with the help of stored data at the end of contract (June 1983).

5. Conclusion

With these means, the L.A.A.S. can study in depth the behaviour of SOPHOCLE generators and their evolution. New solutions to problems of watertightness and insulation are tested in order to define a reliable system.

Parallel works which may lead to a second generation of concentrating system are conducted by P.I.R.S.E.M. and L.A.A.S. to develop :
- cells with better conversion efficiency going up to 18 % ;
- dichroic mirror techniques :

- semi-passive cooling by Freon which allows :
  . cooler structure to be lightened (44 % of material costs at present);
  . higher concentration to be reached.
These advances will lead to a decrease of the cost of the system per peak watt.

LIST OF SYMBOLS
SOPHOCLE : Solar Photovoltaïc with Limited Concentration of Energy
C.N.R.S. : Scientific Research National Center (France)
P.I.R.D.E.S. : Research and Development for Solar Energy Program
L.A.A.S. : CNRS-Laboratory for Automatism and Analysis of Systems

W_p : peak watt C : sunlight concentration ratio
I^p : current intensity V : voltage
W_d : direct solar radiation power W_{gh} : total horizontal radiation power
MPPT: Maximum Power Point Tracking

REFERENCES
(1) D. ESTEVE, G. VIALARET, A. ACHAIBOU, D. FOLLEA, "Design of a photo-
     voltaïc power array with concentrated sunlight."
     Proceedings of the Luxembourg Photovoltaïc Solar Energy Conference
     p. 360, Septembre 1977

(2) Centrale Photovoltaïque 1 kW à concentration SOPHOCLE
     LAAS document n° 1891 : Avril 1979

(3) Utilisation de convertisseurs DC/DC pour la poursuite du point de
     puissance maximum dans les installations photovoltaïques.
     COMES report n° 79.121
     LAAS document n° 2234 : Novembre 1980

(4) O. SOUMAORO, "Etude des cellules solaires au Silicium, au GaAs et
     de leur couplage optique au moyen d'un miroir dichroïque".
     Thesis (3rd Cycle), Mars 1982

(5) J.P. FORTEA, "Etude des différents procédés de refroidissement des
     photopiles dans les centrales photovoltaïques à concentration."
     Thesis (Docteur-Ingénieur), Avril 1981

Photo n° 5
expansion membrane
on SOPHOCLE modules

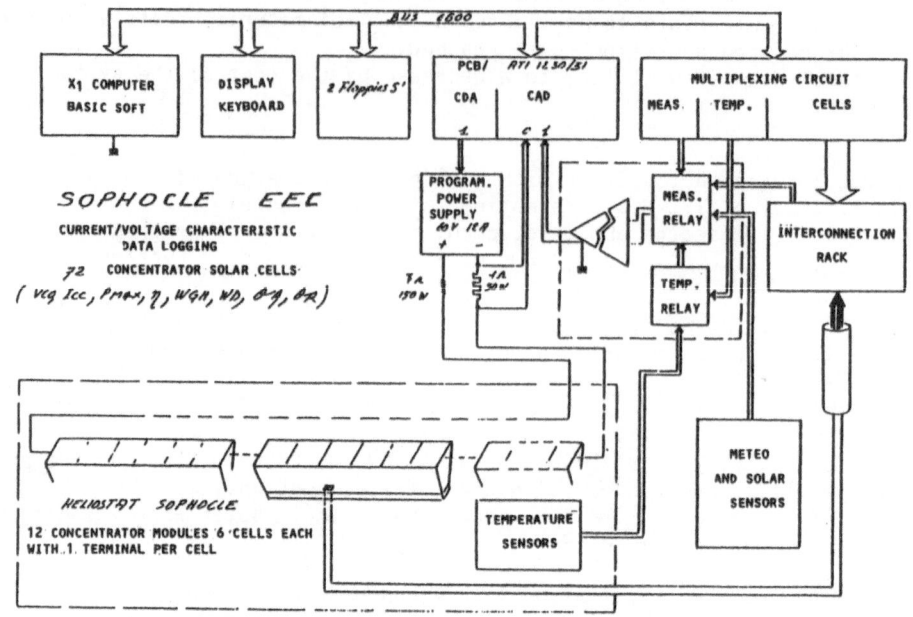

FIG. 1

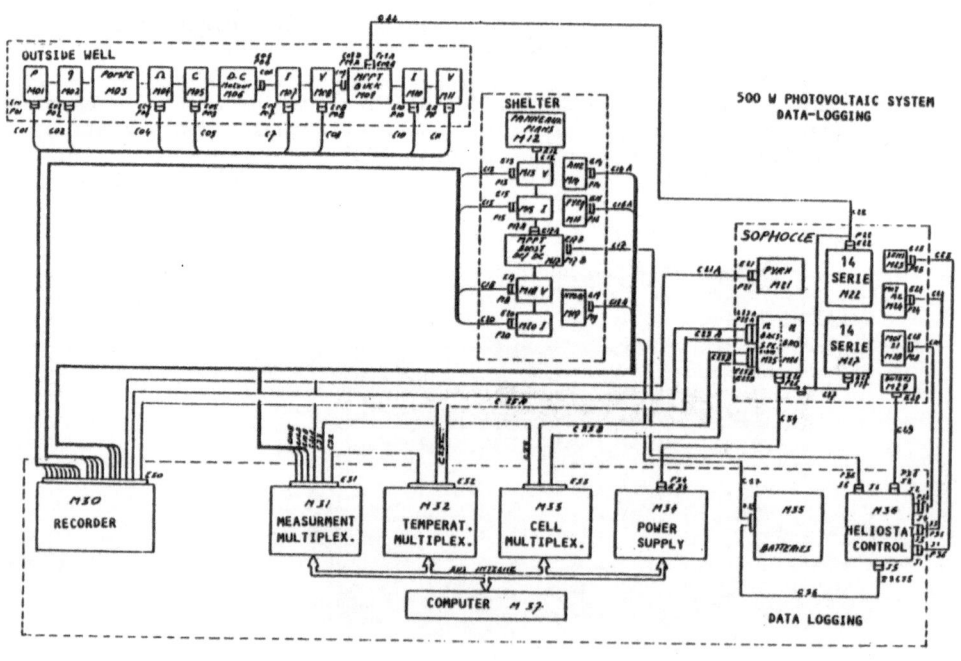

FIG. 2

- 264 -

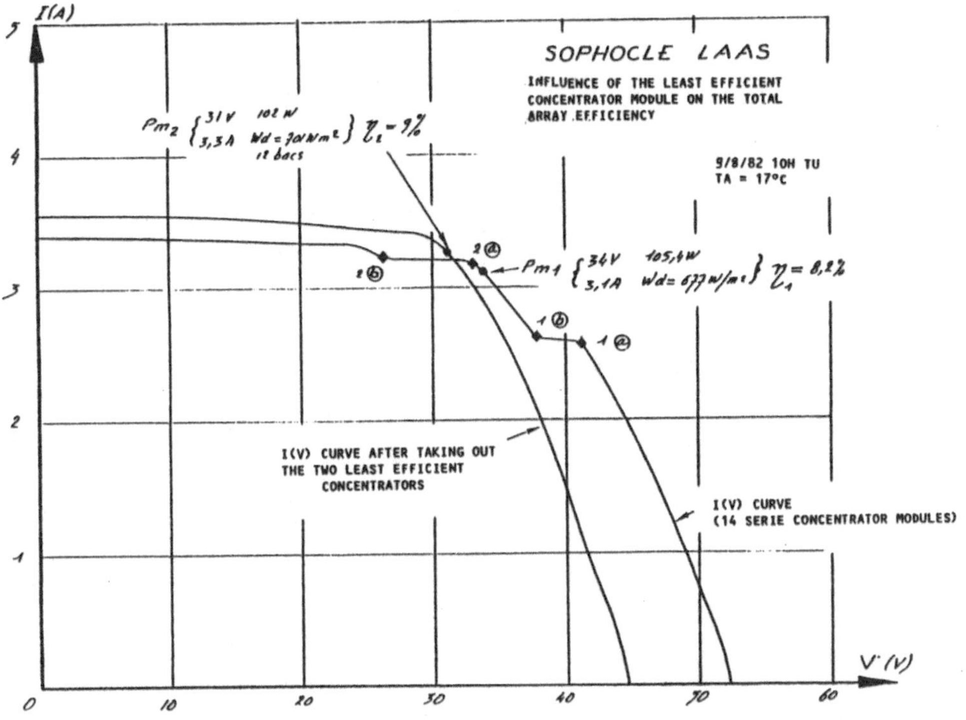

SOPHOCLE LAAS
INFLUENCE OF THE LEAST EFFICIENT CONCENTRATOR MODULE ON THE TOTAL ARRAY EFFICIENCY
9/8/82 10H TU
TA = 17°C
I(V) CURVE AFTER TAKING OUT THE TWO LEAST EFFICIENT CONCENTRATORS
I(V) CURVE (14 SERIE CONCENTRATOR MODULES)
I(A)
V (V)

FIG. 3

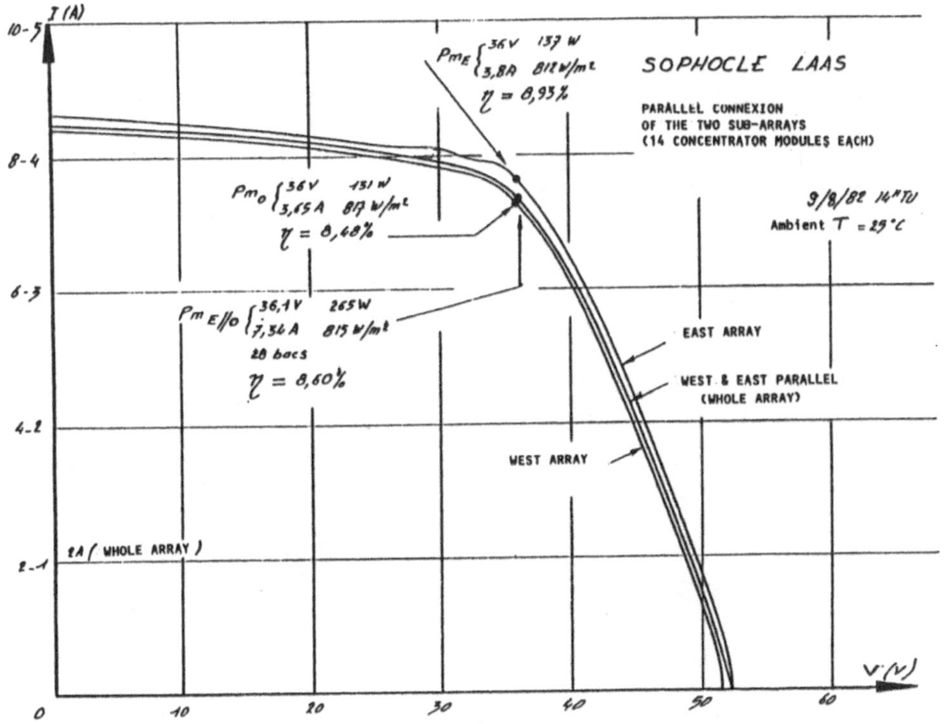

SOPHOCLE LAAS
PARALLEL CONNEXION OF THE TWO SUB-ARRAYS (14 CONCENTRATOR MODULES EACH)
EAST ARRAY
WEST & EAST PARALLEL (WHOLE ARRAY)
WEST ARRAY
2A (WHOLE ARRAY)
I (A)
V (V)

FIG. 4

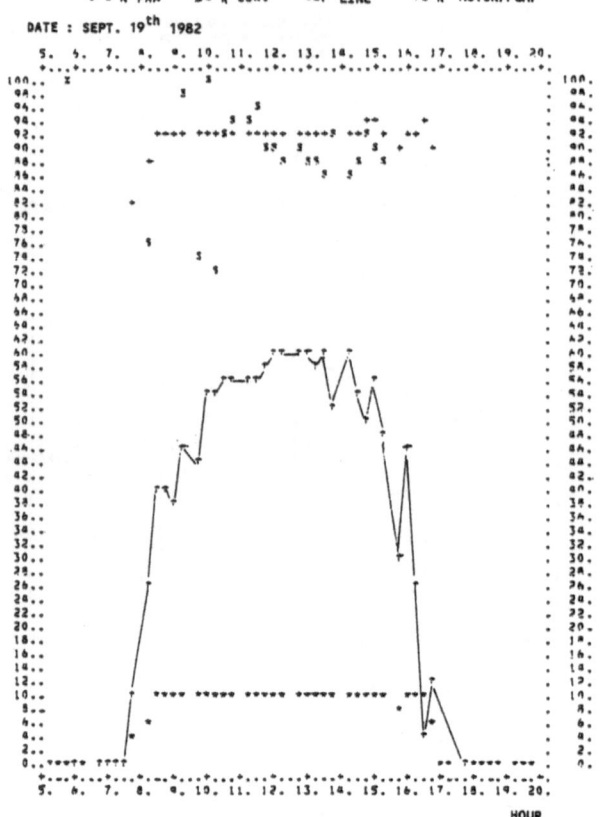

FIG. 5

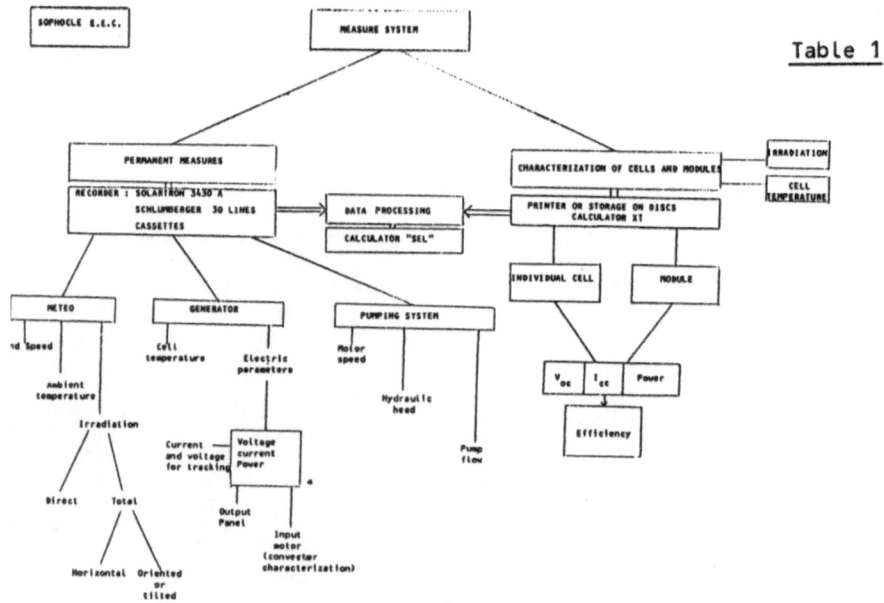

Table 1

CONVERSION OF SOLAR ENERGY USING FLUORESCENT COLLECTORS:
INSTALLATION OF A TEST COLLECTOR TO DELIVER SEVERAL WATTS POWER

Authors   : H.R. WILSON, V. WITTWER

Contract number : ESC-R-037 D

Duration   : 24 months
                    1 July 1981 - 30 June 1983

Head of project : Prof. Dr. A. Goetzberger, Dr. V. Wittwer, ISE

Contractor   : Fraunhofer-Institut für Solare Energiesysteme

Address   : Fraunhofer-Institut für Solare Energiesysteme
                    Oltmannsstr. 22
                    D-7800 Freiburg

Summary

A 1 m² fluorescent collector has been built which, with Si solar cells, is already capable of delivering 6 W at maximum insolation (1 kWm^{-2}), and has the potential for output powers exceeding 20 W, when optimized according to presently known methods to increase the amount of sunlight absorbed. With GaAs or GaAlAs cells, the efficiency could be doubled. The results of outdoor measurements from this collector are supplemented by those from longer-term measurements of two smaller triangular collectors and a reference solar cell. These results show the importance of having solar cells which are intended for use in concentrated light, and the necessity of continued research on the interactions in dye-matrix systems. As an aid to the prediction of the performance of fluorescent collectors, continuous measurements of the solar spectrum have been carried out, and these results, compared with the measured collector output, are also presented here.

Introduction

Progress has been made in the development of the fluorescent collector, such that it is now possible to obtain an output power of several watts from a single collector. Long-term, open-air tests have recently begun on such a collector, which has an area of 1 m². Results from this collector are supplemented by those from two triangular collectors with a hypotenuse of 400 mm. These have been continuously measured on the roof of the Fraunhofer-Institut für Solare Energiesysteme over a period of five months.

An overview of recent experimental and theoretical results from the fluorescent collector was presented at the Fourth E.C. Photovoltaic Solar Energy Conference in Stresa (Heidler, Goetzberger, Wittwer, 1982). As determined by the fluorescent dyes, with which the plexiglass collectors are doped, each collector absorbs only within a narrow spectral band. Thus the overall efficiency can be increased when a stack of two or more different collectors is used. A stack of two triangular collectors, with reflectors on the two shorter edges (making the triangles optically equivalent to squares having 400 mm sides) yielded an electrical efficiency of 3 % when measurements were made with Si solar cells. With GaAs cells, an electrical efficiency of 4 % has been measured, due to the better matching of the cell's absorption to the emitted light, and its superior performance with concentrated light. These are significant improvements on the values of electrical efficiency around 1.5 % which were measured with earlier collectors at the beginning of the contract period.

The long-term stability of the collectors has also been improved by the introduction of new dyes. After illumination of 1000 kWh, corresponding approximately to the annual insolation for Central Europe, the amount of fluorescent light is 60 - 80 % of the original value, depending on the size of the collector. This reduction is not caused by the deterioration of the dye itself, but by the formation of reaction products between the dye and the matrix which reabsorb the fluorescent light.

As the absorption band of each collector is relatively narrow (wavelength range <100 nm), it is necessary to know not only the total amount of insolation, but also its spectral distribution, for the evaluation and prediction of collector efficiency. Thus, the solar spectrum for wavelengths between 400 and 1200 nm has been measured concurrently with the output from the collectors and a reference solar cell.

In the following report, the new 1 m² collector will be described, and results for it, two triangular collectors and the associated spectral measurements will be presented.

Description

The basis of the new collector is a 1 m x 1 m x 3 mm sheet of plexiglass doped with a fluorescent dye, which has its absorption and emission maxima at 525 and 580 nm respectively. At the doping concentration in the collector, this dye absorbs about 15 % of sunlight, which, when the optical losses are taken into account, leads to a concentration factor of about 4 for the light emitted at the collector edges. Two sets of 24 solar cells (each cell 20 x 5 mm²) connected in parallel are mounted on each edge; the eight sets are then connected in series. The collector and solar cells are supported by an aluminium frame, and are protected by a transparent cover. At first a clear plastic film was used, but this was later replaced by a sheet of glass. One of the triangular collectors, the "yellow" one, is doped with the same dye as the 1 m² collector; the other, the "red" collector, is doped with a dye which has its absorption and emission maxima at longer wavelengths, namely 582 and 660 nm. All collectors and measurement instruments were directed towards the south and tilted 45°.

Results

Figure 1 shows the spectral distribution of the total insolation measured for each month from January to October 1982. It can be seen that not only the total amount of sunlight varies from month to month, but also the relative amount at each wavelength changes; for example, April, a very dry month, has markedly higher values than normal at long wavelengths, where otherwise significant absorption by water vapour occurs. By convoluting these spectra with the absorption spectra of the solar cell alone and the two dyes considered, the proportion of available light used by each device can be predicted. This is shown in the upper part of figure 2, in which the integrated value of the convoluted spectra has been divided by the total amount of light received, as measured with a pyranometer and indicated in the accompanying bargraph. The line graphs have been normalized by the values from April, as the average position of the sun during this month corresponds most closely to the AM 1.5 value commonly taken for evaluation of spectrally selective devices. It can be seen that the silicon solar cell, with its broadband absorption, remains relatively constant in its response. By contrast, the narrow absorption range of the fluorescent collector leads to greater variation. As the path length of the sunlight through the atmosphere increases, the proportion of light at wavelengths shorter than 600 nm decreases due to Rayleigh scattering.

The actual output from the reference solar cell and the two triangular collectors has been monitored from June through to October, and is shown in figure 3. Each value has again been divided by the measured value of global radiation. As expected, the solar cell response remained pratically constant. For the collectors, however, the response to spectral variation is not observable due to the instability of the dyes used.

In order to see the effect of spectral variation, it is therefore more worthwhile to examine the values measured and predicted for a single day. Figure 4 compares these values for the 1 m² collector, as well as the triangular collectors and the reference solar cell. The predicted trend is clearly followed by the reference cell and the red triangular collector. In the results for the other two collectors however, other effects appear in the experimental results. The output from the 1 m² collector is lower than expected during the first half of the day, because water had condensed on the plastic cover film and on the plexiglass sheet itself. This effect has since been eliminated by using a glass cover plate. In the case of the triangular collector, the dip over midday is due to the inability of the solar cell to operate satisfactorily with concentrated light. This type of cell deviates from a linear response in the short circuit current at light intensities of 1500 Wm^{-2}, whereas the intensity of the concentrated light leaving the collector edges is in the range 3000 - 4000 Wm^{-2} for several hours around midday on a fine day.

It is clear that this type of cell is not ideal for use with a concentrating device, but more appropriate cells were not available to us in sufficient quantities when this series of measurements was begun. Recently it has been possible to obtain Si cells from ASEC, Sandia and the Université Catholique de Lourain, and GaAs cells from CISA in Milan, all of which improve in efficiency with increased intensity at least up to the concentrations achieved with the fluorescent collector. The curves shown in figure 5 correspond to the predictions and measurements of figure 4, with the difference that in this case, ASEC cells were used. The ability of this type of cell to operate efficiently in concentrated light is shown in the much improved agreement between the predicted and measured results. These measurements with an ASEC cell on the 1 m² collector indicate that with solar radiation of 1000 Wm⁻², the power obtainable from the collector would exceed 6.7 W. When it is considered that 85 % of the light is transmitted through this collector and is available for other purposes (e.g. illumination in dwellings or greenhouses), this is a satisfying result. By increasing the amount of light absorbed either by increasing the concentration of dye in the collector, or by introducing a second collector which absorbed in a different spectral region, a total power output of 30 W could be achieved at present. With solar cells having higher efficiency (> 35 %, compared to the 18 % of those used here) and improved dyes, it is conceivable that power outputs approaching 100 W could be achieved in the future.

Even if an efficiency of 2 % is assumed for the 1 m² collector, the peak watt price for the collector with solar cells mounted is less than that from standard solar cell panels currently available. As Table I shows, the price per peak watt is DM 11, which could be reduced to DM 3 if the cells were mass produced and the plexiglass plate were incorporated as the coverplate in a hybrid system. This is to be compared with DM 25/pW for standard solar cell panels.

Table I

Efficiency %	Plexiglass DM/m²	Mounting Costs DM/m²	Solar Cells DM/m²	Total Price DM/pW
2	60	80	6000	11
2	-- (transparent hybrid system)	40	2000 (mass production)	3

Conclusion

A test fluorescent collector has been built which is presently able to deliver 6 W when the solar radiation density is 1000 Wm⁻². Increasing the concentration of this dye, or co-doping the collector with another dye, would readily allow values up to 30 W to be reached. With better solar cells and optimized dyes, still higher power outputs could be expected. The importance of having appropriate solar cells which are efficient at high light intensities has been demonstrated by long-term tests on the triangular collectors. These have also underlined the need for further improvements in the dye stability. Finally, the possibility and necessity of using spectral measurements for predicting the performance of narrow-band devices has been shown.

Reference

K. Heidler, A. Goetzberger, V. Wittwer
Fluorescent Planar Concentrators (FPC): Monte-Carlo Computer Model; Limit Efficiency and Latest Experimental Results
Fourth E.C. Photovoltaic Solar Energy Conference. Stresa, 1982.

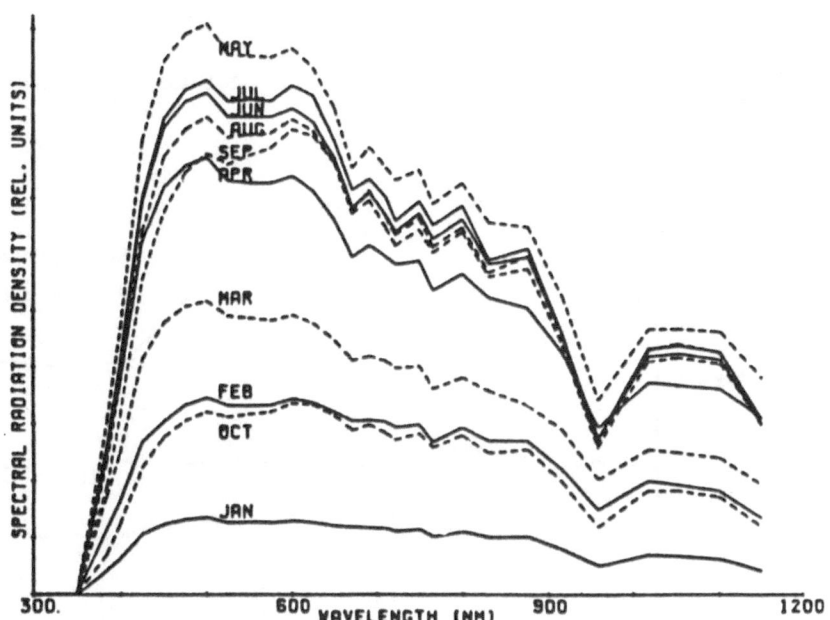

Fig. 1: Monthly solar spectra.

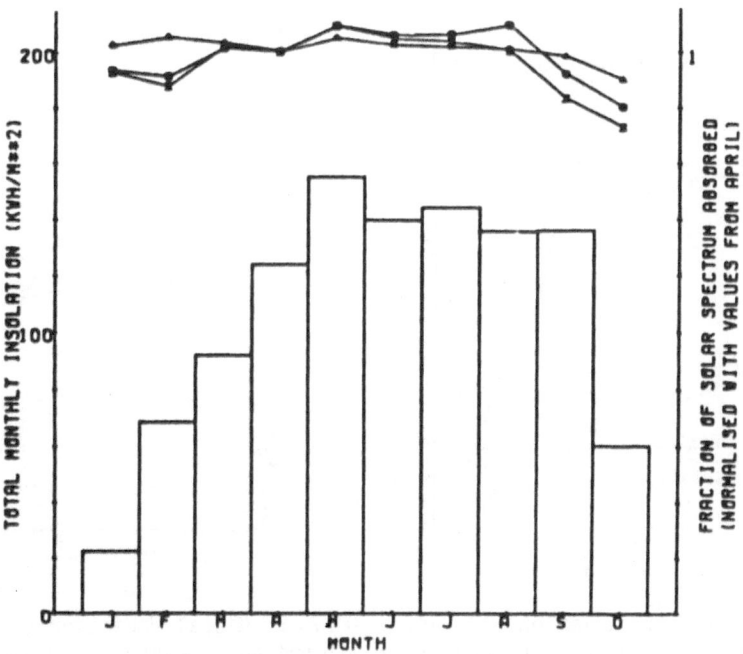

Fig. 2: Monthly global radiation (bargraph) and predicted response of the yellow collector (x-x), the red collector (o-o) and the solar cell alone (Δ - Δ).

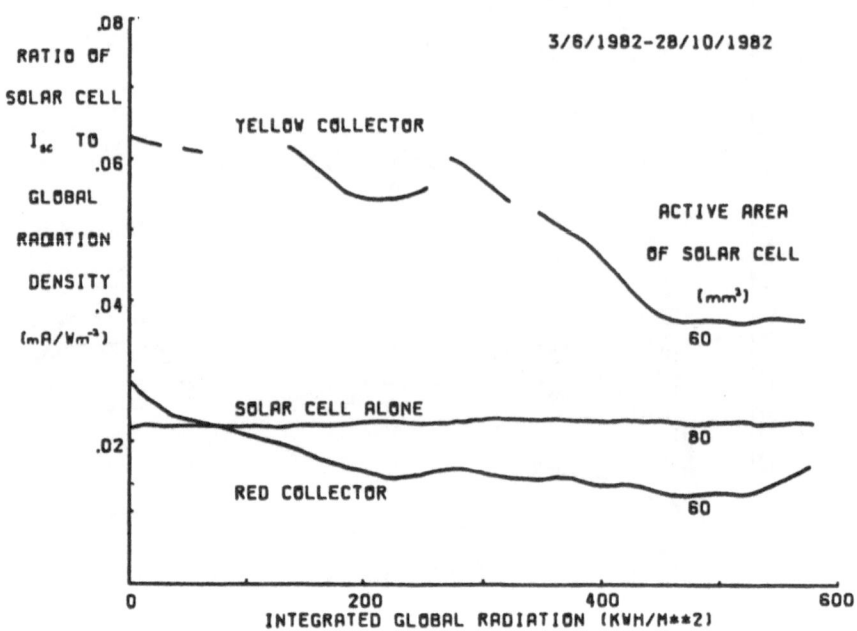

Fig. 3:   Measured response of the solar cell alone and on
          the yellow and red collectors.

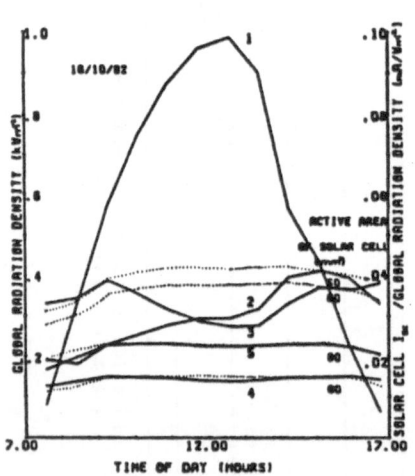

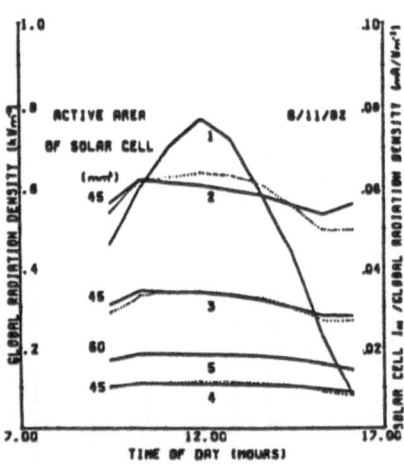

Fig. 4: Global radiation (1), and
predicted and measured responses
of the large collector (2), the
yellow (3), and red (4) collectors, and the reference solar
cell (5) on 16/10/82.

Fig. 5: Global radiation (1) and predicted and measured responses of the
large collector (2), the yellow (3)
and red (4) collectors, and the reference solar cell (5) on 6/11/82.

HOLOGRAPHIC THIN FILM SYSTEM FOR MULTIJUNCTION SOLAR CELLS

Authors   : W.H. BLOSS, M. GRIESINGER, E.R. REINHARDT

Contract number : ESC-R-38-D (B)

Duration   : 36 months   1 July 1980 - 30 June 1983

Head of project : Dr.-Ing. E.R. Reinhardt

Contractor   : Universitaet Stuttgart, Institut fuer
                   Physikalische Elektronik

Address   : Institut fuer Physikalische Elektronik,
                   Universitaet Stuttgart,
                   Boeblinger Str. 70
                   D-7000 Stuttgart 1, F.R.G.

Summary

The efficiency of photovoltaic generators based on
different semiconductor materials with optimized band
gaps can achieve considerably higher values than those
obtained with single junction solar cells. For the required splitting of the solar spectrum into various wavelength regions, phase volume holograms are used, which
act as dispersive concentrators (DISCO). The spectral
imaging properties of these holograms can be optimized
in respect with the requirements for solar applications.
An essential tool for this optimization process is a
theoretical and numerical description of volume phase
holograms. For this purpose, two theories are available
which allow calculation of diffraction efficiency of
volume holograms under various conditions. Together with
a theoretical model of an ideal solar cell, the total
conversion efficiency of the system has been computed
and optimized by varying the dispersive features of the
hologram and the band gaps of the solar cells. For a
simple system consisting of a holographic diffraction
grating and two solar cells with optimized band gaps
the calculated conversion efficiency amounts 30 %.

1. Introduction

The efficiency in photovoltaic power generation can be increased by the use of a system consisting of solar cells with different band gaps. One method, theoretically treated by Gokcen and Loferski /1/ makes use on an arrangement of solar cells with decreasing band gaps positioned behind each another (stack system).

An alternative method is based on spectral spitting of the incoming light. This approach allows the use of spatially separated solar cells, but requires wavelength selective optical elements like dichroic mirrors /2/ or dispersive Fresnel lenses /3/.

A new type of optical elements presented here are volume phase holograms, which allow both, high concentration and splitting of the spectral regions. Due to the dispersive and concentrating features, these optical systems are called DISCO systems.

For realization of a DISCO system, two principles are available /4,5/:

- The imaging features of volume phase holograms can be used to concentrate light on solar cells. For example fig. 1 represents a holographic lens system consisting of two different off-axis Fresnel zone plates with different focus locations.

- A method of non-imaging concentration is obtained by coupling light into a waveguide by a high efficient phase hologram (fig. 2). By this procedure the incident light can be collected over a large area and concentrated on relatively small planes. From a technical point of view it is advantageous to mount the solar cells directly at the side of the hologram plates, which leads to economic compact systems.

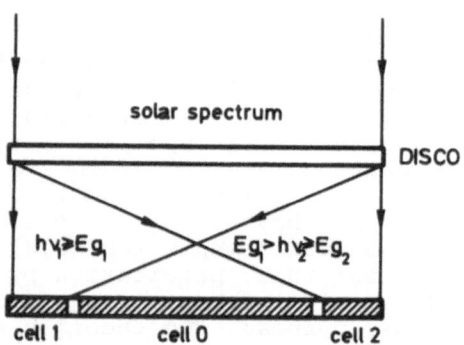

Fig. 1: Holographic lens system with two spatially separated image planes.

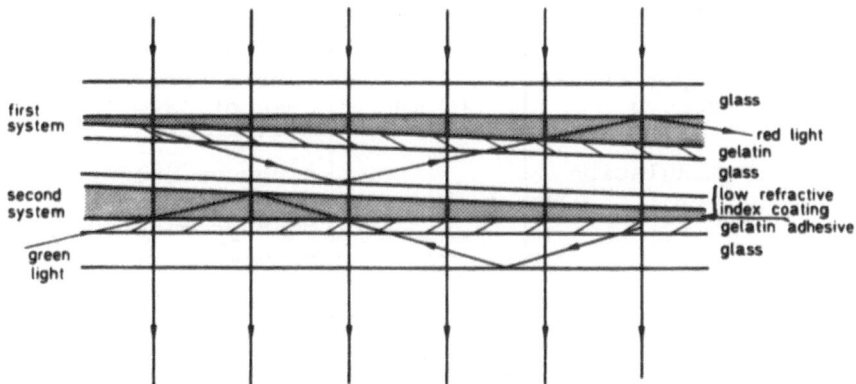

Fig. 2: Holographic waveguide system composed of two elements matched for two different spectral regions.

2. Diffraction efficiency of volume phase holograms

Theoretical optimization of photovoltaic conversion efficiency of a solar cell system working with a DISCO-element requires a computation of diffraction efficiency of volume phase holograms. This procedure is based on two theories:

- The coupled wave theory of Kogelnik /6/ offers a very fast method to calculate the diffraction efficiency as a function of wavelength for single exposed holograms. Holograms diffracting in higher orders and multiply exposed holograms cannot be treated with this theory.

- The theory of Alferness /7/ is much more general and overcomes these limitations, but requires higher expenses due to the non-analytical form of the solution. This method can also be extended to the case of inclined light incidence which is important for studying the influence of sun motion on diffraction efficiency and focus geometry.

The holographic recording method offers the possiblity to generate two or more holograms in one layer. Depending on the recording geometry, this can be used for example to extend the spectral bandwith by diffracting two different wavelengths on the same focus as demonstrated in fig. 3 as well as to seperate two different wavelength regions spatially and to focus them on two solar cells, as shown in fig. 4 /5/.

To influence the dispersive and imaging features of volume phase holograms, a variety of parameters is available, as shown in table I. These parameters have been studied extensively to find optimum conditions for the adaption of holograms to the spectral sensitivity of solar cells.

Parameter	typ. values	depending on
layer thickness d	15 μm	fabrication process of layer
modulation Δn	0.015	exposure energy E, development process
angle between recording beams α	30°	
recording wavelength λ	450-514 nm	spectral sensitivity of DCG, available lasers
maximum diffraction efficiency η	95 %	modulation of refractive index, film thickness
spectral selectivity Δλ	≤120 nm	film thickness, recording geometry
exposure energy E	60 $\frac{mJ}{cm^2}$	spectral sensitivity

Table 1:   Example of a set of parameters for single exposed holograms in dichromated gelatin (DCG).

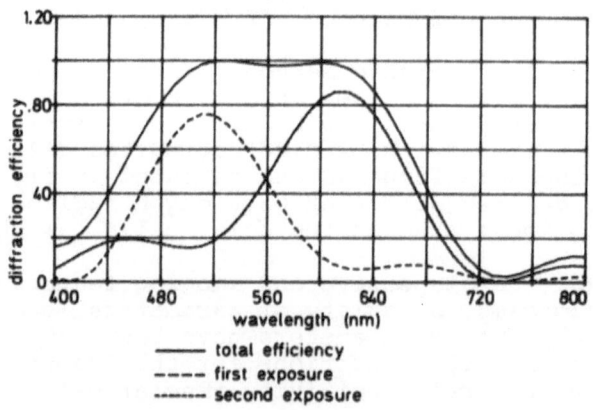

Fig. 3:   Calculated diffraction efficiency of a doubly exposed grating, optimized for extension of spectral bandwith of diffracted light in one direction.

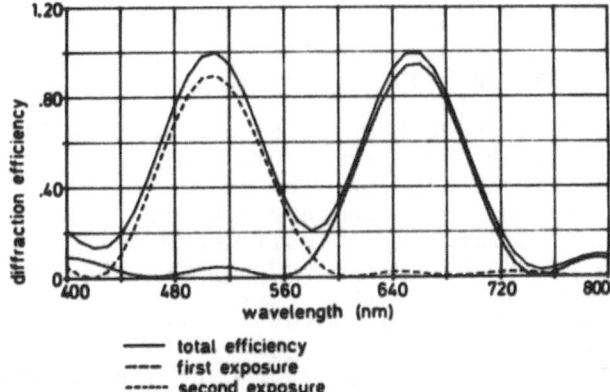

Fig. 4: Calculated diffraction efficiency of a doubly exposed grating, optimized for spatial seperation
of two wavelength regions.

3. Photovolatic conversion efficiency of a tandem system
based on DISCO-elements
   In a first step theoretical investigations have been
applied to a system consisting of a holographic diffraction
grating and two solar cells, as shown in fig. 5.

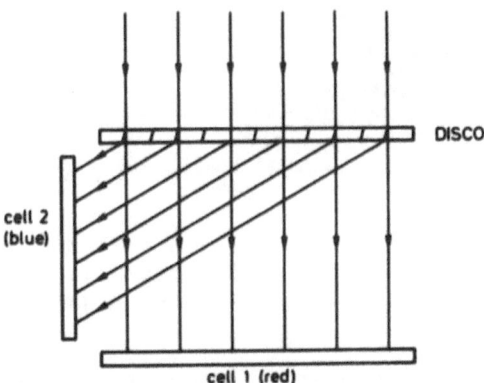

Fig. 5: Tandem system, consisting of a diffraction grating
and two cells.

Calculation of the conversion efficiency of the cells are
based on a theoretical model proposed by Gokcen and Loferski
/1/. The I-V-characteristics herein were assumed to be given
by

$$I = I_O \; (e^{q \cdot V/AkT} - 1)$$

with
$$I_O = K \cdot e^{-E_g/BkT}$$

and
$$A = B = 1, \quad T = 300 \text{ K}$$

The calculations have been performed for two different solar spectra, the AMO spectrum of Thekaekara /9/ and the AM 1.5 spectrum listed in /10/. The solar spectra are split into two parts according to the experimentally derived wavelength dependence of diffraction efficiency of the grating, and for each part the optimized bad gap and the total conversion efficiency is computed.

Fig. 6 shows the AM 1.5 energy spectrum, which is divided into two spectral regions according to the wavelength dependence of a diffraction grating with a spectral bandwidth of 150 nm and maximum of diffraction efficiency at 500 nm. The shaded part of the spectrum is converted by a cell with a band gap E_{g_1} of 1.2 eV, whereas the dotted part which corresponds to the diffracted light illuminates a cell with a band gap of 2.1 eV.

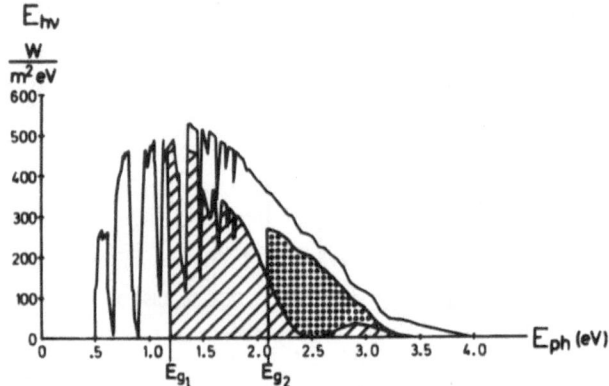

Fig. 6:   AM 1.5 solar spectrum, divided into two spectral
          regions by a holographic diffraction grating and
          converted by two cells with band gaps of 1.2 eV
          and 2.1 eV.

Fig. 7 demonstrates the total conversion efficiency as a function of wavelength of maximum diffraction efficiency λ_{max} and spectral bandwidth of the grating $\Delta\lambda$. The solid lines are values which can be obtained by a single exposed grating, the dashed lines require an extension of spectral bandwidth by doubly exposure. For a spectral bandwidth of more than 250 nm the conversion efficiency decreases again.

Table II gives an overview over the computed results, compared with the values obtained for the stack system of Gokcen and Loferski /1/.

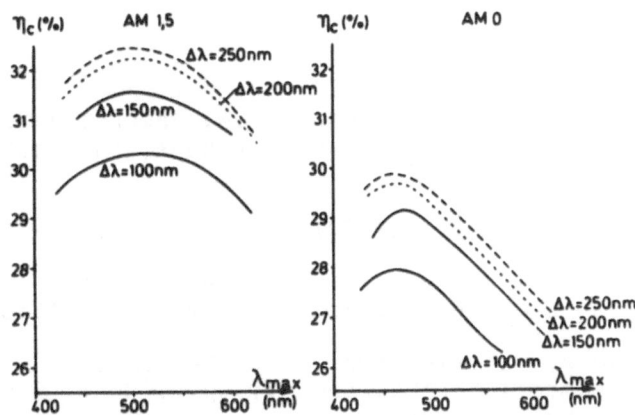

Fig. 7: Total conversion efficiency for a two cell system
as a function of wavelength of maximum diffraction
efficiency and spectral bandwidth.

Single cell	Eg_1	η_c	tandem system	Eg_1	Eg_2	η_c
AM 1.5	1.4 eV	26.3 %	AM 1.5	1.9 eV	1.2 eV	32.4 %
AM 0	1.5 eV	23.7 %	AM 0	2.0 eV	1.2 eV	29.9 %
AM 0 /1/	1.5 eV	23.2 %	AM 0/1/	2.0 eV	1.2 eV	32.5 %

Table II: Optimized band gap and theoretical conversion
efficiency for a single cell and a tandem system.

The DISCO-system, consisting of a doubly exposed grating
and two solar cells with band gaps of 2.0 eV and 1.2 eV shows
for the AM O spectrum a theoretical increase of conversion
efficiency from 23.7 % to 29.9 % and lies only about 3 %
under the conversion efficiency of the stack system, although
the two spectral ranges are not sharply seperatred compared
with the stack system, where a sharp truncation of the spectrum
is performed at the photon energy of the higher band gap.
    A further improvement can be obtained by concentrating
the two spectral ranges on the cells.

References:

/1/   N.A. Gokcen, J.J. Loferski
      Sol. Energy $\underline{1}$, 271 (1979)

/2/   H.A. Vander Plas, R.L. Moon, L.W. James,T.O. Yep,R.R.Fulks
      Proc. 2nd Photovol. Sol. Energy Conf. 515 (1979)

/3/   G. Sassi
      Sol. Energy 24, 451 (1980)

/4/   W.H. Bloss, M. Griesinger, E.R. Reinhardt
      Proc. 3nd E.C. Photov. Sol. Energy Conf. (1980)

/5/   W.H. Bloss, M. Griesinger, E.R. Reinhardt
      Conf. Record 16th IEEE Photov. Spec. Conf. San Diego
      1982 (to be published)

/6/   H. Kogelnik
      Bell Syst. Techn. Journ. $\underline{48}$, 2909 (1969)

/7/   R. Alferness, S.K. Case
      Apll. Phys. $\underline{7}$, 29, (1979)

/8/   W.H. Bloss, M. Griesinger, E.R. Reinhardt
      Appl. Opt. 21, 3739, (1982)

/9/   M.P. Thekaekara
      Proc. Int. Conf., Heliotechnique and Development Dhahran,
      Nov. 2 - 6, 1975, p.47

/10/ Standard Procedures for Terrestrial Photovoltaic
      Performance Measurements
      C.E.C. Specification No. 101 (1979).

HIGH CONCENTRATION P.V. 100 Wp MODULE MAKING USE OF SPECTRUM SPLITTING AND Si-GaAlAs COUPLED CELLS

Authors   : E. FANETTI, C. FLORES, G. GUARINI, F. PALETTA

Part 2 : Modules and Systems

Contract n°   : ESC-R-039-I

Duration   : 18 months, from 1.6.80 to 31.12.81

Total budget   : 240 000 000 lire

CEC contribution   : 120 000 000 lire

Head of project   : ing. C. MALAGUTI

Contractor   : ENEL

Address   : Via G.B. Martini, 3 - 00198 Roma

Subcontractor   : CISE SpA

Address   : Via Reggio Emilia, 39 - 20090 Segrate (Milano)

Summary

In the present report the activity concerning the design and fabrication
of a small size, high concentration module using GaAs and GaAlAs cells
is described.
The aim of the work was that of checking the maturity of GaAs technology
for solar cells, making experience on very high concentration and demonstrating the feasibility of multi-cell solutions using spectrum splitting
optical filters.
In theory, the bandgap combination (1.7 e V - 1,1 e V) corresponding to
GaAs and Si solar cells is nearly the optimum, but no high efficiency Si
cells specifically designed and optimized for spectrum split applications
could be supplied by other laboratories. The combination of GaAs (1,4 e V)
and GaAlAs (1,7 e V) was therefore chosen and high efficiency cells
designed and fabricated in CISE were used.
The module includes six Fresnel lenses, six solar cell couples with
spectrum splitting filters, a full tracking structure, an active cooling
circuit and a maximum power tracking circuit. All the principal components
such as cells and optical elements, have been individually characterized
up to 1000 suns concentration ratios.
For Fresnel lenses, the energy distribution as a function of the distance
from the lens, have been measured. The transmission curves of the
splitting filters have been measured as well.
All the individual cells have been characterized in full spectrum
demonstrating an average conversion efficiency at 800 suns AM1.5 of
18% for the GaAs cells and 16.5% for the GaAlAs cells. Due to lack of
sun during the winter season in Milan, only a preliminary evaluation of
the module performances could be carried on. On the base of the first
measurement sets a total peak power of $\sim$ 120 W (at P_{sun} = 100 mW/cm^2) and
an overall efficiency of 14.1% resulted for the module working at a $\sim$ 800
suns concentration.

1. INDOOR CELL MEASUREMENTS

Concentration I-V characteristics are, of course, of particular interest. Measurements up to effective concentration ratios of about 100 suns (AM 1.5) could be performed by using the sun simulator combined with a Fresnel lens.

Higher concentrations, typically 800 suns, were obtained with flash technique widely described in (1). The circuit can perform the scanning of the cell characteristics during the constant illumination time of the flash lamp. After the lamp turn-on a triggering signal is sent to the pulse generator, whose output is applied to the transistor base circuit. This pulse, smoothed by capacitor C, gradually drives the transistor from the initial cut off to a saturation condition where the collector current I_c equals the cell short-circuit current I_{sc}. The current signal accross a 0.1Ω resistance and the cell voltage are recorded by a digital storage oscilloscope as a function of time. The dependence of the current on the cell voltage, i.e. the I-V curve, can also be displayed. The advantages of the method are related to the possibility of achieving very high (2000x) concentrations without heating the cells.

In fig. 1 a I-V characteristic measured by the flash technique is shown on the screen of a storage oscilloscope.

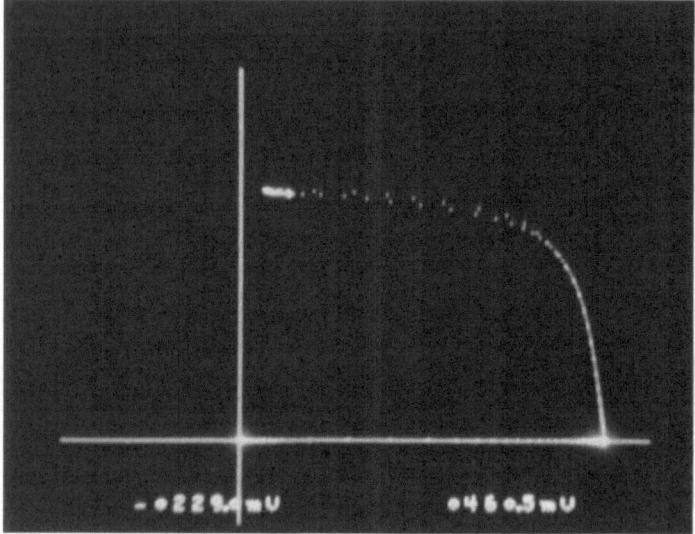

Fig. 1 - Oscilloscope photograph of a flash I-V curve, I_{sc} = 15.5A; V_{oc} = 1.35 V

In Table I the efficiencies measured by the flash technique are reported. The values do not correspond to the same concentration ratio due to a different light intensity from test to test.

TABLE I - GaAs and GaAlAs CELL EFFICIENCIES. AM 1.5 WAS TAKEN AS REFERENCE.

TYPE CELL	CONCENTRATION	FF	η(%)
170	830	0.71	18.4
171	820	0.68	18.1
172	945	0.70	17.1
173	880	0.66	18.0
175	875	0.70	18.1
177	910	0.62	17.3

TABLE I (continued)

TYPE	CELL	CONCENTRATION	FF	η(%)
	T-56	980	0.80	16.7
	T-57	950	0.81	16.0
	T-58	965	0.75	17.0
	T-59	900	0.73	16.2
	T-61	915	0.77	16.4
	T-62	890	0.78	16.0

The efficiencies values reported above and measured by the flash technique refer to optimal conditions (room temperature and uniform illumination of the cells). Outdoor conditions will be therefore somewhat less favourable.

2. ACTIVE COOLING

In the design of the 100 Wp module using spectrum splitting, an active cooling has been preferred both because it is more effective and also more promising for instance for hybrid power P V systems.

The cooling device that will be included in the module design is schematically represented in fig. 2.

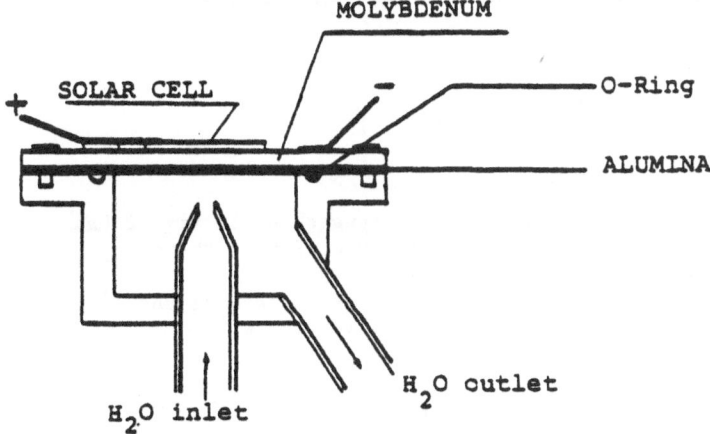

Fig. 2 - Scheme of the "jet impinge-ment" active dissi-pator

The cooling liquid passes through a very thin nozzle and impinges at high velocity against the metal base supporting the solar cell.
The base supporting the cell is hermetically blocked on the dissipator by means of screws and O-ring.
A defective cell can be easily replaced without changing the whole dissipator.

In the active cooling version the electrical isolation problem is much more critical than for passive cooling. In fact the cooling fluid is water, which is a very bad insulator. The electrical insulation of the cells is achieved by plasma deposition of $\sim$ 150μm alumina on the back of the mounting molybdenum base. The same deposition is performed on the bottom of the p-contact copper tab; the n-contact is the mounting base itself. A simplified drawing of the mounting is given in fig. 3. The metal base was a molybdenum disk, with a thermal expansion coefficient similar to that of GaAs. In this way high temperature tests could also be carried on without introducing thermal stresses in the cells. The cell and the copper tabs were attached in the Mo disk using the same epoxy resin previously described.

A few cells after 1 month outdoor operation showed a delamination of the antireflection coating. Therefore the cells used in the spectrum split module were protected with a silicon resin or a glass directly attached to the cell.

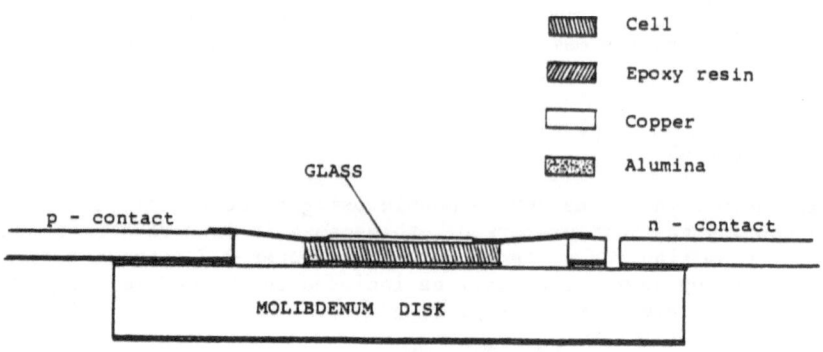

Fig. 3 - *Active cooling solar cell mounting.*

3. THE COOLING CIRCUIT

The "jet impingement" dissipator of the type shown in fig. 2 has been characterized by a set of measurements for different thermal powers and two water flow rates.

The temperature difference between the upper surface of the dissipator and that of the input water, is reported in fig. 4.

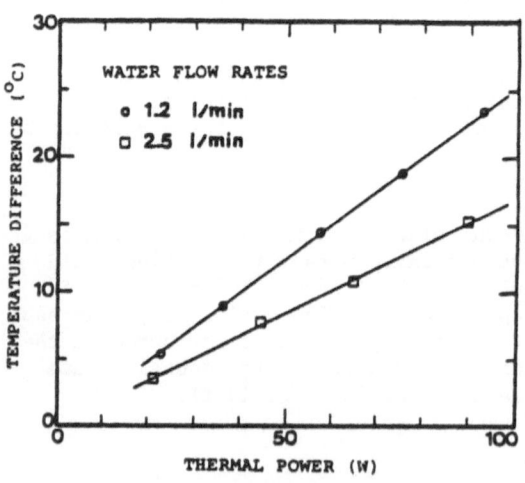

Fig. 4 - *Temperature difference as a function of the thermal power for two flow rates.*

As it can be seen even for higher power levels the temperature differences were low. For example, the thermal resistance at 100 W power was 0.17°C/W for a 2.4 l/min flow and 0.25°C/W for a 1.2 l/min flow.

The hydraulic circuit of the module is schematically shown in fig. 5. It is composed of three parallel branches, each cooling four cells (two GaAlAs and two GaAs cells). The heated water enters an air exchanger made of a series of finned tubes. The exchanger is designed in order to keep the temperature difference between input and output < 2°C, while at a room temperature of 30°C the water maximum temperature should be 38°C.

The circulation of the coolant is forced by a pump. A flow switch is placed in series to the hydraulic circuit and regulated for a minimum flow under which it drives a relais which operates on the tracking system defocusing the module. An overheating of the cells due to a misfunctioning of the pump or of the circuit, will so be avoided. The coolant will be a mixture of water and ethylene glycol in order to obtain a lower freezing point.

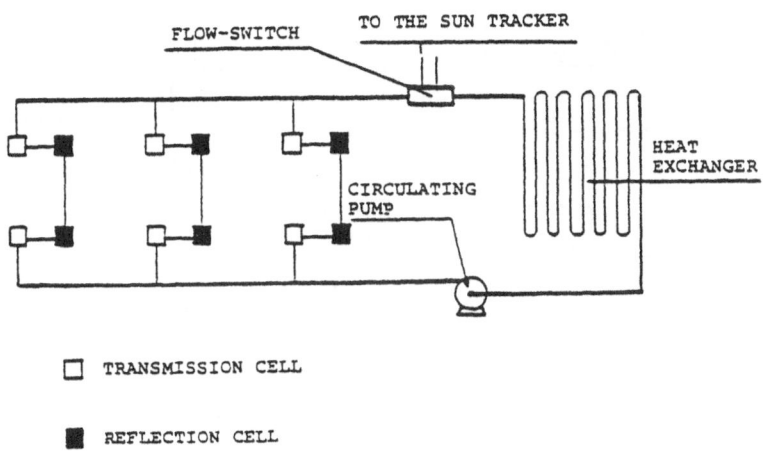

Fig. 5 - Hydraulic circuit layout for the active cooling.

4. MECHANICAL STRUCTURE AND TRACKING SYSTEM

A scheme of the module structure is shown in fig. 6. It is composed by a fork supporting an horizontal axis around which a tank including lenses and cells, can rotate. On the top of the tank, two rows of three Fresnel lenses each, are placed while, in the inside, the adjustable supports for cells and filters can ensure the correct positioning. The heat exchanger is hooked to the back of the tank, while the circulating pump is placed on the fork base.

The whole structure was designed to resist to winds up to 100 km/h allowing a correct performance up to 30 km/h.

The sun tracking system is composed by a solar light sensor and by a logic cricuit driving the alt-azimuthal movements of the structure.

The sensor is made of 4 photodevices (2 for the azimuthal and 2 for the zenithal movement), separated by a cross shaped screen projecting balanced or unbalanced shadows on the 4 photodevices. When the system is exactly directed towards the sun the outputs of the 4 elements are the same and no logic signal drives the structure.

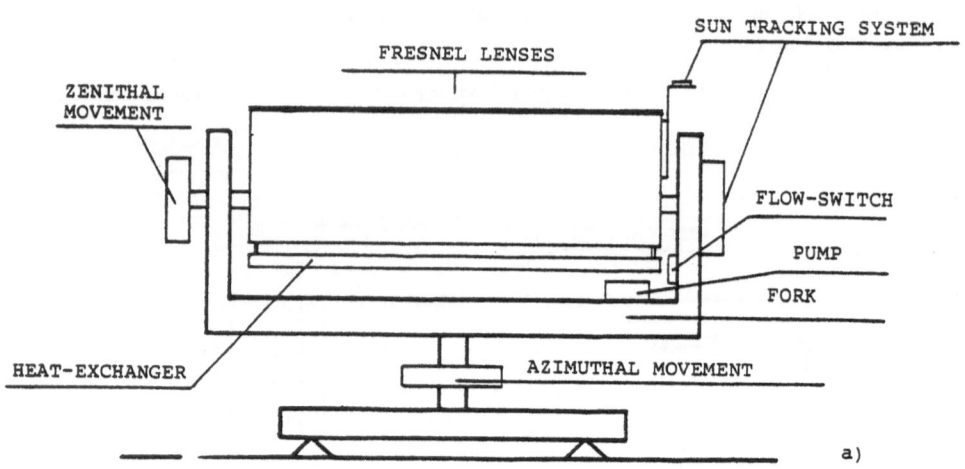

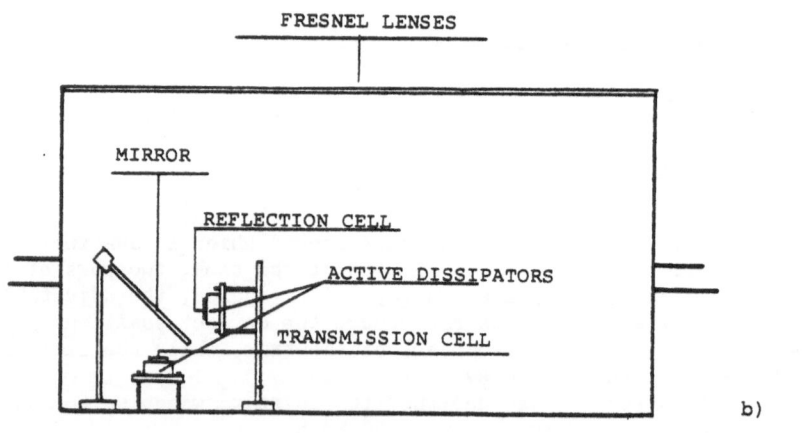

Fig. 6 - a) Simplified structure of the module;
b) Scheme of the spectrum splitting arrangement (not in scale).

On the other side, if the sun is not exactly tracked, the four outputs result unbalanced and the module moves, looking for a new balance. The tracking precision is < 3 mrad.

Some other characteristics of the system are :
- after the sunset the tracker returns to the morning position;
- in the absence of sun, the zenithal position is maintained while the azimuthal position varies with a speed approximately equal to the apparent velocity of the sun.

Because the sensor has a wide field of view, when the sun appears again it is quickly realigned.

5. MAXIMUM POWER TRACKER (M.P.T.)

This instrument dynamically tracks the maximum power point of the solar cells array in a 2 - 20 A current range (2). In order to perform this function the M.P.T. calculates the instantaneous module output power which is displayed. The array voltage and the current signal across the R_{sense} resistance are measured by two operational amplifiers and sent to an analogic block performing the product of the two signals and giving an output proportional to the instantaneous electrical power.

This signal enters a peak detector whose output is a continuous voltage proportional to the maximum array power. The peak detector output and the multiplier signal attenuated by a $\sim$0.95 factor are processed by a comparator driving a flip-flop which, in turn, switches the load on and off.

When the circuitry is correctly regulated a self sustaining cycle is established which keeps the array at the maximum power point.
The system linearity is better than 1%.
The power is measured at the peak detector output and displayed on a 3 digit voltmeter directly calibrated in watts.

6. PRELIMINARY MODULE TESTS

Due to the lack of sun during the winter months only three of the six GaAlAs - GaAs cell couples could be tested outdoor. These three couples were mounted on the module structure and the cells were indivi- dually measured under concentrated sunlight, with full tracking and active cooling.

Two types of splitter filters have been used with cut-off wavelength respectively $\lambda o = 0.73\mu m$ and $\lambda o = 0.79\mu m$.

The GaAs cell was used in reflection while the GaAlAs cell received the light transmitted by the filter.
The I-V curve was measured for each separated cell (3GaAs and 3 GaAlAs cells).
Typical characteristics measured in the conditions described above for the GaAs and the GaAlAs cells are reported in fig. 7. The position of the cell with repect to the lens and the filter was adjusted to achieve the maximum output.
The effective concentration on each cell (measured without the spectrum splitting filter) was 800 suns. For the 0.73μm filters, the short circuit current values were nearly the same for the GaAs and the GaAlAs cells.
This would greatly simplify the interconnection problem; in fact all the cells could be series connected without any heavy penality for the total power output.

Table II summarizes the cell performances 3+3 cells measured with 0.73μm splitting filters.

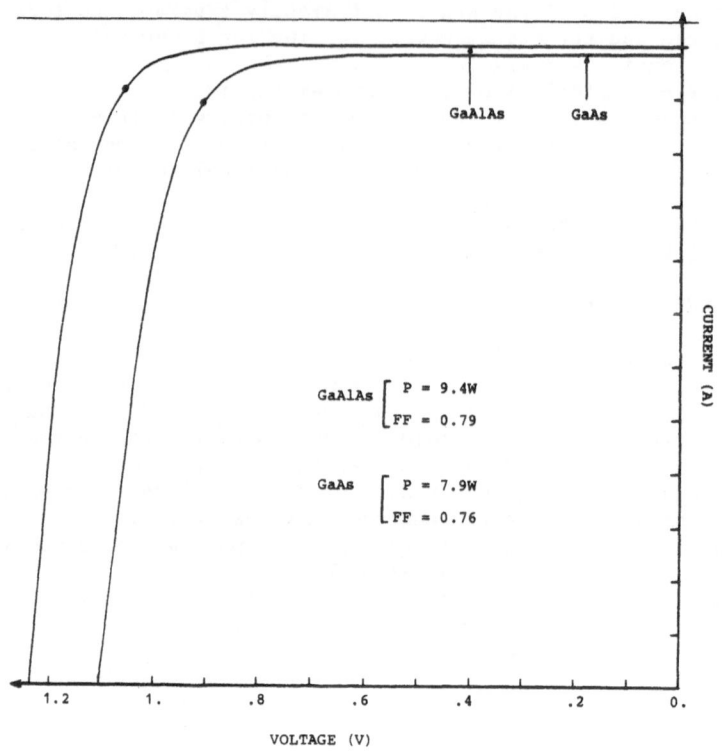

*Fig. 7 - Spectrum split I-V curves for a λ_o= 0.73µm filter.
Current scale : 0.8 A/div.*

TABLE II - CELL PERFORMANCES AT 800 SUNS, AM 1.5, 45°C, WITH 0.73µm FILTERS

CELL TYPE	CELL N	Isc (A)	Voc (V)	FF	P (W)	η(%)
[GaAs	170	9.5	1.1	0.76	7.9	6.45
[GaAlAs	T-56	9.6	1.24	0.79	9.4	7.66
[GaAs	171	9.35	1.11	0.75	7.8	6.36
[GaAlAs	T-58	9.5	1.25	0.78	9.3	7.58
[GaAs	175	9.4	1.09	0.76	7.8	6.36
[GaAlAs	T-61	9.5	1.24	0.78	9.2	7.5

The square brackets in table II indicate how the single cells are coupled together.

The above efficiencies were derived as the ratio between the maximum power output of a cell and the sun power incident on the lens (the lens active area was 0.146 m^2). The values reported in table II therefore include both filter and lens losses. (15% for the filter and 25% for the lenses).

From this values, the following average characteristics can be reasonably assumed for the GaAs-GaAlAs couples employed in the module, including filter and lens losses, at 800 suns AM 1.5 :

Cell coupling efficiency 14% (6.5 + 7.5)
Cell couple power output 17.5 W (8 + 9.5).

The above efficiency value would be of about 19% if the lens were ideal. The result reported above refer to solar cells with exposed area of 1.37 cm^2. Much higher combined efficiencies have been measured for some couples of small (0.36 cm^2) of GaAs and GaAlAs cells with particularly good individual performances. Moreover, due to their smaller dimensions the series resistance for these cells were much lower than for the larger ones. 23% overall efficiencies have been measured at about 750 suns, AM 1.5, 45°C including filter losses. If also 20% lens losses are included, this best efficiency value, is reduced to 18.4%. The average efficiency of 14% measured for the coupled cells mounted on the module was much lower than the correspondent 18.4% measured on small area best cells. The principal reasons for this difference may be listed as follows :
- LPE epitaxial technique actually does not guarantee a perfect uniformity on large areas, so that small area regions can be more easily found with optimal material properties.
- A small area cell, intrinsically has a smaller series resistance.
- The small area cell measurements have been made with "just arrived" lenses with 80% transmission while, because of dust the transmission was only 75% in the other case.

The module structure is finally shown in Fig. 8

BIBLIOGRAPHY

1) E. FANETTI : "Flash technique for GaAs concentrator solar cell measurement". Electronics letters, 1981, 17, (13) pp. 469-470.
2) D.M. DI VAN, M.M. HASAN : "Maximisation of operating efficiency of solar cells". Energy Conversion, 1977, Vol. 17, pp. 183-188.

LIST OF PARTICIPANTS

ALLISON, J.
University of Sheffield
Dept. of Electronic & Electrical Engineering
Mappin street
UK - SHEFFIELD S1 3JD
Tel. (0742) 785 55
Telex

BELOUET, C.
Laboratoires de Marcoussis
CGE
Route de Nozay
F - 91460 MARCOUSSIS
Tel. 449 12 37
Telex

BEYER, W.
KFA Jülich
Postfach 1913
D - 5170 JUELICH
Tel. (02461) 61 41 28
Telex

BLOSS, W.
Universität Stuttgart
Boeblingerstr. 70
D - 7000 STUTTGART 1
Tel. (0711) 66 54 11
Telex

BOBO, J.C.
Laboratoires de Marcoussis
CGE
Route de Nozay
F - 91460 MARCOUSSIS
Tel. 449 11 76
Telex

BONNET, D.
Battelle Institut
Am Roemerhof 35
D - 6000 FRANKFURT 90
Tel. (0611) 79082631
Telex 0411966

BRESLIN, G.
Commission of the European Communities
DG Information Market and Innovation
P.O.Box 1907 - JMO B4/072
L - 1019 LUXEMBOURG
Tel. 43011 x 3163
Telex 3423/3446/3476 comeur lu

BUFFET, P.

Société européenne de Propulsion
Forêt de Vernon
B.P. 802
F - 27207 VERNON
Tel. (32) 51 31 21
Telex sepvern 770019 f

CHABOT, B.

AFME (Agence française pour la
maîtrise de l'énergie)
Route des Lucioles
F - 06565 VALBONNE CEDEX
Tel. (93) 74 79 79
Telex 461357 f

CHAPMAN, M.H.

Electricity Supply Board
Commercial Department
Fitzwilliam Street
IRL - DUBLIN
Tel. 76 58 31 x 7364
Telex

CHEEK, G.

Laboratory ESAT
Katholieke Universiteit Leuven
Kardinaal Mercierlaan 94
B - 3030 HEVERLEE
Tel. (016) 220 931
Telex 25941 elekul

COM-NOUGUE', J.

Laboratoires de Marcoussis
CGE
Route de Nozay
F - 91460 MARCOUSSIS
Tel. (6) 449 10 00
Telex

COWACHE, P.

Lab. d'électrochimie analytique et appliquée
Ecole nationale sup. de chimie de Paris
11, rue Pierre et Marie Curie
F - 75231 PARIS CEDEX 05
Tel. 336 25 25
Telex

CUMBERBATCH, T.

Thorn-Emi plc
Central Research Laboratories
Trevor Road
UK - HAYES, Middlesex UB3 1HH
Tel. (01) 573 3888 x 2880
Telex 22417

DEMONGEOT, F.

Photowatt International SA
6, rue de la Girafe
B.P. 6069
F - 14002 CAEN CEDEX
Tel. (31) 95 09 46
Telex rtc 170333 f

DONON, J.   Photowatt International SA
                              6, rue de la Girafe
                              B.P. 6069
                              F - 14002 CAEN CEDEX
                              Tel. (31) 95 09 46
                              Telex rtc 170333 f

EQUER, B.   PIRSEM
                              CNRS
                              282 Boulevard St. Germain
                              F - 75007 PARIS
                              Tel. (1) 705 77 15
                              Telex

EVANGELISTI, F.   Istituto di Fisica G. Marconi
                              Università di Roma "La Sapienza"
                              P.le Aldo Moro 2
                              I - 00185 ROMA
                              Tel. (06) 49 76 385
                              Telex

EVERTH, T.   JMC Kunststoff GmbH
                              Adolf-Flöring-Str. 22
                              D - 5632 WERMELSKIRCHEN
                              Tel. (02194) 822 99
                              Telex

FABRE, E.   Photowatt
                              131, Route de l'Empereur
                              F - 92500 RUEIL MALMAISON
                              Tel. (1) 708 05 05
                              Telex 202084 f

FALLY, J.   Laboratoires de Marcoussis
                              CGE
                              Route de Nozay
                              F - 91460 MARCOUSSIS
                              Tel. (6) 449 12 41
                              Telex

GALLONI, R.   CNR
                              Istituto LAMEL
                              Via Castagnoli 1
                              I - 40126 BOLOGNA
                              Tel. (051) 51 95 93
                              Telex 511350

GRASSI, G.   Commission of the European Communities
                              D.G. Science, Research and Development
                              200, rue de la Loi - SDME 03/18
                              B - 1049 BRUSSELS
                              Tel. 235 68 01
                              Telex 21877 comeu b

GRIESINGER, M.   Universität Stuttgart
                        Institut für Physikalische Elektronik
                        Boeblinger Str. 70
                        D - 7000 STUTTGART 1
                        Tel. (0711) 665 375
                        Telex

GRUBER, E.   RESART-IHM AG
                        Gassnerallee 40
                        D - 6500 MAINZ
                        Tel. (06131) 671001
                        Telex

HERTLEIN, H.   BMFF
                        P.O.Box 20 07 06
                        D - 5300 BONN
                        Tel. (0228) 59 35 41
                        Telex

KALBITZER, S.   Max-Planck-Institut
                        für Kernphysik
                        Saupfercheckweg 1
                        D - 6900 HEIDELBERG
                        Tel. (06221) 5161
                        Telex

KAUT, W.   Commission of the European Communities
                        D.G. Energy
                        200, rue de la Loi - GUIM 03/8
                        B - 1049 BRUSSELS
                        Tel. 235 39 70
                        Telex 21877 comeu b

KOUROGENIS, C.N.   Ministry of Research and Technology
                        Academias 42
                        GR - ATHENS
                        Tel. 3641 596
                        Telex

KREBS, K.-H.   Commission of the European Communities
                        Joint Research Centre
                        Ispra Establihsment
                        I - 21020 ISPRA (VARESE)
                        Tel. (0039) 332 / 78 99 02
                        Telex 380042 eur i

LAURENT, B.   LAAS (CNRS)
                        Lab. d'Automatique et d'Analyse des Systèmes
                        7, avenue du Colonel Roche
                        F - 31400 TOULOUSE
                        Tel. (61) 25 21 47
                        Telex

LINCOT, D.   Lab. d'électrochimie analytique et appliquée
                              ENSCP
                              11, rue Pierre et Marie Curie
                              F - 75231 PARIS CEDEX 05
                              Tel. 336 25 25 x 3833
                              Telex

MAKIOS, V.   Electromagnetics Laboratory
                              University of Patras
                              GR - PATRAS
                              Tel. (061) 991 902
                              Telex 312347 efap gr

McCARTHY, S.   Microelectronics Research Centre
                              University College
                              Lee Maltings str., Prospect Row
                              IRL - CORK
                              Tel. (021) 26 871 x 2694
                              Telex

McINALLY, I.   THORN-EMI
                              Central Research Laboratories
                              Trevor Road
                              UK - HAYES, Middlesex UB3 1HH
                              Tel. (01) 573 38 88 x 2880
                              Telex 22471

MELCHIOR, B.   JMC-Jmchemie Kunststoff GmbH
                              Richard-Wagner-Str.
                              D - 5632 WERMELSKIRCHEN
                              Tel. (02196) 822 99
                              Telex 08513382

MERTENS, R.   Katholieke Universiteit Leuven
                              Kardinaal Mercierlaan 94
                              B - 3030 HEVERLEE
                              Tel. (016) 22 09 31
                              Telex 25941 elekul

MULLER, J.-C.   Centre de Recherches nucléaires, Groupe de
                              Physique et Applications des Semiconducteurs
                              23, rue du Loess
                              F - 67037 STRASBOURG CEDEX
                              Tel. (88) 29 90 33 x 6607
                              Telex

NYLANDSTED LARSEN, A.   Institute of Physics
                              University of Aarhus
                              Ny Munkegade
                              DK - 8000 AARHUS
                              Tel. (06) 12 88 99
                              Telex

PALZ, W.   Commission of the European Communities
                         D.G. Science, Research and Development
                         200, rue de la Loi, SDME 03/19
                         B - 1049 BRUSSELS
                         Tel. 235 69 22
                         Telex 21877 comeu b

RICARD, J.   Pechiney Ugine Kuhlmann
                         Centre de Recherches de Grenoble
                         F - 38560 JARRIE
                         Tel. (76) 68 16 11
                         Telex 320709

RICAUD, A.M.   France-Photon
                         Zone Industrielle Les Agriers
                         F - 16015 ANGOULEME
                         Tel. (45) 95 70 66
                         Telex 790244 f

RICKUS, E.   Battelle Institut E.V.
                         Am Roemerhof 35
                         D - 6000 FRANKFURT 90
                         Tel. (0611) 7908 2886
                         Telex 0411966

SARIS, F.   FOM - Institute for Atomic and
                         Molecular Physics
                         Kruislaan 407
                         NL - 1098 SJ AMSTERDAM
                         Tel. (31) (20) 94 67 11
                         Telex

SAVELLI, M.   Université des Sciences et Techniques
                         du Languedoc
                         F - 34060 MONTPELLIER CEDEX
                         Tel. (67) 54 23 19
                         Telex

SCHMID, J.   Fraunhofer Institute
                         for Solar Energy Systems
                         Oltmannsstr. 22
                         D - 7800 FREIBURG
                         Tel. (0761) 40 50 46
                         Telex

SINKE, W.   FOM - Institute for Atomic
                         and Molecular Physics
                         Kruislaan 407
                         NL - 1098 SJ AMSTERDAM
                         Tel. (020) 946711
                         Telex

SIRTL, E.   Heliotronic GmbH
                           Postfach 1129
                           D - 8263 BURGHAUSEN
                           Tel. (08677) 83 25 80
                           Telex 56923 (wacker chemitronic)

SONNEVILLE, R.P.M.   Holecsol Systems BV
                           P.O.Box 2300
                           NL - 5600 CH EINDHOVEN
                           Tel. (040) 524 655
                           Telex 59030 hosol nl

SØRENSEN, G.   Institute of Physics
                           University of Aarhus
                           Ny Munkegade
                           DK - 8000 AARHUS
                           Tel. (06) 12 88 99
                           Telex

SPINDLER, O.   Wacker Chemitronic
                           Postfach 1129
                           D - 8263 BURGHAUSEN
                           Tel. (08677) 83 25 80
                           Telex 56923

TREBLE, FC.   Consultant
                           43, Pierrefondes Avenue
                           UK - FARNBOROUGH, Hant., GU14 8PA
                           Tel. (0252) 54 20 37
                           Telex

TURNER, D.P.   Sheffield University
                           Dept. of Electronic and Electrical Engineering
                           Mappin Street
                           UK - SHEFFIELD S1 3JD
                           Tel. (07427 78 555
                           Telex

VAN GYSEL, M.   IDE
                           Parc Industriel
                           B - 5430 ROCHEFORT
                           Tel. (084) 21 37 71
                           Telex 41969 soleil b

VAN OVERSTRAETEN, R.   ESAT Laboratory
                           Katholieke Universiteit Leuven
                           Kardinaal Mercierlaan 94
                           B - 3030 HEVERLEE
                           Tel. (016) 22 09 31
                           Telex 25941 elekul

VIEUX-ROCHAZ, L.   Centre d'études nucléaires de Grenoble
                          LETI/CE
                          85 X
                          F - 38041 GRENOBLE CEDEX
                          Tel. (76) 97 41 11
                          Telex

VOSS, B.   Fraunhofer Institut für
                          Solare Energiesysteme
                          Oltmannsstrasse 22
                          D - 7800 FREIBURG
                          Tel. (0761) 40 50 46
                          Telex

WELSCHEN, J.M.M.   Holec Solar Energy Systems
                          HOLECSOL SYSTEMS
                          Lagedijk 26 - P.O.Box 30
                          NL - 5700 AA HELMOND
                          Tel. (04920) 48 886
                          Telex 51686 hocom nl

WILSON, H.R.   Fraunhofer Institut für
                          Solare Energiesysteme
                          Oltmannsstrasse 22
                          D - 7800 FREIBURG
                          Tel. (0761) 40 50 46
                          Telex

WITTWER, V.   Fraunhofer Institut für
                          Solare Energiesysteme
                          Oltmannsstrasse 22
                          D - 7800 FREIBURG
                          Tel. (0761) 40 50 46
                          Telex